Communications in Computer and Information Science 1915

Rationale

The CCIS series is devoted to the publication of proceedings of computer science conferences. Its aim is to efficiently disseminate original research results in informatics in printed and electronic form. While the focus is on publication of peer-reviewed full papers presenting mature work, inclusion of reviewed short papers reporting on work in progress is welcome, too. Besides globally relevant meetings with internationally representative program committees guaranteeing a strict peer-reviewing and paper selection process, conferences run by societies or of high regional or national relevance are also considered for publication.

Topics

The topical scope of CCIS spans the entire spectrum of informatics ranging from foundational topics in the theory of computing to information and communications science and technology and a broad variety of interdisciplinary application fields.

Information for Volume Editors and Authors

Publication in CCIS is free of charge. No royalties are paid, however, we offer registered conference participants temporary free access to the online version of the conference proceedings on SpringerLink (http://link.springer.com) by means of an http referrer from the conference website and/or a number of complimentary printed copies, as specified in the official acceptance email of the event.

CCIS proceedings can be published in time for distribution at conferences or as post-proceedings, and delivered in the form of printed books and/or electronically as USBs and/or e-content licenses for accessing proceedings at SpringerLink. Furthermore, CCIS proceedings are included in the CCIS electronic book series hosted in the SpringerLink digital library at http://link.springer.com/bookseries/7899. Conferences publishing in CCIS are allowed to use Online Conference Service (OCS) for managing the whole proceedings lifecycle (from submission and reviewing to preparing for publication) free of charge.

Publication process

The language of publication is exclusively English. Authors publishing in CCIS have to sign the Springer CCIS copyright transfer form, however, they are free to use their material published in CCIS for substantially changed, more elaborate subsequent publications elsewhere. For the preparation of the camera-ready papers/files, authors have to strictly adhere to the Springer CCIS Authors' Instructions and are strongly encouraged to use the CCIS LaTeX style files or templates.

Abstracting/Indexing

CCIS is abstracted/indexed in DBLP, Google Scholar, EI-Compendex, Mathematical Reviews, SCImago, Scopus. CCIS volumes are also submitted for the inclusion in ISI Proceedings.

How to start

To start the evaluation of your proposal for inclusion in the CCIS series, please send an e-mail to ccis@springer.com.

Christopher L. Buckley · Daniela Cialfi ·
Pablo Lanillos · Maxwell Ramstead · Noor Sajid ·
Hideaki Shimazaki · Tim Verbelen ·
Martijn Wisse
Editors

Active Inference

4th International Workshop, IWAI 2023
Ghent, Belgium, September 13–15, 2023
Revised Selected Papers

Springer

Editors
Christopher L. Buckley
University of Sussex
Brighton, UK

Pablo Lanillos
Donders Institute for Brain, Cognition and Behaviour
Nijmegen, The Netherlands

Noor Sajid
University College London
London, UK

Tim Verbelen
VERSES Research Lab
Los Angeles, CA, USA

Daniela Cialfi
La Sapienza University of Rome
Rome, Italy

Maxwell Ramstead
VERSES Research Lab
Los Angeles, CA, USA

University College London
London, UK

Hideaki Shimazaki
Kyoto University
Kyoto, Japan

Martijn Wisse
Technische Universiteit Delft
Delft, The Netherlands

ISSN 1865-0929 ISSN 1865-0937 (electronic)
Communications in Computer and Information Science
ISBN 978-3-031-47957-1 ISBN 978-3-031-47958-8 (eBook)
https://doi.org/10.1007/978-3-031-47958-8

This Springer imprint is published by the registered company Springer Nature Switzerland AG
The registered company address is: Gewerbestrasse 11, 6330 Cham, Switzerland

Paper in this product is recyclable.

Preface

The 4th International Workshop on Active Inference (IWAI) took place in Ghent, Belgium on September 13–15, 2023. In contrast to the previous editions (2020–2022), which were held in conjunction with the European Conference on Machine Learning and Principles and Practice of Knowledge Discovery in Databases (ECML-PKDD), this was the first standalone edition. We gathered around 70 active inference researchers from academia and industry in the beautiful scenery of the Saint Peter's Abbey, with a three-day program packed with two tutorials, three keynotes, a poster session, and six sessions of paper presentations.

This volume presents the 17 full papers that were accepted and presented at the workshop. Out of 34 submissions, 17 full papers were selected through a double-blind review process. The papers are clustered in the six sections as presented at the workshop. These sections cover a wide range of domains that find applications of active inference, ranging from robotics, decision-making and control, psychology, representation learning, to theoretical advancements of learning and inference as well as active inference implementations.

The IWAI 2023 organizers would like to thank the Program Committee for their valuable review work, all authors for their contributions, Noor Sajid and Ajith Anil Meera for their excellent tutorials, Tetsuya Ogata, Antonella Maselli and Karl Friston for their inspiring keynotes, and of course all the attendees. We would also like to thank our sponsors VERSES AI, Ghent University (IDLab), imec and the AI Flanders Research Program, which made this event possible.

Group picture taken at IWAI 2023, September 14, Ghent, Belgium.

September 2023

Christopher L. Buckley
Daniela Cialfi
Pablo Lanillos
Maxwell Ramstead
Noor Sajid
Hideaki Shimazaki
Tim Verbelen
Martijn Wisse

Organization

General Chair

Tim Verbelen	VERSES, USA

Local Organisation Chairs

Bart Dhoedt	Ghent University, Belgium
Toon Van de Maele	Ghent University, Belgium

Organizing Committee

Christopher L. Buckley	University of Sussex, UK
Daniela Cialfi	Institute of Complex Systems (CNR); La Sapienza University of Rome, Italy
Pablo Lanillos	Donders Institute for Brain, Cognition and Behaviour, The Netherlands
Maxwell Ramstead	VERSES, USA; and University College London, UK
Noor Sajid	University College London, UK
Hideaki Shimazaki	Kyoto University, Japan
Tim Verbelen	VERSES, USA
Martijn Wisse	Delft University of Technology, The Netherlands

Program Committee

Anjali Bhat	University College London, UK
Christopher Buckley	University of Sussex, UK
Ozan Catal	VERSES, USA
Daniela Cialfi	Institute of Complex Systems (CNR); La Sapienza University of Rome, Italy
Lancelot Da Costa	Imperial College London, UK
Cedric De Boom	Statistiek Vlaanderen, Belgium
Bart Dhoedt	Ghent University, Belgium
Daniel Friedman	University of California, USA

Karl Friston	University College London, UK
Conor Heins	Max Planck Institute of Animal Behavior, Germany
Alex Kiefer	VERSES, USA
Brennan Klein	Northeastern University, USA
Pablo Lanillos	Donders Institute for Brain, Cognition and Behaviour, The Netherlands
Christoph Mathys	Aarhus University, Denmark
Pietro Mazzaglia	Ghent University, Belgium
Ajith Anil Meera	Donders Institute for Brain, Cognition and Behaviour, The Netherlands
Thomas Parr	University College London, UK
Corrado Pezzato	TU Delft, The Netherlands
Maxwell Ramstead	VERSES, USA; and University College London, UK
Noor Sajid	University College London, UK
Dalton Sakthivadivel	VERSES, USA
Eli Sennesh	Northeastern University, USA
Panos Tigas	Oxford University, UK
Alexander Tschantz	VERSES, USA
Hideaki Shimazaki	Kyoto University, Japan
Ruben van Bergen	Radboud University, The Netherlands
Toon Van de Maele	Ghent University, Belgium
Tim Verbelen	VERSES, USA
Martijn Wisse	Delft University of Technology, The Netherlands

Contents

From Theory to Implementation

Learning Representations for Active Inference

Theory of Learning and Inference

Active Inference and Robotics

Contextual Qualitative Deterministic Models for Self-learning Embodied Agents

Jan Lemeire[1,2(✉)], Nick Wouters[2], Marco Van Cleemput[1], and Aron Heirman[2]

[1] Department of Industrial Sciences (INDI), Vrije Universiteit Brussel (VUB), Pleinlaan 2, 1050 Brussels, Belgium
jan.lemeire@vub.be

[2] Department of Electronics and Informatics (ETRO), Vrije Universiteit Brussel (VUB), Pleinlaan 2, 1050 Brussels, Belgium

Abstract. This work presents an approach for embodied agents that have to learn models from the least amount of prior knowledge, solely based on knowing which actions can be performed and observing the state. Instead of relying on (often black-box) quantitative models, a qualitative forward model is learned that finds the relations among the variables, the contextual relations, and the qualitative influence. We assume qualitative determinism and monotonicity, assumptions motivated by human learning. A learning and exploitation algorithm is designed and demonstrated on a robot with a gripper. The robot can grab an object and move it to another location, without predefined knowledge of how to move, grab or displace objects.

Keywords: Autonomous robots · Developmental learning · Open-ended learning · Qualitative Models

1 Introduction

We believe that self-learning capabilities are crucial for fully autonomous agents. With the right learning architecture, agents will be able to adapt to new, unseen, and uncontrolled, open environments. They can discover knowledge autonomously, with no need for external supervision, while also having the capability to redefine their own behavior in case of unexpected perturbations.

We developed a qualitative approach, that allows for a high level of abstraction. It is therefore effective, data-efficient, relates to symbolic reasoning, and provides explainability. The idea of self-learning agents in general, is to start with an empty brain that doesn't contain any prior knowledge, apart from a generic learning architecture. With this developmental learning philosophy in mind, our agent aims to achieve the following goals. Firstly it wants to formalize the effect of its actions on the world in a forward model. It does so by interacting with the world, and relating its observations to its actions. Secondly, it will exploit the learned model to make effective action plans to achieve desired goal states. In other words, the agent will learn to manipulate its environment through its directly controllable actuators. In order to do this, our agent's learning architecture needs algorithms for exploration, learning, and exploitation.

C. L. Buckley et al. (Eds.): IWAI 2023, CCIS 1915, pp. 3–13, 2024.
https://doi.org/10.1007/978-3-031-47958-8_1

As opposed to monolithic black box models, such as neural networks, our approach is based on explicitly modeling the structure and qualitative properties of the system. Some examples of qualitative relations that our system will learn are the following: "positive motor input gives an increase of x- and y-coordinate if the robot's orientation is North", "if I touch a wall, I cannot move further, but can move in the other direction", or "If I touch an object that is located North of me, and if I move North, the object's position will change". Those examples illustrate the plausibility of the assumption of qualitative determinism on which our approach relies: in our world, things happen roughly deterministically. Certainly, if we only look at the direction of changes in variables and not the quantity of the changes.

The contributions of this work are the explicit modeling of context, the identification of the context and the graph describing the relations among the variables, and the exploitation algorithm based on the graph and context.

First, the related work is discussed. The three assumptions are given in the subsequent section. Then we present the experimental setup before defining the model class. In the last 2 sections, the learning and exploitation algorithms are given and the experimental results are shown.

2 Related Work

There is extensive work on qualitative models [3,6], in which the advantages over pure quantitative models are extensively discussed and proven. Bratko et al. [2,13] proposed qualitative models for robotic control. They are based on a small set of variables, while we try to learn networks over a large set of variables. Also, with our representation of context, we enable high-level reasoning. The work of Mugan et al. [10] tackles the same problem as in this paper. They also employ qualitative dynamic Bayesian networks. But they use a probabilistic approach and turn the networks into MDPs to use the model for solving tasks. In Sect. 7, we show how that model can directly be used to determine effective actions. Mugan's quantitative space allows more qualitative values than just the sign. The space is partitioned by so-called landmarks, which resemble our contextual partitioning.

There is quite extensive work on algorithms for causal structure learning, such as the PC algorithm and its variants [11]. However, these algorithms heavily rely on the probabilistic nature of the relations since they rely on some type of faithfulness, which is violated in the presence of deterministic relations [9].

The advantage of explicitly adding context to the models has been studied by [12]. Applied to Bayesian networks this results in context-specific independencies [1], which come from our contextual edges. Our representation is based on the work of [5].

The field of *Active Inference* [7] is also concerned with the self-learning of embodied agents. The theoretical foundations are based on probabilistic models, while we challenge their necessity for the robotic settings on which we focus. Many approaches for active inference are based on (deep) neural networks, e.g., Çatal et al. [4]. These are monolithic black-box models, while the presented approach is based on *explicit modeling of the qualitative properties*, which can then be exploited for reasoning about plans and linking them to symbolic approaches.

3 Assumptions

We assume that the system under study can be described by a *limited set of piece-wise deterministic monotonic functions* defined over variables that are observed or derived:

- *piece-wise monotonic functions*: the relations among variables are primarily monotonous. At certain points/boundaries/constraints (called landmarks by [10]), the monotonicity might be 'broken' and a new 'tone' starts. We say that the state space is divided into **subspaces**.
- *limited*: almost every continuous mathematical function can be split into pieces of monotonicity, but we assume that for the functions of our models, the number of pieces is limited.
- we assume that the set of observed variables is sufficient to characterize the state of the system. This will also become possible by relying on *derived variables*. Derived variables are defined over observed variables or other derived variables.
- we assume *qualitative determinism*. Although this will be relaxed later. We plan to add a *don't know*-value for variables and function outputs. This value can also be used in regions where the variable's sign changes and there is some uncertainty.

4 Experimental Setup

We will perform experiments on a simulated robot. The robot has 4 motors: one for the right wheel (m_R), one for the left wheel (m_L), one to close the gripper ($close$), and one to lift the gripper ($lift$). The wheel motors can only turn in one direction (to drive forward). The robot knows its position (x and y) and its orientation (or). It also has a camera for the position of an object (obx, oby and obz) and a sensor for detecting whether an object is held ($hold$).

The goal is that the robot explores the effect of its actions and learns a model such that it can grab an object and move it to another location. This setup is similar to the setup used by Mugan and Kuipers [10].

5 Contextual Qualitative Deterministic Causal Models

Here we define the proposed model class.

5.1 Problem Definition

Similar to the Markov Decision Processes (MDPs), the problem is defined by a tuple $\langle S, A, T\rangle$, where S is a set of states that are observed, A a set of actions, and T is a transition model, which, in our approach, is a deterministic function $S' = T(S, A)$. As opposed to MDPs, we do not have a reward signal R; the agent will learn a model for T by intrinsic motivation.

5.2 Model Definition

A model is defined over the action variables from A, the previous state $\_S$ (the underscore is used for previous state variables) and the new state S. State variables are directly observed or calculated from other variables. The latter we call **derived variables**. For each state variable, we add the derived variable $ds = s - \_s$, which measures the change of variable s and can be regarded as an approximation of the time derivative. Note that we use capitalized names for vectors and small letters for single variables.

The model consists of two parts: the description of the relations among the variables and the nature of these relations. Similar to a dynamic Bayesian network, the relations among the variables are modeled by a Directed Acyclic Graph (DAG), which we will interpret causally: the orientation of the edges represents the causal influence. Variables A and $\_S$ are input or root variables. Only the variables of S have incoming edges. The parents of variable s are denoted with $Pa(s)$.

By assuming determinism, all (dependent) state variables could be expressed as a function of the (free) input variables and the previous state. However, since we want to have simple relations, we want to find an order in which all dependent variables can be calculated from input or other dependent variables such that the relations are 'simple': each variable has a minimum of parents and the relations are basic qualitative functions based on the monotonicity assumption.

Once the DAG is established, the dependence of each state variable on its parents has to be established. As we are only interested in the qualitative relation, the function returns the sign of the variable.

We denote the sign of variable v by $\mathcal{Q}(v)$, which has three possible outcomes: PLUS, MINUS and ZERO, also denoted with +, − and 0. When applied to a vector, $\mathcal{Q}(V)$ returns the vector containing the signs of the vector elements. To each state variable s, a deterministic qualitative function $\mathcal{Q}(s) = QF(Pa(s))$ is determined. For the moment, we allow two types of qualitative functions. In the first case, the qualitative value of s can be determined by the qualitative values of the parent variables only, while in other cases, the quantitative values are needed. The first type is a function that only depends on the qualitative value of the independent variable and can thus be written as $\mathcal{Q}(s) = QF(\mathcal{Q}(Pa(s)))$. The function can be described by a ternary truth table. The second type is a function $\mathcal{Q}(s) = QF(Pa(s))$ that can be described by a monotone *decision function* DF such that $\mathcal{Q}(s) = Sign(DF(Pa(s)))$. The function separates the positives from the negatives, as shown in Fig. 1. When the left motor is actuated more than the right motor, the robot turns to the left (the change of orientation is positive). Otherwise to the right, except if both actuations are equal, then the robot drives straight.

5.3 Representing Context

So far, our model is able to model deterministic monotonic functions qualitatively. These functions are only valid in parts of the state space, which are defined by the context. For some of the state variables, different qualitative functions might apply according to the **context**, which depends on some action or state variables. Here we

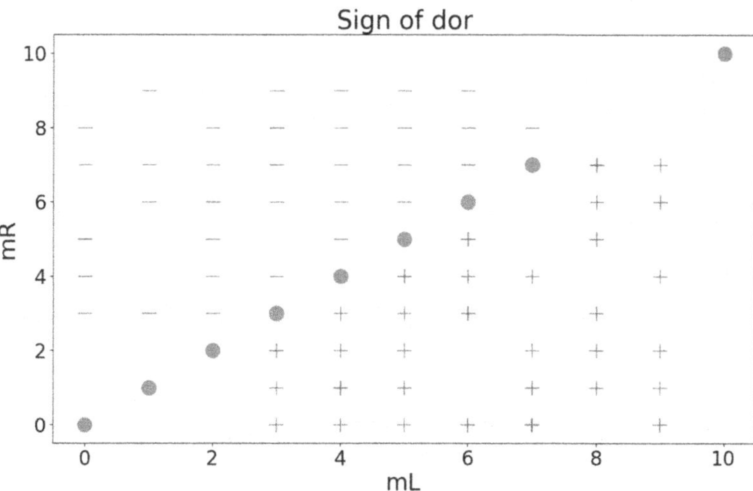

Fig. 1. How the left and right motor determine the sign of the change of the robot's orientation. This is a type 2 function.

limit the context of a specific function to a specific range of one variable (which we call the **context variable**): the total range of that variable is partitioned into two or more contexts.

Definition 1. *Variable c is a context variable of state variable s if its range can be subdivided into regions in which the qualitative function can be written as a truth table (Type 1) or with a decision function (Type 2).*

In each context of c, s might depend on another set of parents, or it is just the qualitative function that starts another 'tone'. An example of the latter is shown in Fig. 2. The orientation (expressed in degrees) determines the sign of the change of the x-coordinate. Between 90 and $-90°$, x changes positively, otherwise negatively. Driving forward is determined by the minimum of the left and right motor actuation ($minLR$) since the difference between both actuations results in a turn of the robot.

The edge between context variable c and state variable s is called a **context edge**. We augment the DAG with this contextual information and call it the *meta-DAG*. If some edges towards s depend on the context defined by c, they are called **contextual edges**. When drawing the DAG, we point the context edge towards these edges. The current state makes the contextual edges active or inactive. If a context edge only determines the qualitative function of s, we point it towards s. An example is given in the next section.

Figure 3 shows the meta-DAG for the example robot defined in Sect. 4. A few additional derived variables are added. $distx$ and $disty$ represent the distance of the robot and the object's x and y coordinate respectively. $mdist$ is the maximal value of $distx$ and $disty$.

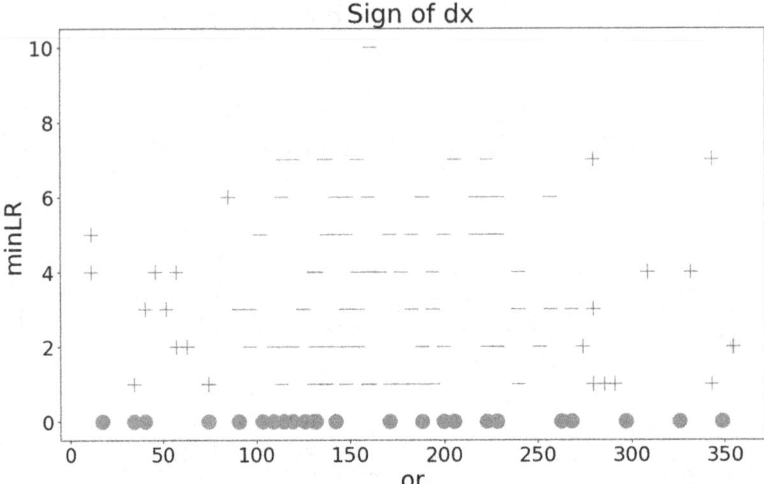

Fig. 2. The change of the x-coordinate depends on the robot's orientation or. The orientation is a context variable. It also depends on $minLR$ which is the minimum of the left and right motor actuation.

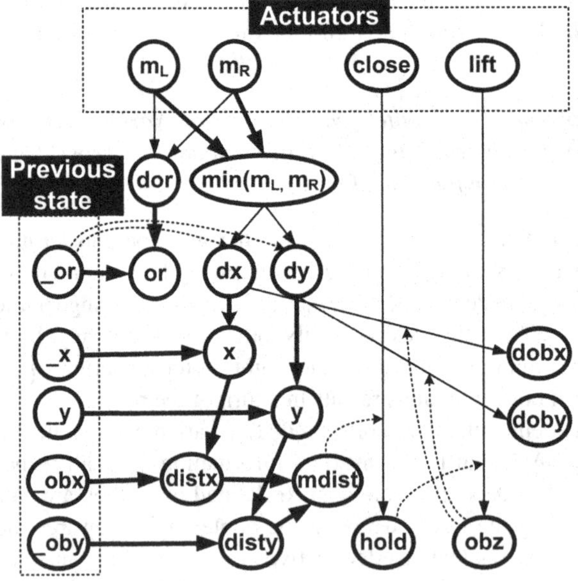

Fig. 3. The contextual qualitative model of the example robot. Contextual edges are shown with dashed lines. The thick edges are known relations coming from derived variables.

6 The Learning

In this section we give the algorithm for learning the model defined in Sect. 5.

6.1 The Tests

The algorithm is based on analyzing the relations among the variables by applying the following tests the observed data.

- Function **depStr** measures the dependency strength between two variables. Pearson's correlation coefficient is used for this.
- Function **condDepStr** measures the conditional dependency strength between two variables conditional on some others with Pearson's partial correlation coefficient.
- Function **isQDet** tests whether a variable can be written as a deterministic function given a set of other variables. The test checks in each context the two types of qualitative functions that are allowed:
  Type 1: the data is arranged according to the truth table (all possible sign combinations of the independent variables). A conflict in a cell happens when there are samples with different signs for the dependent variable.
  Type 2: a monotone decision function is fit on the data to separate the PLUS for the MINUS values in the space defined by the independent variables. Then it is checked whether the decision function can effectively separate the PLUS from the MINUS values. A support vector machine is trained with a linear kernel. To test the separation, we ignore the ZERO values and take a margin of 5 percent.
- Function **isContext** for context identification: data is filtered according to the range of the proposed context variable, and the test for determinism is applied. For Type 1 functions, the truth table is gradually filled by gradually enlarging the context's range. As soon as a conflict occurs, a new context starts. For Type 2 functions, the classification is retrained with a conflict to check whether this can annul the conflicts.

6.2 The Learning Algorithm

The goal is to construct the model: identification of the relations that form the DAG and parameterization of $QF(Pa(s))$ of each variable s. However, this is not necessary for derived variables since their functions are known by their definition. An exception is the variables indicating the change of state variables denoted with the prefix 'd'. These variables are added to the unknown variables, called target variables, while the corresponding state variables are considered to be known by their relation $s = ds + \_s$. The known edges are shown with thick lines in Fig. 3.

The algorithm has to find for each target variable a set of parents that qualitatively determine the target variable. By choosing a set of potential parents for a target variable, the tests of Sect. 6.1 are used to determine whether it results in a possibly-contextual function of type I or type II. The potential parents are chosen in order of the correlation and partial correlation coefficients until a deterministic function is found. We start with the target variable having the highest correlation with one of the action variables. Then, additional action variables or variables from the previous state are added according to their partial correlation. If no deterministic function is found, the target variable will be reconsidered at a later stage (when other target variables have been resolved). Once a state variable is resolved, it is added to the list of action variables to select the next target variable. As such, it can serve as a parent of the other target variables.

6.3 Exploration and Learning

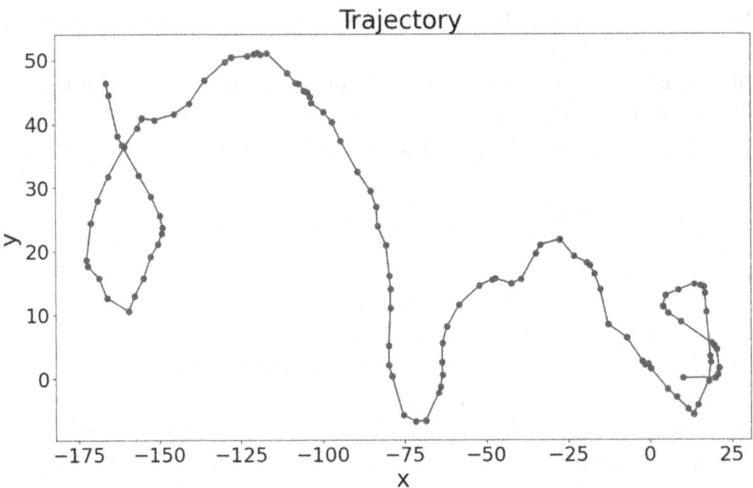

Fig. 4. The exploratory trajectory of the robot with random driving.

During exploration, the robot gathers data that will be used for learning the model. At first, random inputs are given for the motors (a so-called motor 'babbling') by which the driving will be learned. This is the upper part of the model, controlling the robot's position. The exploratory trajectory, which contains 150 data points, is shown in Fig. 4.

For learning how to grab and move an object, there must be data acquired in which the object is accidentally grabbed and displaced. Therefore, the robot is, during its random exploration, regularly oriented towards the object and the gripper is regularly closed to make the chance of grabbing possible. The second exploratory trajectory, by which the lower part of the model involving the object is learned, is shown in Fig. 5. This trajectory contains 350 points during which the object was successfully grabbed and moved 4 times.

With the collected data of 500 points, the learning algorithm correctly learns the model of Fig. 3 with the correct qualitative functions.

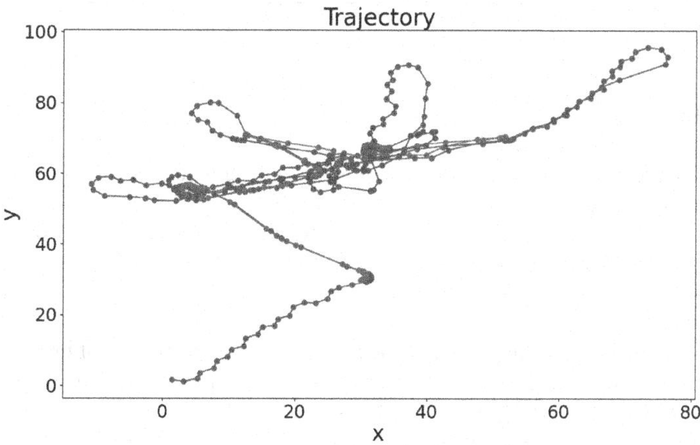

Fig. 5. The exploratory trajectory of the robot (in blue) when trying to grab the object. The trajectory of the object is shown in red. (Color figure online)

7 Exploitation

A task is defined by a goal state in which some state variables should attain certain values. The robot has to take actions in a control loop such that the goal state is reached effectively. Algorithm 1 describes how the qualitative model is used to choose the actions to achieve the goals. Applied to our case, the robot has to travel through 5 **subspaces**: turn -> drive to object -> grab -> lift -> move. This chain is calculated backward: to move an object, the context indicates that the object should be lifted, then to lift an object, it should be grabbed, etcetera.

A hierarchical plan is created: at the higher level, a path across subspaces is sought; at the lower level, a path within a subspace is calculated through a simple control loop. This corresponds to most top-down approaches for robot control [8]. Here it follows naturally from our bottom-up approach.

With the model learned in Sect. 6, the robot, starting from position $(0, 0)$, is assigned to grab an object at position $(30, 30)$, bring it to $(10, 40)$ and return home. In Fig. 6, the paths of the robot and the grabbed object are shown.

The path of the robot might not be a straight line to the object. This is because of our qualitative approach. It makes calculations and reasoning simpler at the expense of accuracy. By accuracy, we mean that we do not calculate the exact command the robot has to drive to reach the desired goal. The qualitative information the robot is using is shown in Fig. 2. It tells him in which range the orientation should be to travel in the direction of the object, but it does not tell him exactly how much the orientation should be to travel straight to the object. Once the robot is in the desired orientation, it drives straight. By doing this, the robot gets closer to the object. There comes a point in time when the robot stops getting closer to the object because of this non-exact path.

Algorithm 1. ReachGoal(Goal G, State S)

while $G \neq S$ **do**
  $dGS \leftarrow G - S$
  for all dgs in dGS **do**
    if dgs is non-zero **then**
      construct a list of tuples of contexts and action signs such that $\mathcal{Q}(dgs) = \mathcal{Q}(ds)$ is attained in the model
      rank all tuples on amount of context values that have to be changed with respect to the current state (lower is better)
    end if
  end for
  search for the simplest (according to the sum of ranks) combination of tuples of the lists (one tuple per dgs) so that the contexts and action signs are the same for all tuples.
  if context $\neq$ current state S **then**
    recursively execute function ReachGoal with Goal set to the wanted context
  end if
  estimate values with the right sign for the action variables (controller)
  execute actions
  update S with the new state
end while

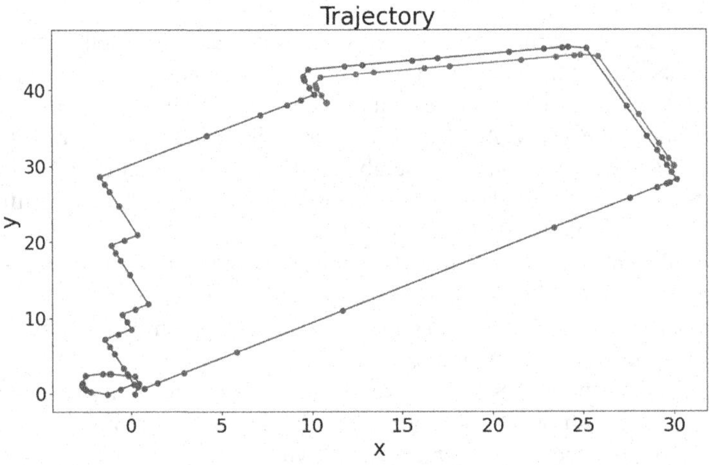

Fig. 6. The trajectory of the robot (in blue) for bringing the object from position (30, 30) to position (10, 40) and returning home. The trajectory of the object is shown in red. (Color figure online)

The robot creates a new subgoal at this point. This subgoal is changing the orientation to another quadrant so that the robot once again is able to get closer to the object by driving straight. It keeps doing this until the desired goal is reached.

8 Conclusions

This work is part of the quest for the 'first principles' that allows self-learning. With the human example in mind, we put some thought-provoking ideas on the table. In everyday situations, probabilistic models are not needed except for the notion that there are things we don't know (yet). Modeling qualitative properties explicitly enables reasoning, makes the link with top-down symbolic approaches, and is easier to learn than quantitative approaches that often require 1000s of samples. The algorithms presented in this paper showed that it is possible and resulted in promising results. Important remaining challenges are the autonomous identification of useful derived variables, effective curiosity-driven exploration and incremental learning.

References

1. Boutilier, C., Friedman, N., Goldszmidt, M., Koller, D.: Context-specific independence in Bayesian networks. In: Uncertainty in Artificial Intelligence, pp. 115–123 (1996)
2. Bratko, I.: An assessment of machine learning methods for robotic discovery. J. Comput. Inf. Technol. **16**, 247–254 (2008)
3. Bratko, I., Suc, D.: Learning qualitative models. AI Mag. **24**, 107–119 (2004)
4. Çatal, O., Wauthier, S.T., De Boom, C., Verbelen, T., Dhoedt, B.: Learning generative state space models for active inference. Frontiers Comput. Neurosci. **14** (2020)
5. Corander, J., Hyttinen, A., Kontinen, J., Pensar, J., Väänänen, J.: A logical approach to context-specific independence. Ann. Pure Appl. Logic **170**(9), 975–992 (2019)
6. Forbus, K.D.: Chapter 9 qualitative modeling. In: Foundations of Artificial Intelligence. Handbook of Knowledge Representation, vol. 3, pp. 361–393. Elsevier, January 2008
7. Friston, K., Kilner, J., Harrison, L.: A free energy principle for the brain. J. Physiol. **100**(1–3), 70–87 (2006)
8. Konidaris, G., Kaelbling, L., Lozano-Perez, T.: Constructing symbolic representations for high-level planning. Proc. AAAI Conf. Artif. Intell. **28**(1) (2014)
9. Lemeire, J., Meganck, S., Cartella, F., Liu, T.: Conservative independence-based causal structure learning in absence of adjacency faithfulness. Int. J. Approx. Reasoning **53**(9), 1305–1325 (2012)
10. Mugan, J., Kuipers, B.: Autonomous learning of high-level states and actions in continuous environments. IEEE Trans. Auton. Ment. Dev. **4**(1), 70–86 (2012)
11. Spirtes, P., Glymour, C., Scheines, R.: Causation, Prediction, and Search. 2nd edn. Springer, New York (1993). https://doi.org/10.1007/978-1-4612-2748-9
12. Tikka, S., Hyttinen, A., Karvanen, J.: Identifying causal effects via context-specific independence relations. In: Advances in Neural Information Processing Systems, vol. 32 (2019)
13. Zabkar, J., Bratko, I., Mohan, A.C.: Learning qualitative models by an autonomous robot. In: 22nd International Workshop on Qualitative Reasoning, pp. 150–157 (2008)

Dynamical Perception-Action Loop Formation with Developmental Embodiment for Hierarchical Active Inference

Kanako Esaki(✉), Tadayuki Matsumura, Shunsuke Minusa, Yang Shao, Chihiro Yoshimura, and Hiroyuki Mizuno

Center for Exploratory Research, Hitachi, Ltd., Tokyo, Japan
kanako.esaki.oa@hitachi.com

Abstract. To adapt an autonomous system to a newly given cognitive goal, we propose a method to dynamically combine multiple perception-action loops. Focusing on the fact that humans change their embodiment during development, the perception-action loops associated with each body part are combined. Applying the method to an end-effector movement task with a robot arm shows that the joints necessary to accomplish the target task are selectively moved in practical time. The result suggests that the robot adapts to the newly given cognitive goal and that developmental embodiment is an essential component in the design of an autonomous system.

Keywords: active inference · embodiment · robot · cognitive goal

1 Introduction

Active inference is a mathematical description of the rules that organisms should obey. Organisms select their actions based on their beliefs about their environment and attempt to minimize their free energy. This allows the organism to reach a preferable state while minimizing uncertainty about the environment [8,12,13,29]. Active inference has been found to explain a variety of human characteristics [2,9,11,14,15,28]. However, when active inference is used to reveal characteristics of organisms or to construct autonomous systems, the methods for designing generative models are not yet fully understood [6,34,35].

In recent years, many studies have used deep learning techniques to learn generative models in active inference. Ueltzhöffer [33] proposed to implement the generative models with neural networks, called "deep active inference". The proposed methods performed as well as conventional reinforcement learning on the toy problems [7,25,33,37]. By modeling the above probabilities with neural networks, existing learning methods can be used to infer generative models even when the state and action space is multidimensional. Wei et al. [35] also proposed a method for learning generative models from human demonstration behavior

C. L. Buckley et al. (Eds.): IWAI 2023, CCIS 1915, pp. 14–28, 2024.
https://doi.org/10.1007/978-3-031-47958-8_2

and evaluated it on car driving behavior. The experiments show that the proposed method is able to mimic human driving behavior on highways. Other studies have also applied active inference to robots [20,22,24,26,31,32]. These studies have revealed an important aspect of active inference: acting to realize the predicted outcome of sensory input makes complex inverse kinematics models unnecessary. However, all of these studies assumed that the state and action spaces for a given cognitive goal were given. In other words, the Markov Blanket (see Sect. 2.1) for a given cognitive goal must be designed in advance. The Markov Blanket extracts from the world the observations and actions required for each cognitive goal. It is virtually impossible to predesign these Markov Blankets for all cognitive goals that would be given in an autonomous system.

To adapt to a newly given cognitive goal, organisms, including humans, are said to have multiple Markov Blankets dynamically [27,29]. The Markov Blanket can be applied at different scales, such as separating the outside of the brain from the whole brain, and separating the self from others [23]. In addition, multiple Markov Blankets can be nested within each other [3,10]. Given the role of Markov Blanket as an interface to the world, flexible combinations of Markov Blanket require developmental embodiment. Developmental embodiment is essential for organisms to adapt autonomously to different cognitive goals [4,5].

In this paper, we propose a method to adapt to a newly given cognitive goal through dynamically formed Markov Blankets with developmental embodiment. The proposed method defines two types of Markov Blankets: primitive Markov Blankets, which are preconfigured according to the system's embodiment, i.e., the hardware configuration, such as joints and sensors, and meta cognitive Markov Blankets, which are created as a higher level of the primitive Markov Blankets when a cognitive goal is given. Active inference is performed in each Markov Blanket, and in addition, the meta cognitive Markov Blanket develops embodiment by selecting the necessary primitive Markov Blankets according to the cognitive goal (hereafter "attention"). In the process of adapting to the cognitive goal, the primitive Markov Blankets that are the attention targets are gradually determined. By dynamically combining multiple Markov Blankets, the system can adapt to a newly given cognitive goal.

Dynamical Markov Blanket formation is implemented and validated using a robot task. A robot is one of the solutions to realize an embodied system. However, robots currently used in industry, as seen in robots used in factories, tend to focus on efficiency and are superior as "automated" systems, but still have many problems as "autonomous" systems. In the future, robots will be used in everyday spaces where there may be more human-robot interaction. In such situations, the robot is expected to autonomously perform complex tasks in which new goals of the tasks are unexpectedly assigned or the operating environment is constantly changing. The proposed method is validated using the basic robot arm task, end-effector movement. The validation will be an important first step for autonomous systems to adapt to more complex tasks.

2 Method

2.1 Markov Blanket

Markov Blanket [10,19,27,30] determines the appropriate observations and actions for the system to adapt to the newly given cognitive goal. Determining observations and actions means that no other information is considered. For example, if a person tries to pick up a cup within reach while sitting, he or she will observe the position of the cup's handle (observation) and move his or her hand (action), but will not observe the color of the curtain behind him or her, nor will he or she move his or her toes. Given a cognitive goal, the organism uses the Markov Blanket to determine the necessary information.

In the free energy principle underlying active inference, the determination of observations and actions by the Markov Blanket results in the setting of the generative model inside the system and the generative process in the environment. The ultimate goal of the system is to approximate the generative process with the generative model. The closer the generative model is to the generative process, the more appropriately the system observes and acts on the environment.

2.2 Generative Model

Given the world as a discrete space, the generative model can be described in linear algebraic form [29]. Observations o_τ, hidden states s_τ, and actions π_τ are all assumed to be categorical variables. Under the assumption, the generative model is decomposed into likelihood $P(o_\tau \mid s_\tau)$, transition probability $P(s_{\tau+1} \mid s_\tau, \pi)$, preference $P(o_\tau \mid C)$, and prior belief $P(s_1)$, represented by matrices **A**, **B**, **C**, and **D**, respectively:

$$\begin{aligned} P(o_\tau \mid s_\tau) &= Cat(\mathbf{A}) \\ P(s_{\tau+1} \mid s_\tau, \pi) &= Cat(\mathbf{B}_{\pi\tau}) \\ P(o_\tau \mid C) &= Cat(\mathbf{C}_\tau) \\ P(s_1) &= Cat(\mathbf{D}) \end{aligned} \tag{1}$$

The free energy is minimized by updating the generative model. The update rule for each generative model is derived by transforming the equation that minimizes the variational free energy:

$$\begin{aligned} \mathbf{a} &= a + \sum_\tau o_\tau \otimes \mathbf{s}_\tau \\ \mathbf{b}_{\pi\tau} &= b_{\pi\tau} + \sum_\tau \mathbf{s}_{\pi\tau} \otimes \mathbf{s}_{\pi\tau-1} \\ \mathbf{c} &= c + \sum_\tau o_\tau \\ \mathbf{d} &= d + \mathbf{s}_1 \end{aligned} \tag{2}$$

where a to d are the elements of matrices **A** to **D**, respectively.

2.3 Action Selection

Actions with smaller expected free energy $\mathbf{G}$, weighted by precision γ_G, are selected with higher probability. The belief about policy $P(\pi)$ is as follows:

$$\begin{aligned} P(\pi) &= Cat(\boldsymbol{\pi}_0) \\ \boldsymbol{\pi}_0 &= \sigma(\ln \mathbf{E} - \gamma_G \mathbf{G}) \end{aligned} \tag{3}$$

where $\mathbf{E}$ is the habit term. Precision in general is defined as the inverse of the variance of the probability distribution and represents the confidence in the probability distribution. The higher the precision, the higher the confidence in that probability distribution. Precision is discussed in relation to attention to the information stream conveyed as a probability distribution. Adjusting precision higher leads to attention to the information stream, while adjusting precision lower diverts attention away from the information stream.

In this study, the precision of each primitive Markov Blanket $\gamma_{\mathbf{G}}^{prim(i)}$ is adjusted according to the preference of each primitive Markov Blanket $\mathbf{C}_{\tau}^{prim(i)}$ for the action of meta cognitive Markov Blanket π_{τ}^{meta}. In everyday space, multiple ways of achieving a cognitive goal are expected. That is, the preference of each primitive Markov Blanket $\mathbf{C}_{\tau}^{prim(i)}$ for achieving the action of the meta cognitive Markov Blanket π_{τ}^{meta} have variance. A large variance in the preference of each primitive Markov Blanket $\mathbf{C}_{\tau}^{prim(i)}$ is interpreted as not having to "stick" to the action of that primitive Markov Blanket $\pi_{\tau}^{prim(i)}$, while a small variance means that the action of that primitive Markov Blanket $\pi_{\tau}^{prim(i)}$ is indispensable. Based on the above, the precision of each primitive Markov Blanket $\gamma_{\mathbf{G}}^{prim(i)}$ is adjusted by the precision (i.e., the inverse of the variance) of the preference of each primitive Markov Blanket $\gamma_{\mathbf{C}}^{prim(i)}$.

$$\begin{aligned} \gamma_G^{prim(i)} &= \begin{cases} 1(\gamma_{\mathbf{C}}^{prim(i)} > \theta^{prim(i)}) \\ 0(\gamma_{\mathbf{C}}^{prim(i)} \leq \theta^{prim(i)}) \end{cases} \\ \gamma_{\mathbf{C}}^{prim(i)} &= 1/Var(Cat(\mathbf{C}_{\tau}^{prim(i)})) \end{aligned} \tag{4}$$

If the precision of the preference of each primitive Markov Blanket $\gamma_{\mathbf{C}}^{prim(i)}$ is higher than the threshold $\theta^{prim(i)}$, then the action with the smaller expected free energy of that primitive Markov Blanket $\mathbf{G}_{\tau}^{prim(i)}$ will be selected with higher probability as shown in Eq. (3). On the other hand, if the precision of the preference of each primitive Markov Blanket $\gamma_{\mathbf{C}}^{prim(i)}$ is less than the threshold $\theta^{prim(i)}$, the habit term $\mathbf{E}^{prim(i)}$ is preferred. In this study, the habit term $\mathbf{E}^{prim(i)}$ was set so that the action that preserves the current state of each primitive Markov Blanket $s_{\tau}^{prim(i)}$ (i.e., "do nothing") is selected. Thus, adjusting the precision of each primitive Markov Blanket $\gamma_{\mathbf{G}}^{prim(i)}$ determines which of the primitive Markov Blankets is given attention.

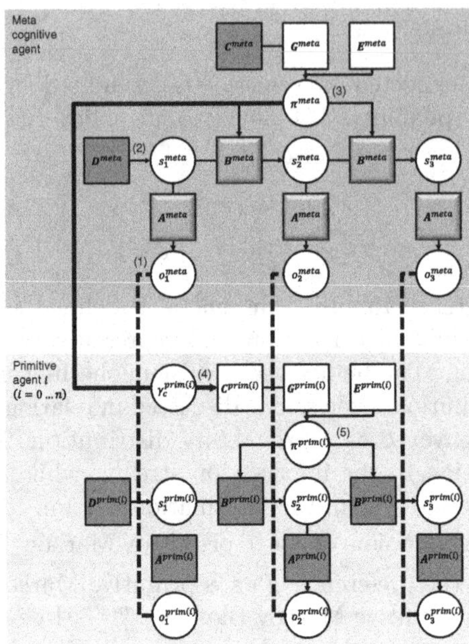

Fig. 1. Dynamical Markov Blanket formation. Meta cognitive and primitive agents perform active inference, respectively. These are connected by transformations of observations and actions between primitive and meta cognitive agents. The generative models of the meta cognitive agent $\mathbf{C}^{meta}$ and $\mathbf{D}^{meta}$ and the generative models of the primitive agent $\mathbf{A}^{prim(i)}$, $\mathbf{B}^{prim(i)}$, and $\mathbf{D}^{prim(i)}$ are assumed to be known, and the generative models of the meta cognitive agent $\mathbf{A}^{meta}$ and $\mathbf{B}^{meta}$ are updated.

2.4 Dynamical Markov Blanket Formation Process

The dynamical Markov Blanket formation process consists of active inference, generative model updating, and attention updating by agents in the system, determined by Markov Blankets. In the following, the agent corresponding to a primitive Markov Blanket is called a primitive agent and the agent corresponding to a meta cognitive Markov Blanket is called a meta cognitive agent. The causal graphs of primitive and meta cognitive agents are shown in Fig. 1. Since primitive agents are associated with embodiment of the system such as joints and sensors, determining the hardware configuration (usually at the time the system is shipped) means that generative models of the primitive agents $\mathbf{A}^{prim(i)}$, $\mathbf{B}^{prim(i)}$, and $\mathbf{D}^{prim(i)}$ is given. When a new cognitive goal is given to the system, a new instance of the meta cognitive agent is created. At this time, only generative models $\mathbf{C}^{meta}$ and $\mathbf{D}^{meta}$ are given, and for generative models $\mathbf{A}^{meta}$ and $\mathbf{B}^{meta}$, only categorical variables are given.

In active inference, after acquiring observations and inferring the hidden state, the action is selected based on the expected free energy. First, the observations of n primitive agents $o_{\tau}^{prim(i)} (i = 0 \ldots n)$ are transformed into observation of the meta cognitive agent o_τ^{meta} using the hardware configuration of the system, e.g. kinematics (Fig. 1(1)). Next, the meta cognitive agent infers the hidden state s_τ^{meta} from the observation o_τ^{meta} (Fig. 1(2)). The meta cognitive agent then computes the expected free energy $\mathbf{G}_\tau^{meta}$ and probabilistically selects an action π_τ^{meta} so that the expected free energy $\mathbf{G}_\tau^{meta}$ becomes smaller (Fig. 1(3)). The action of meta cognitive agent π_τ^{meta} are then translated into preferences of n primitive agents $\mathbf{C}_\tau^{prim(i)}(i = 0 \ldots n)$ (Fig. 1(4)). The primitive agent then calculates its expected free energy $\mathbf{G}_\tau^{prim(i)}$ and selects its action $\pi_\tau^{prim(i)}$according to the Eq. (3) (Fig. 1(5)).

In generative model updating, only the likelihood $\mathbf{A}^{meta}$ and transition probability $\mathbf{B}^{meta}$ of the meta cognitive agent are updated according to the Eq. (2). Since meta cognitive agents are created according to cognitive goals, the initial likelihood $\mathbf{A}^{meta}$ and transition probability $\mathbf{B}^{meta}$ are assumed to be uniformly distributed, and preferences $\mathbf{C}^{meta}$ and prior distributions $\mathbf{D}^{meta}$ are known.

In attention updating, the precision of each primitive agent $\gamma_{\mathbf{G}}^{prim(i)}$ is updated according to the precision of preference of each primitive agent $\gamma_{\mathbf{C}}^{prim(i)}$. Since primitive agents are associated with hardware, the initial attention targets are all primitive agents, i.e., $\gamma_{\mathbf{G}}^{prim(i)} = 1(i = 0 \ldots n)$. If the precision of preference of each primitive agent $\gamma_{\mathbf{C}}^{prim(i)}$ becomes below its threshold $\theta^{prim(i)}$, the precision of the primitive agent $\gamma_{\mathbf{G}}^{prim(i)}$ is switched to zero, as shown in Eq. (3).

3 Results and Discussion

3.1 Experimental Setup

Using the robot task of moving the position of an end-effector with a robot arm, we validated that the robot adapts to a newly given cognitive goal through dynamical Markov Blanket formation. Figure 2(a) shows the Universal Robotics UR5e robot arm used for the validation. The robot arm has a base, shoulder, elbow, wrist 1, wrist 2, and wrist 3 joints, and a Robotiq 2F-140 Adaptive Gripper as an end-effector. Each joint angle of the robot arm was controlled using ROS Melodic Morenia installed on Ubuntu 18.04 LTS. The dynamical Markov Blanket formation was implemented in Python using pymdp [17].

The cognitive goal of the system in our validation is a robot task that moves the end effector of a robot arm from one position to another. Figure 2(b) shows the three robot tasks used in this validation. The FC task is to move the position of the robot arm's end-effector from position F (front) to position C (center) to confirm the basic performance of the dynamical Markov Blanket formation. The FCR task is to move the position of the robot arm's end-effector from position F to position C and then to position R (right). The FCR task involves two cognitive goals: moving from position F to C and moving from position C to R to validate the advantage of attention. The CR-CL task has two phases. In the first phase, the robot repeatedly moves the end-effector from position C to position R, forming a meta cognitive Markov Blanket. After the generative

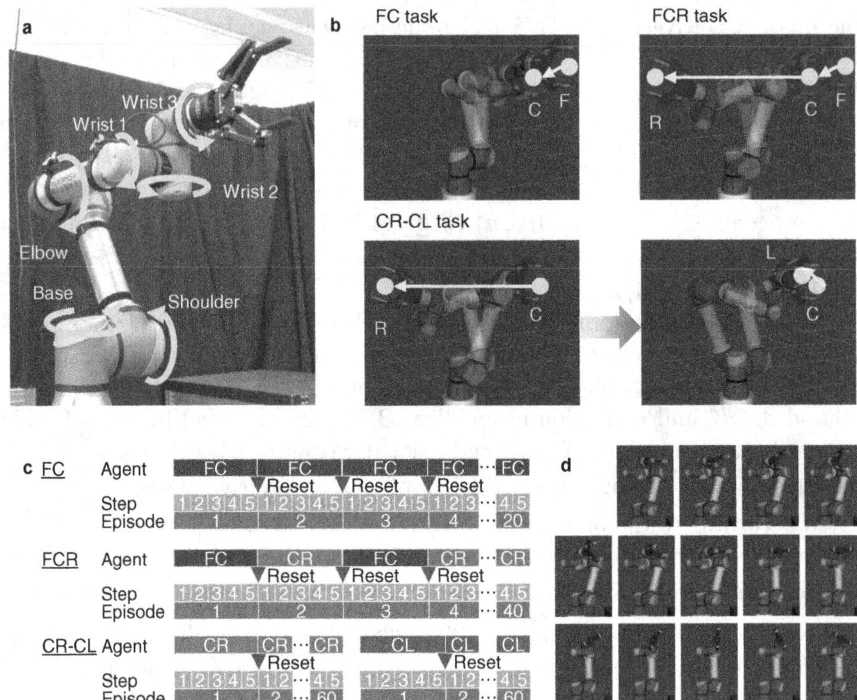

Fig. 2. Validation Setting. (a) Universal Robots UR5e 6-axis robot arm used for validation. (b) Cognitive goals. The cognitive goal in this validation is the task of moving the end-effector of the robot arm. (c) Sequence in each task. Each episode contains five steps, and in each step, each agent performs active inference, generative model updating, and precision updating. At the beginning of each episode, the end-effector position is reset to the initial position. (d) Orientation patterns of the end-effector at position C for the FC and FCR tasks.

models and the number of primitive Markov Blankets that are the attention targets have converged sufficiently, the second phase is executed. In the second phase, the robot repeatedly moves the end-effector from position C to position L (left) to form a Markov Blanket. The CR-CL task compares adaptation to cognitive goals with a single Markov Blanket and that with dynamically formed multiple Markov Blankets. The FC and FCR tasks were validated on the real machine, while the CR-CL task was validated by simulation.

In our validation, the primitive agents correspond to each joint of the 6-axis robot arm, and the meta cognitive agents correspond to each robot task. Each agent is assumed to be in a discrete space. Table 1 lists the observations, hidden states, and actions of each agent. For primitive agents, joint angles are discretized. For the meta cognitive agent, both obsesrvation and hidden states were set to the end-effector positions possible in the task. Since the information

Table 1. Observation, hidden state, and action in validation.

Agent	Primitive	Meta cognitive (FC case)
Observation o_τ	Current joint angle [deg]: $\{10x \mid \lvert x\rvert \leq 27, x \in \mathbb{Z}\}$	Current end-effector position: Position F/Position C
Hidden state s_τ	Current joint angle [deg]: $\{10x \mid \lvert x\rvert \leq 27, x \in \mathbb{Z}\}$	Current end-effector position: Position F/Position C
Action π_τ	Move to [joint angle]/Stop Joint angle [deg]: $\{10x \mid \lvert x\rvert \leq 27, x \in \mathbb{Z}\}$	Move to C/Stop

used by each agent has a very simple structure, the hidden state was identical to the observation and thus observable.

Each task was repeated for 20 episodes (FC and FCR tasks) or 60 episodes (CR-CL task), with 5 steps per episode for each agent. Figure 2 (c) shows sequence in each task. In each episode, the end-effector position was first reset to its initial position. In each step, the primitive and meta cognitive agents performed active inference, and then the meta cognitive agent updated the likelihood and transition probability and updated attention to each primitive agent (precision $\gamma_{\mathbf{C}}^{prim(i)}$). In the following, all episodes are consistently represented in terms of time steps. For example, time step 7 is step 2 in episode 2.

To simplify the implementation of precision-based attention updating, variance, the inverse of precision, was used. The preference of the primitive Markov Blanket corresponds to the angle pattern of each joint that achieves the target position of the end-effector. For the threshold $\theta^{prim(i)}$ in Eq. (4), $1/\theta^{prim(i)} = 4[deg^2]$. Figure 2(d) shows the orientation patterns at position C in the FC and FCR tasks. The robot tasks in our validation differs from robot tasks commonly used in factories, where various orientations are allowed at the target position. Similarly, at position R in the FCR task, 14 different orientation patterns are allowed. In the CR-CL task, however, only one orientation pattern is allowed at all positions C, R and L in order to eliminate the influence of attention.

The transformation from primitive agent observations to meta cognitive agent observations, and from meta cognitive agent actions to the preferences of each primitive agent, used a kinematics database to correspond to discretized joint angles. The kinematics database maps the aforementioned orientation patterns, i.e. the set of joint angles, to the positions of the end-effector. In the transformation from the observation of a primitive agent to that of a meta cognitive agent, the end-effector position corresponding to the joint angles observed by the primitive agent were obtained from the kinematics database and used as the observations of the meta cognitive agent. In addition, the transformation from the meta cognitive agent's action to each primitive agent's preference requires the conversion of the meta cognitive agent's action to the end-effector position. If the meta cognitive agent's action is "Move to C/R/L", it is converted to the end-effector's position C/R/L. If the meta cognitive agent's action is "Stop", it is converted to the current end-effector position. The joint angles corresponding to the converted end-effector position were randomly sampled from the kinematic database and used as the primitive agent's preferences.

3.2 Adaptation to Newly Given Cognitive Goal

To confirm that the robot arm adapts to the newly given cognitive goal, an FC task was performed with the robot arm. Figure 3(a) shows the transition in the number of primitive agents that were the attention targets in the five trials of the FC task. Attention was pruned from time step 3 to 8, and the number of the attention targets converged to 3 at time step 8 for all trials. This was because the minimum number of dimensions required at position C was 3. Among the six dimensional variables indicating position and orientation, only the three dimensional variable indicating position was uniquely specified at position C in the FC task. Figure 3(b) shows the transition of each joint angle and orientation of the robot arm from time step 1 to 5 and from time step 16 to 20 for trial 2 in Fig. 3(a). From time step 1 to 5, the multiple orientation patterns in Fig. 2(d) were attempted to move the end-effector to position C, because more than five primitive agents were the attention targets. In contrast, from time step 16 to 20, the end-effector reached position C by moving only the shoulder, elbow, and wrist 1 joints and executing only one orientation pattern, because only the primitive agents associated with those joints were the attention targets.

The results suggest that by updating attention and refining the primitive agent combination, the robot arm adapts to the newly given cognitive goal. The decrease in the number of orientation patterns attempted seems to correspond to the phenomenon that humans, when given a new cognitive goal, initially act with hesitation and then gradually become more confident in their actions and adapt to the new goal. In addition, the finding that only half of the joints of the robot arm were moved at later time steps is considered equivalent to the phenomenon that humans adapt by moving only the necessary body parts in order to reduce energy costs [1,16,21].

3.3 Attention Switching

To confirm the advantage of attention, the FCR task was performed with the robot arm. The FCR task consists of multiple cognitive goals of moving from position F to C and moving from position C to R. Attention was switched between these movements so that the appropriate joints were moved for each movement. Figure 4 shows the variance transition of each primitive agent's preference. The closer to blue, the higher the variance, i.e., the lower the precision. At time step 1, immediately after the FC agent was created, and at time step 6, immediately after the CR agent was created, the variance was zero because each primitive agent experienced only one preference pattern. As the time step progressed, the variance of the primitive agents differed between the FC and CR agents. During the time steps of the FC agent, the variances of the shoulder, elbow, and wrist 1 joints, which contribute significantly to pitch rotation,

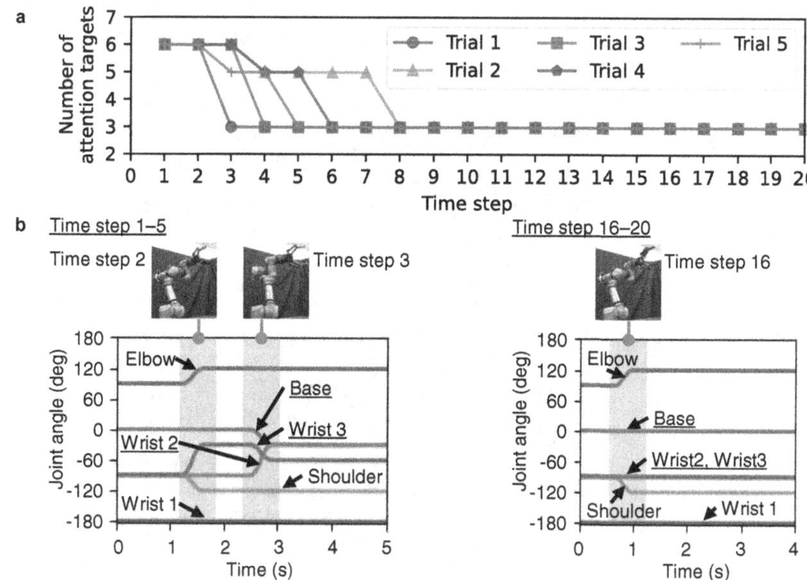

Fig. 3. Attention transitions. (a) Transitions in the number of primitive agents that are attention targets for five trials. Since the same number of the attention targets were maintained after time step 20, the plot was not shown. (b) Transition of joint angles from time step 1 to 5 and from time step 16 to 20.

remained small, while those of the base, wrist 2, and wrist 3 joints became larger. During the time steps of the CR agent, the variances for the base, shoulder, elbow, and wrist 2 joints, which contribute significantly to the yaw rotation, remained small, while the variances for the wrist 1 and wrist 3 joints became larger. By time steps 11 and 16, the unnecessary attentions of the FC and CR agents, respectively, were pruned. In later time steps, the respective attention targets of both FC and CR agents remained unchanged.

Attention switching would enable the system to adapt to complex cognitive goals. Complex cognitive goals are generally assumed to be decomposable, either spatially or temporally, into simpler cognitive goals. By setting up meta cognitive Markov Blankets for each decomposed cognitive goal, the system will adapt to complex cognitive goals within a single framework of dynamical Markov Blanket formation. Adapting to complex cognitive goals with multiple Markov Blankets also means that the system has the potential to respond flexibly to changes in the environment. It has already been suggested that the system is able to respond to context switching by switching the generative models [18]. Our proposed method would enable the system to respond to unexpected context switching. An example of an unexpected context switching is when an obstacle appears in the path of the robot arm and the robot arm must avoid it. Specifically, it can be handled by replacing some of the previously formed Markov Blanket sequences, or by adding meta cognitive Markov Blankets to the Markov Blanket sequences.

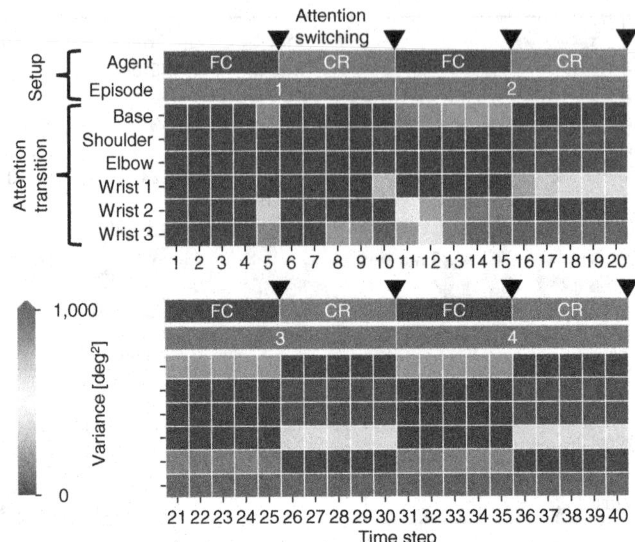

Fig. 4. Transition of attention. The attention targets changed gradually and then, starting at time step 21, clearly switched between the FC and the CR agent's steps.

3.4 Comparison with Single Markov Blanket

Comparison with the single Markov Blanket in the CR-CL task confirms the advantages of dynamical Markov Blanket formation. The single Markov Blanket here is a task independent meta cognitive Markov Blanket. In the CR-CL task, according to the proposed method, the Markov Blanket corresponding to the CR task is formed in the first phase, and that corresponding to the CL task is formed in the second phase. The single Markov Blanket, on the other hand, has at least the current end-effector positions "position C", "position R", and "position L" as observation and hidden states, and "Move to position R", "Move to position L", and "Stop" as actions.

Dynamically formed Markov Blanket showed higher learning performance than the single Markov Blanket. Table 2 shows the average, minimum and maximum time steps of the five trials required for the expected free energy to converge. The convergence of the expected free energy, i.e. the learning of the generative model, took longer for the single Markov Blanket than for the dynamically formed Markov Blanket, because the number of dimensions for all observations, hidden states, and actions is higher in the single Markov Blanket. Table 3 shows examples of generative models of the single Markov Blanket **A** and **B** that failed to learn. In some cases, generative models became unexpected, even when the expected free energy converged. The single Markov Blanket sometimes fell into local solutions due to the high dimensionality of the observations, hidden states, and actions, which made learning unstable. In contrast, dynamical Markov Blanket formation only requires generative models with the minimum number of

Table 2. Convergence time step.

Values	Dynamical	Single
Average	125.2	176
Minimum	111	154
Maximum	144	201

Table 3. Failed examples of generative models.

Type	Truth	Failure
$\boldsymbol{A}: p(o_\tau \mid s_\tau)$ (columns s_τ: C R L; rows o_τ: C R L)	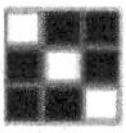	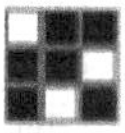
$\boldsymbol{B}: p(s_{\tau+1} \mid s_\tau, \pi = \text{"Move to R"})$ (columns s_τ: C R L; rows $s_{\tau+1}$: C R L)		
$\boldsymbol{B}: p(s_{\tau+1} \mid s_\tau, \pi = \text{"Move to L"})$ (columns s_τ: C R L; rows $s_{\tau+1}$: C R L)		
$\boldsymbol{B}: p(s_{\tau+1} \mid s_\tau, \pi = \text{"Stop"})$ (columns s_τ: C R L; rows $s_{\tau+1}$: C R L)	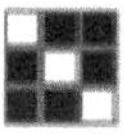	N/A

dimensions of observations, hidden states, and actions for a newly given cognitive goal, and learning was more stable.

The results suggest the importance of dynamically forming Markov Blankets in acquiring the adaptive capabilities of an autonomous system. The proposed method is inspired by changes in human embodiment during development. Humans are said to be able to respond to newly given cognitive goals by gradually accumulating what they can do during development [36]. What would happen if we had adult bodies at birth? The single Markov Blanket addresses just such an assumption. Just as the learning performance of the single Markov Blanket was lower than that of the dynamically formed Markov Blanket, humans with an adult body at birth would not be able to adapt well to cognitive goals because of the lack of the developmental embodiment. We believe that the developmental embodiment is an essential part of the design of an autonomous system.

4 Conclusion

We proposed dynamical Markov Blanket formation to adapt an autonomous system to a newly given cognitive goal, focusing on human embodiment during development. Applying the method to an end-effector movement task with a robot arm showed that the joints necessary to accomplish the target task are selectively moved. Furthermore, the learning performance of dynamically formed Markov Blankets was better than that of the single Markov Blanket. The results suggest that the robot adapts to the newly given cognitive goal and that developmental embodiment is essential for designing an autonomous system. Future work includes scaling the precision adjustment that determines the attention to more than just the joint angle.

References

1. Chai, J., Hayashibe, M.: Motor synergy development in high-performing deep reinforcement learning algorithms. IEEE Rob. Autom. Lett. **5**(2), 1271–1278 (2020). https://doi.org/10.1109/LRA.2020.2968067
2. Cittern, D., Nolte, T., Friston, K., Edalat, A.: Intrinsic and extrinsic motivators of attachment under active inference. PLOS ONE **13**(4), 1–35 (2018). https://doi.org/10.1371/journal.pone.0193955
3. Da Costa, L., Parr, T., Sajid, N., Veselic, S., Neacsu, V., Friston, K.: Active inference on discrete state-spaces: a synthesis. J. Math. Psychol. **99**, 102447 (2020). https://doi.org/10.1016/j.jmp.2020.102447
4. Esaki, K., Matsumura, T., Ito, K., Mizuno, H.: Sensorimotor visual perception on embodied system using free energy principle. In: Kamp, M., et al. (eds.) Machine Learning and Principles and Practice of Knowledge Discovery in Databases, vol. 1524, pp. 865–877. Springer, Cham (2021). https://doi.org/10.1007/978-3-030-93736-2_62
5. Esaki, K., Matsumura, T., Yoshimura, C., Mizuno, H.: Extended-self recognition for autonomous agent based on controllability and predictability. In: 2022 IEEE Symposium Series on Computational Intelligence (SSCI), pp. 1036–1043 (2022). https://doi.org/10.1109/SSCI51031.2022.10022161
6. Ferraro, S., Van de Maele, T., Mazzaglia, P., Verbelen, T., Dhoedt, B.: Disentangling shape and pose for object-centric deep active inference models. In: Buckley, C.L., et al. (eds.) Active Inference, pp. 32–49. Springer, Cham (2023). https://doi.org/10.1007/978-3-031-28719-0_3
7. Fountas, Z., Sajid, N., Mediano, P., Friston, K.: Deep active inference agents using Monte-Carlo methods. In: Larochelle, H., Ranzato, M., Hadsell, R., Balcan, M., Lin, H. (eds.) Advances in Neural Information Processing Systems, vol. 33, pp. 11662–11675. Curran Associates, Inc. (2020). https://proceedings.neurips.cc/paper/2020/file/865dfbde8a344b44095495f3591f7407-Paper.pdf
8. Friston, K.: The free-energy principle: a unified brain theory? Nat. Rev. Neurosci. **11**(2), 127–138 (2010). https://doi.org/10.1038/nrn2787
9. Friston, K.: The Bayesian savant. Biol. Psychiat. **80**(2), 87–89 (2016). https://doi.org/10.1016/j.biopsych.2016.05.006
10. Friston, K.: A free energy principle for a particular physics (2019). https://arxiv.org/abs/1906.10184
11. Friston, K., Adams, R., Perrinet, L., Breakspear, M.: Perceptions as hypotheses: saccades as experiments. Frontiers Psychol. **3** (2012). https://doi.org/10.3389/fpsyg.2012.00151
12. Friston, K., FitzGerald, T., Rigoli, F., Schwartenbeck, P., Pezzulo, G.: Active inference: a process theory. Neural Comput. **29**(1), 1–49 (2017). https://doi.org/10.1162/NECO_a_00912
13. Friston, K., Kilner, J., Harrison, L.: A free energy principle for the brain. J. Physiol. Paris **100**(1), 70–87 (2006). https://doi.org/10.1016/j.jphysparis.2006.10.001. Theoretical and Computational Neuroscience: Understanding Brain Functions
14. Friston, K.J., et al.: Dopamine, affordance and active inference. PLOS Comput. Biol. **8**(1), 1–20 (2012). https://doi.org/10.1371/journal.pcbi.1002327
15. Friston, K.J., Stephan, K.E., Montague, R., Dolan, R.J.: Computational psychiatry: the brain as a phantastic organ. Lancet Psychiatry **1**(2), 148–158 (2014). https://doi.org/10.1016/S2215-0366(14)70275-5

16. Hayashibe, M., Shimoda, S.: Synergetic learning control paradigm for redundant robot to enhance error-energy index. IEEE Trans. Cogn. Dev. Syst. **10**(3), 573–584 (2018). https://doi.org/10.1109/TCDS.2017.2697904
17. Heins, C., et al.: pymdp: a Python library for active inference in discrete state spaces. J. Open Source Softw. **7**(73), 4098 (2022). https://doi.org/10.21105/joss.04098
18. Hesp, C., Smith, R., Parr, T., Allen, M., Friston, K.J., Ramstead, M.J.D.: Deeply felt affect: the emergence of valence in deep active inference. Neural Comput. **33**(2), 398–446 (2021). https://doi.org/10.1162/neco_a_01341
19. Kirchhoff, M., Parr, T., Palacios, E., Friston, K., Kiverstein, J.: The Markov blankets of life: autonomy, active inference and the free energy principle. J. R. Soc. Interface **15**(138), 20170792 (2018). https://doi.org/10.1098/rsif.2017.0792
20. Lanillos, P., et al.: Active inference in robotics and artificial agents: survey and challenges (2021). https://arxiv.org/abs/2112.01871
21. Liu, L., Ballard, D.: Humans use minimum cost movements in a whole-body task. Sci. Rep. **11**(1), 20081 (2021). https://doi.org/10.1038/s41598-021-99423-5
22. Matsumoto, T., Ohata, W., Benureau, F.C.Y., Tani, J.: Goal-directed planning and goal understanding by extended active inference: evaluation through simulated and physical robot experiments. Entropy **24**(4) (2022). https://doi.org/10.3390/e24040469
23. Matsumura, T., Esaki, K., Mizuno, H.: Empathic active inference: active inference with empathy mechanism for socially behaved artificial agent. In: ALIFE 2022: The 2022 Conference on Artificial Life (2022). https://doi.org/10.1162/isal_a_00496,18
24. Meo, C., Franzese, G., Pezzato, C., Spahn, M., Lanillos, P.: Adaptation through prediction: multisensory active inference torque control. IEEE Trans. Cogn. Dev. Syst. **15**(1), 32–41 (2023). https://doi.org/10.1109/TCDS.2022.3156664
25. Millidge, B.: Deep active inference as variational policy gradients. J. Math. Psychol. **96**, 102348 (2020). https://doi.org/10.1016/j.jmp.2020.102348
26. Oliver, G., Lanillos, P., Cheng, G.: An empirical study of active inference on a humanoid robot. IEEE Trans. Cogn. Dev. Syst. **14**(2), 462–471 (2022). https://doi.org/10.1109/TCDS.2021.3049907
27. Palacios, E.R., Razi, A., Parr, T., Kirchhoff, M., Friston, K.: On Markov blankets and hierarchical self-organisation. J. Theor. Biol. **486**, 110089 (2020). https://doi.org/10.1016/j.jtbi.2019.110089
28. Parr, T., Friston, K.J.: Active inference and the anatomy of oculomotion. Neuropsychologia **111**, 334–343 (2018). https://doi.org/10.1016/j.neuropsychologia.2018.01.041
29. Parr, T., Pezzulo, G., Friston, K.J.: Active Inference: The Free Energy Principle in Mind, Brain, and Behavior. The MIT Press (2022). https://doi.org/10.7551/mitpress/12441.001.0001
30. Pearl, J.: Probabilistic Reasoning in Intelligent Systems: Networks of Plausible Inference. Morgan Kaufmann (1988). https://dl.acm.org/doi/10.5555/534975
31. Pezzato, C., Corbato, C.H., Bonhof, S., Wisse, M.: Active inference and behavior trees for reactive action planning and execution in robotics. IEEE Trans. Rob. **39**(2), 1050–1069 (2023). https://doi.org/10.1109/TRO.2022.3226144
32. Taniguchi, T., et al.: World models and predictive coding for cognitive and developmental robotics: frontiers and challenges. Adv. Rob. **37**(13), 780–806 (2023). https://www.tandfonline.com/doi/full/10.1080/01691864.2023.2225232
33. Ueltzhöffer, K.: Deep active inference. Biol. Cybern. **112**(6), 547–573 (2018). https://doi.org/10.1007/s00422-018-0785-7

34. Wauthier, S.T., Vanhecke, B., Verbelen, T., Dhoedt, B.: Learning generative models for active inference using tensor networks. In: Buckley, C.L., et al. (eds.) Active Inference, pp. 285–297. Springer, Cham (2023). https://doi.org/10.1007/978-3-031-28719-0_20
35. Wei, R., et al.: World model learning from demonstrations with active inference: application to driving behavior. In: Buckley, C.L., et al. (eds.) Active Inference, pp. 130–142. Springer, Cham (2023). https://doi.org/10.1007/978-3-031-28719-0_9
36. Weng, J., Zhang, Y.: Developmental robots-a new paradigm. Technical report, Michigan State University, East Lansing, Department of Computer Science (2005). https://apps.dtic.mil/sti/citations/ADA437286
37. Catal, O., Verbelen, T., Nauta, J., Boom, C.D., Dhoedt, B.: Learning perception and planning with deep active inference. In: ICASSP 2020–2020 IEEE International Conference on Acoustics, Speech and Signal Processing (ICASSP), pp. 3952–3956 (2020). https://doi.org/10.1109/ICASSP40776.2020.9054364

Decision-Making and Control

Towards Metacognitive Robot Decision Making for Tool Selection

Ajith Anil Meera[1(✉)] and Pablo Lanillos[1,2]

[1] Donders Institute, Department of Artificial Intelligence, Radboud University Nijmegen, Nijmegen, The Netherlands
ajith.anilmeera@donders.ru.nl

[2] Cajal International Center for Neuroscience, Spanish National Research Council, Madrid, Spain

Abstract. The capability to self-asses our performance before doing a task is essential for the decision making process, e.g., when selecting the most suitable tool for a given task. While this form of awareness has been identified in humans as metacognitive performance (thinking about the performance), robots still lack this cognitive ability. This awareness has a potential to enhance their embodied decision power, robustness and safety. Here, we take a step in this direction by proposing a novel synthetic model that unites active inference with some ideas from metacognition. We (mathematically) identify three main components that contribute to the agent's self-evaluation when making a decision: i) its performance for task completion, ii) its control effort towards task completion, and, very importantly and novel, iii) its self-confidence about the decision. We further show that these quantities are seamlessly balanced inside the free energy objective. As a proof of concept, we framed our theoretical account within the tool selection problem as a use case. Results show that the agent is able to select the best tool—modelled as spring-mass-damper systems—given three types of control tasks: attain a goal position, velocity and acceleration. Interestingly, the proposed tool selection criteria prioritises the performance during a hard task, and self-confidence during an easy task. Furthermore, we discuss how our mathematical framework can be generalized for tool/model optimization and invention.

Keywords: Active Inference · Tools · Metacognition · Robotics

1 Introduction

One crucial difference in decision making between humans and machines is the human's capacity to self-evaluate their performance before (predictively) and after (postdictively) doing the task. This ability of thinking about their performance—or metacognitive performance [4]—provides a powerful second order decision making where confidence plays a primary role. This confidence monitoring has been described as an "independent" cognitive process that encodes how good you think you are at performing a task, and that affects

C. L. Buckley et al. (Eds.): IWAI 2023, CCIS 1915, pp. 31–42, 2024.
https://doi.org/10.1007/978-3-031-47958-8_3

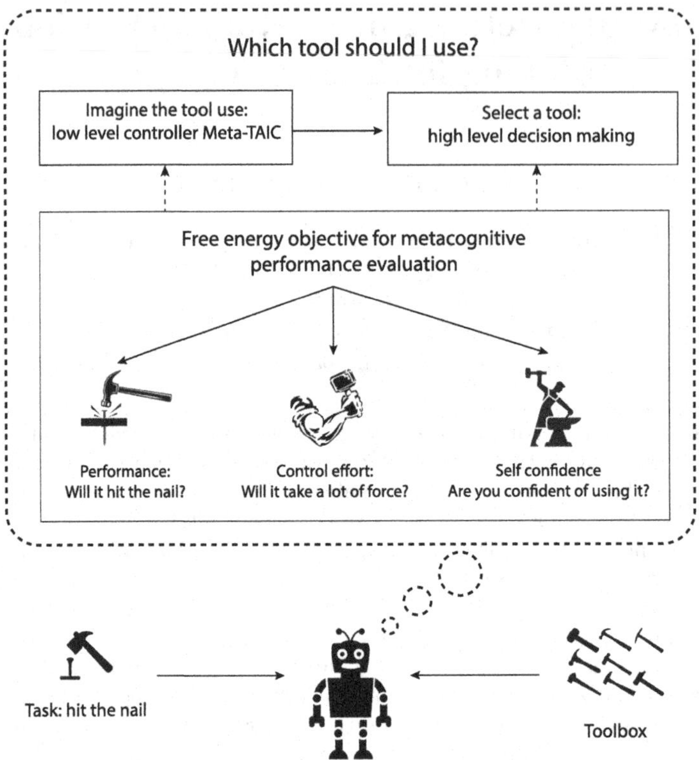

Fig. 1. An intuitive illustration of the proposed tool selection scheme based on the metacognitive performance evaluation. The agent selects a tool from a given tool-set to complete the task, based on its imagined performance towards task completion, the required control effort and its self confidence in using the tool. In this work, we consider the tools as spring mass damper systems for modelling. The illustrative example is only for conceptual clarity.

the decision made [3]. For example, given two routes with the same travel distance, the route that you are more confident might be selected, as this decision offers less uncertainty in reaching the destination. Conversely, in robot decision making (e.g., optimal control), with rare exceptions, task performance (or task completion) is only taken into account. For instance, state of the art controllers like LQR optimise the sum of weighted quadratic cost[1] for states and control input [2]. This cost is purely performance driven and does not take the confidence levels of the control signal into account. We propose that incorporating metacognitive capabilities within robot control and decision making is of prime importance for the development of brain-inspired robot controllers [11]. Such agents will be able to solve complex cognitive tasks using self-assesment as a proxy for both high-level and low-level decisions.

[1] $J = \int_0^\infty (x^T Qx + u^T Ru + 2x^T Nu)dt$ where Q, R and N are weights.

With the aim of stepping closer to a metacognitive robot decision making, we contribute with a decision making model that can balance between performance, control effort and, crucially, self-confidence, grounded on the fundamentals of the Free Energy Principle (FEP) [5]. Particularly, we propose i) a low-level (force) controller design for task completion, based on continuous Active Inference [9], ii) a closed form solution to compute the agent's self-confidence, and iii) a high-level decision making criteria that balances between the performance, control effort and self-confidence of the low-level controller—e.g, for selecting which tool is the best. While there have been other attempts to model metacognition in discrete active inference [7] this is, to the best of our knowledge, the first model in continuous state and action space that is able to incorporate self-confidence evaluation within the low level control and the high-level decision making.

As a practical use case to validate our proposal, we focus on the tool selection problem, where the agent has to select the best tool given a goal. Robots that are aware (or capable to self-evaluate) of the low level control to select the right tool for the given task is a challenging and impactful problem. For example, robots autonomously selecting the right spanner from a tool kit for tightening the bolts is expected to improve the process automation [10]. Besides, addressing tool use may be useful to validate current metacognitive theories about human behaviour.

In this paper, we provide the mathematical description of our proposal to solve the tool selection problem using metacognitive performance capabilities, followed by its evaluation in simulated experiments. The results show how the agent selects the best tool—modelled as a spring-mass-damper (SMD) system—using its self-confidence, under three types of control tasks: attain a goal position, velocity and acceleration. Figure 1 shows the proposed tool selection scheme.

2 Tool Selection Problem

The tool selection problem consists of selecting a tool from a set of p tools $\mathbf{T} = \{T^1, T^2, \ldots, T^p\}$, such that it best completes a task using its controller, by fulfilling the desired goal conditions. We restrict the set of possible tools to those whose dynamics can be modelled using a linear state space system of the form:

$$\dot{x} = Ax + Bu, \;\; y = Cx, \tag{1}$$

where $A \in \mathbb{R}^{n\times n}$, $B \in \mathbb{R}^{n\times r}$ and $C \in \mathbb{R}^{m\times n}$ are the matrices defining the system dynamics, $u \in \mathbb{R}^{r\times 1}$ is the control input to the system, $x \in \mathbb{R}^{n\times 1}$ is the hidden state, and $y \in \mathbb{R}^{m\times 1}$ is the measured output. Hence, every tool is fully described by its matrices $T^i = \{A^i, B^i, C^i\}$. We consider three theoretical types of tasks $\mathbf{\Gamma} = \{\Gamma^1, \Gamma^2, \Gamma^3\}$ in terms of goal condition τ^g: reach a desired i) constant goal state x^g (e.g., reaching task), ii) constant goal state velocity $\dot{x}^g$ (e.g., screw tightening tasks), and iii) constant goal acceleration $\ddot{\mathrm{x}}^g$ (e.g., constant force tasks like lifting, pushing). This paper aims to select the tool T^i with a dynamics given in Eq. 1, for a given task Γ^j, such that the task variable τ best reaches the desired goal τ^g within a tolerance (desired goal covariance) of $\Sigma^{\tau^g} = (P^{\tau^g})^{-1}$.

3 Task Active Inference with Metacognitive Performance

We provide a solution on tool selection using a novel decision making model, hereinafter Meta Task Active Inference (Meta-TAIC), that improves previous continuous AIF controllers [9] by redefining the free energy objective to incorporate self-confidence in control, task performance and control cost. To this end we first introduce a novel low-level controller for task completion, that explicitly connects action optimization to the preferred goal state, thus allowing task completion evaluation. Second, we mathematically formalize a high-level decision making criteria that includes confidence evaluation, which allows the agent, for instance, to select the best tool for a specific task.

3.1 Free Energy Objective for Meta-TAIC

We introduce a novel form of the free energy objective for the Meta-TAIC from first principles aimed at task completion with high performance, minimal control effort and high confidence.

According to Bayes rule, the posterior distribution $p(\theta/y)$ of parameter θ, given the measurement y is given by $p(\theta/y) = p(\theta,y)/p(y)$. Since the computation of $p(y) = \int p(y,\theta)d\theta$ is intractable for large search spaces of θ, variational methods use a recognition density $q(\theta)$ to closely approximate $p(\theta/y)$ by minimizing the Kullback-Leibler (KL) divergence between both the distributions. This procedure results in the minimization of an objective function called free energy, given by [5]:

$$F = \int q(\theta)\ln p(y/\theta)p(\theta)d\theta - \int q(\theta)\ln q(\theta)d\theta. \tag{2}$$

Under the FEP, brain's perception and control follows the minimization of its free energy and active inference agents optimize F to choose the control policy via gradient descent on free energy.

We consider the problem of evaluating the control action u, to perform the task Γ^i, by controlling the task variable τ^i to reach the goal τ^{i^g}, within a desired level of uncertainty or prior covariance $\Sigma^\tau = (P^{\tau^g})^{-1}$. The recognition density is assumed to be a Gaussian distribution of the form $q(u) = \mathcal{N}(u : \mu^u, (\Pi^u)^{-1})$. The notation P is used for the prior precision (or inverse covariance) and Π is used for the conditional precision. We assume a Gaussian prior distribution on u, written as $p(u) = \mathcal{N}(u : \eta^u, (P^u)^{-1})$. The distribution $p(\tau^i/u)$ is assumed to be Gaussian distributed as $p(\tau^i/u) = \mathcal{N}(\tau^i : \tau^{i^g}, (P^{\tau^i})^{-1})$. Using the task variable as the direct measurement $y = \tau^i$ and control action as the unknown parameter $\theta = u$, upon simplification of Eq. 2 reduces the free energy (after dropping the constants) to:

$$F = \underbrace{\frac{1}{2}(\tau^i - \tau^{i^g})^T P^{\tau^i}(\tau^i - \tau^{i^g})}_{\text{performance error } U^g} + \underbrace{\frac{1}{2}(u - \eta^u)^T P^u (u - \eta^u)}_{\text{control effort } U^c} - \underbrace{\frac{1}{2}\ln|\Pi^u|}_{\text{self-confidence } H}. \tag{3}$$

The resulting free energy can be seen as a sum of three terms: i) performance measure U^g (the prior precision weighted deviation of the task variable from the goal), ii) control cost U^c (the prior precision weighted control effort), and iii) confidence measure H (the level of confidence in the chosen control action). When $\eta^u = 0$, minimizing F implies, maximizing performance, minimizing control effort and maximizing confidence. This objective can be used to design the controller.

3.2 Controller Design and Its Self-confidence

We optimize the control actions by gradient descent on free energy objective. Under this scheme, the discrete time update rule for Meta-TAIC is written as a function of the first two gradients of free energy as:

$$u(t+dt) = u(t) + \left(e^{-k^l \frac{\partial^2 F}{\partial u^2} dt} - I\right)\left(\frac{\partial^2 F}{\partial u^2}\right)^{-1} \frac{\partial F}{\partial u}, \tag{4}$$

where k^l is the learning rate. Inspired from the dynamic expectation maximization algorithm [6], we propose a closed form solution for the optimal precision of control action[2], from the second gradient of $U^g + U^c$ (following an analogous mathematical derivation from [1]):

$$\Pi^u = \frac{\partial^2 (U^g + U^c)}{\partial u^2}. \tag{5}$$

Equation 3, 4 and 5 together represent our controller design and its self confidence in action. The presence of the measure of agent's confidence in action, third term in Eq. 3 and its closed form computation (Eq. 5) is the novelty of this work. The agent is not only aware of its decisions u, but also of its second order judgement or confidence in decisions (Π^u), making it metacognitive. In the next section, the agent will be equipped with a metacognitive decision making capability for the tool selection problem.

3.3 Tool Selection Criteria Using Free Energy

In this section, we propose a high-level decision making criteria for tool selection based on the objective function formulated in Eq. 3. The criteria involves selecting the tool T^i that minimizes the free energy integral ($\bar{F} = \int F dt$) for the given task Γ^j, within the stipulated time $T0$ as:

$$T^i = \underset{\mathbf{T^i}}{\arg\min} \int_0^{T0} F(T, \Gamma^j) dt \tag{6}$$

The chosen tool maximises the task performance with minimal control effort and maximum confidence in action, leading to the task completion. In addition to task performance, the agent is now aware of its self confidence in actions while using the tool, making the decision making process metacognitive.

[2] Evaluated by also using a mean field term ($W = \frac{1}{2} trace(\Sigma^u \frac{\partial^2 (U^g + U^c)}{\partial u^2})$) in the free energy in Eq. 3, and differentiating F with Σ^u and equating it to 0. W is omitted from F in this work for mathematical simplicity.

3.4 Task Specific Free Energy Gradients

This section describes the free energy expression and its gradients for the tool dynamics given in Eq. 1 for three tasks. These gradients are necessary for the update rule of the controller in Eq. 4.

Constant Goal State Γ^1. The free energy of an agent that acts to reach a desired goal state ($\tau^g = x^g$) with a precision (inverse covariance) of P^{x^g} can be written as:

$$F = \frac{1}{2}(x - x^g)^T P^{x^g}(x - x^g) + \frac{1}{2}(u - \eta^u)^T P^u(u - \eta^u) - \frac{1}{2}\ln|\Pi^u|. \quad (7)$$

Differentiating it by u yields the gradients of free energy as:

$$\frac{\partial F}{\partial u} = (x - x^g)^T P^{x^g}\frac{\partial x}{\partial u} + u^T P^u, \quad \frac{\partial^2 F}{\partial u^2} = \frac{\partial x}{\partial u}^T P^{x^g}\frac{\partial x}{\partial u} + P^u. \quad (8)$$

Constant Goal State Velocity Γ^2. The free energy of an agent taking actions to reach a desired goal state velocity ($\tau^g = \dot{x}^g$) with precision $P^{\dot{x}^g}$, using a tool with the dynamics $\dot{x} = Ax + Bu$, is:

$$\begin{aligned} F &= \frac{1}{2}(\dot{x} - \dot{x}^g)^T P^{\dot{x}^g}(\dot{x} - \dot{x}^g) + \frac{1}{2}(u - \eta^u)^T P^u(u - \eta^u) - \frac{1}{2}\ln|\Pi^u| \\ &= \frac{1}{2}\Big[(Ax + Bu - \dot{x}^g)^T P^{\dot{x}^g}(Ax + Bu - \dot{x}^g) + (u - \eta^u)^T P^u(u - \eta^u) - \ln|\Pi^u|\Big] \end{aligned} \quad (9)$$

Differentiating it with u yields the two gradients of free energy as:

$$\begin{aligned} \frac{\partial F}{\partial u} &= (Ax + Bu - \dot{x}^g)^T P^{\dot{x}^g}(A\frac{\partial x}{\partial u} + B) + u^T P^u \\ \frac{\partial^2 F}{\partial u^2} &= (A\frac{\partial x}{\partial u} + B)^T P^{\dot{x}^g}(A\frac{\partial x}{\partial u} + B) + P^u \end{aligned} \quad (10)$$

Constant Goal State Acceleration Γ^3. The free energy of an agent trying to reach a desired goal state acceleration ($\tau^g = \ddot{x}^g$) with precision $P^{\ddot{x}^g}$ is:

$$F = \frac{1}{2}(\ddot{x} - \ddot{x}^g)^T P^{\ddot{x}^g}(\ddot{x} - \ddot{x}^g) + \frac{1}{2}(u - \eta^u)^T P^u(u - \eta^u) - \frac{1}{2}\ln|\Pi^u| \quad (11)$$

Substituting $\dot{x} = Ax + Bu$ and $\ddot{x} = A\dot{x} + B\dot{u}$ yields:

$$\begin{aligned} F &= \frac{1}{2}\Big[(A\dot{x} + B\dot{u} - \ddot{x}^g)^T P^{\ddot{x}^g}(A\dot{x} + B\dot{u} - \ddot{x}^g) + (u - \eta^u)^T P^u(u - \eta^u) - \ln|\Pi^u|\Big] \\ &= \frac{1}{2}\Big[(A^2x + ABu + B\dot{u} - \ddot{x}^g)^T P^{\ddot{x}^g}(A^2x + ABu + B\dot{u} - \ddot{x}^g) + \\ &\quad (u - \eta^u)^T P^u(u - \eta^u) - \ln|\Pi^u|\Big] \end{aligned} \quad (12)$$

Differentiating it with u yields:

$$\frac{\partial F}{\partial u} = (A^2x + ABu + B\dot{u} - \ddot{x}^g)^T P^{\ddot{x}^g}(A^2\frac{\partial x}{\partial u} + AB + B\frac{\partial \dot{u}}{\partial u}) + (u - \eta^u)^T P^u$$
$$\frac{\partial^2 F}{\partial u^2} = (A^2\frac{\partial x}{\partial u} + AB + B\frac{\partial \dot{u}}{\partial u})^T P^{\ddot{x}^g}(A^2\frac{\partial x}{\partial u} + AB + B\frac{\partial \dot{u}}{\partial u}) + P^u \tag{13}$$

The Eqs. 8, 10 and 13 along with the update rule in Eq. 4 show that the control action for all three tasks using Meta-TAIC is independent of the confidence term Π^u for a linear state space system. In the next section, we provide a proof of concept for the tool selection criteria using simulation experiments.

4 Simulation Results

This section aims to demonstrate the working of Meta-TAIC and the tool selection criteria, with an SMD as the tool used. The MATLAB code used for the simulation is available at: https://github.com/ajitham123/mTAIC_IWAI2023.

4.1 Task Specific Π^u for SMD as the Tool

This section describes the closed-form computation of the self-confidence tailored for the three different tasks. For the sake of simplicity to provide the proof of concept, we define all tools as Spring Mass Damper Systems (SMDs). The system matrices of an SMD is given by:

$$A = \begin{bmatrix} 0 & 1 \\ -\frac{k}{m} & \frac{-b}{m} \end{bmatrix}, \; B = \begin{bmatrix} 0 \\ \frac{1}{m} \end{bmatrix}, \; C = \begin{bmatrix} 1 \\ 0 \end{bmatrix} \tag{14}$$

Using the mathematical formulations in Sect. 3, the precision of action can be computed specifically for the SMD as a tool, for all the three tasks (refer Appendix A for the derivation):

i) Task 1, constant state position

$$\Pi^u = P^u + \frac{p^{x^g}}{k^2} \tag{15}$$

ii) Task 2, constant state velocity

$$\Pi^u = P^u + p^{\dot{x}^g}\left(\frac{k+b-1}{m}\right)^2 \tag{16}$$

iii) Task 3, constant state acceleration

$$\Pi^u = P^u + p^{\ddot{x}^g}\left(\frac{k-1}{m}\right)^2. \tag{17}$$

From Eqs. 15, 16 and 17, it is evident that Π^u is independent of u for all tasks. Intuitively, this means that the confidence in decisions is independent of the decision itself, which is line with the literature on biological metacognition [8]. This completes the proof for mutual exclusivity of u and Π^u within Meta-TAIC for SMD as the tool.

4.2 Performance Evaluation of the Controller

This section shows the effectiveness of our controller in completing all the three tasks. Figure 2 shows the performance of Meta-TAIC during task 1, using all three tools. Tool 1 (in blue) performs the best by quickly taking the SMD to the constant goal state $x^g = \begin{bmatrix} 0.5 \\ 0 \end{bmatrix}$ (in dashed black), with minimal control effort. Intuitively, the solution to keep an SMD at a constant position is by applying a constant force. Meta-TAIC comes up with this solution for a converging u in Fig. 2.

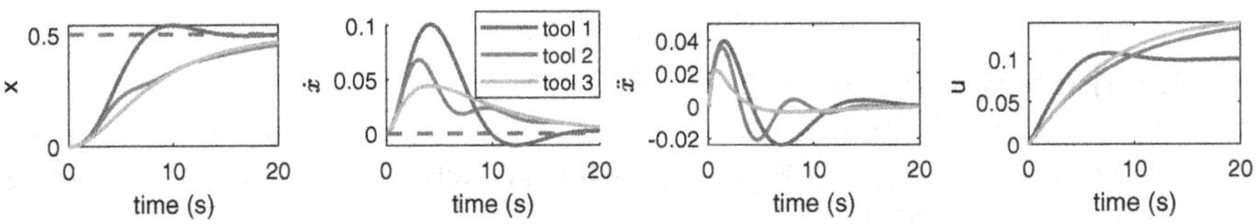

Fig. 2. Meta-TAIC takes the SMD to a constant goal position (task 1) using all tools. All tools reach the goal position (marked by dotted black in Fig. 2a) with a final zero velocity (in Fig. 2b) and a final zero acceleration (in Fig. 2c). (Color figure online)

Similarly, Fig. 3 and 4 shows the success of our controller in completing tasks 2 and 3 using all three tools, by taking the SMD to the constant goal state velocity $\dot{x}^g = \begin{bmatrix} 0.5 \\ 0 \end{bmatrix}$ and constant goal state acceleration $\ddot{x}^g = \begin{bmatrix} 0.5 \\ 0 \end{bmatrix}$ respectively. The solution for an SMD to attain a constant goal velocity is by linearly increasing the force (u in Fig. 3), and to attain a constant goal acceleration is by quadratically increasing the force (u in Fig. 4). This confirms the correct working of our proposed Meta-TAIC controller for all tasks and tools. The details of the simulation setup is given in Appendix B.

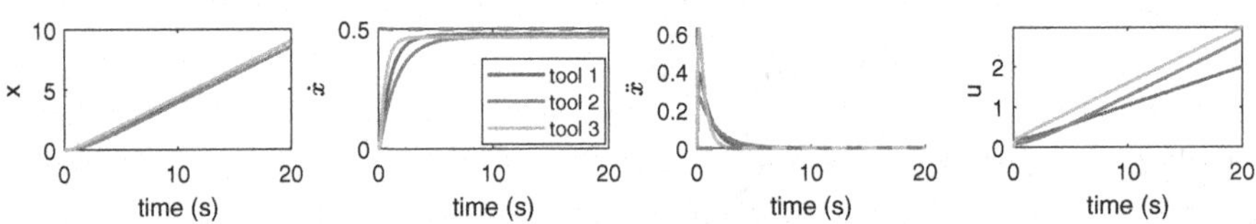

Fig. 3. Meta-TAIC takes the SMD to a constant goal velocity (task 2) using all tools. All tools reach the goal velocity (marked by dotted black in Fig. 3b) with a final zero acceleration (in Fig. 3c).

4.3 Tool Selection

This section aims to use the results of Meta-TAIC from the previous section to demonstrate the functioning of our tool selection criteria introduced in Sect. 3.3.

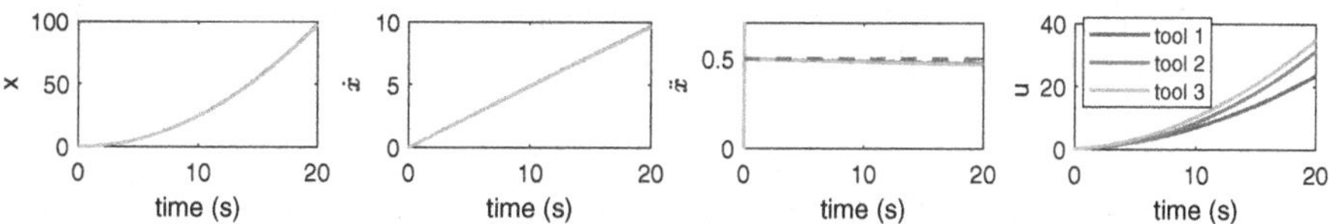

Fig. 4. Meta-TAIC takes the SMD to a constant goal acceleration (task 3) using all tools. The graphs are coinciding for all tools. All tools reach the goal acceleration (marked by dotted black in Fig. 4c).

Table 1 shows the free energy integral ($\bar{F}$) for three tasks when three tools with different parameters were used. The tool selection based on the minimization of $\bar{F}$ results in tool 1 for task 1, tool 2 for task 2 and tool 3 for task 3. The tool selected for the given task with minimum $\bar{F}$ also has the maximum Π^u. Intuitively, this reflects the fact that the agent is more confident about task completion using the selected tool.

Table 1. The free energy integral ($\bar{F}$) and the precision over action for three different tools for three different tasks.

Tool	Tool parameters			$\bar{F}$ for task			Π^u for task		
	k(N/m)	b(Ns/m)	m(kg)	Task 1	Task 2	Task 3	Task 1	Task 2	Task 3
Tool 1	0.2	0.4	0.4	**1384.7**	−678.5	−3678	**.25**	2	40
Tool 2	0.3	0.2	0.3	2194.5	**−1311**	−3984	.11	**3.78**	54
Tool 3	0.3	0.6	0.2	2194.6	−208	**−4792**	.11	1.25	**122**

4.4 Performance vs Confidence

This section aims to illustrate the capability of our tool selection criteria to balance between performance and confidence. The same simulation setup in the previous section was repeated for task 2 under two sets of goal state velocities: i) easy goal with $\dot{x}^g = \begin{bmatrix} .5 \\ 0 \end{bmatrix}$ and ii) hard goal with $\dot{x}^g = \begin{bmatrix} 10 \\ 0 \end{bmatrix}$. Table 2 shows the contribution of performance, control effort and confidence on free energy. With minimal $\bar{F}$, tool 2 is selected for the easy task and tool 1 is selected for the hard task. For the easy task, since all tools perform reasonably well with similar control effort, the confidence plays the dominant role in shaping the decision making for tool selection as per our criteria. However, for a hard task, our criteria prioritises performance over confidence for tool selection. This shows the capability of our tool selection criteria to balance between performance and confidence within the free energy objective.

Table 2. The contribution of free energy components for task 2 under two hardness levels. Tool 2 is selected for the easy task and tool 1 for the hard task.

Tool	Easy task				Hard task			
	$\bar{U}^g$	$\bar{U}^c$	$\bar{H}$	$\bar{F}$	$\bar{U}^g$	$\bar{U}^c$	$\bar{H}$	$\bar{F}$
Tool 1	13.6	.01	−692	−678.5	5442	5.5	−692	**4756**
Tool 2	16	.02	−1327	**−1311**	6391	9	−1327	5073
Tool 3	14.8	.03	−223	−208	5925	12.7	−223	5714

5 Conclusion

The capability of a robot to evaluate its self-confidence in the decisions made is fundamental to the development of brain-inspired agents with metacognitive capabilities. In this work, we proposed a novel controller (Meta-TAIC) that can balance between performance, control action and its confidence in control. Using the free energy formulations, we introduced a closed form expression for an agent's confidence in decisions. We used it to propose a high-level decision making criteria with the capability of metacognitive performance for task completion, and applied it to the tool selection problem. Through simulation experiments on a spring damper system, we showed that our controller achieved the goals for the given tasks. The tool selection criteria selected different tools for different tasks by balancing between performance, control action and confidence in control. Interestingly, the framework could be easily extended for tool optimization—to find an optimal tool for a given task. One of the limitations of our approach is the restriction of tools with linear dynamics. Future research will focus more complex tool dynamics and using self-confidence as a proxy for optimizing new tools.

Acknowledgements. This work was supported by the Metatool project, European Innovation Council through the Pathfinder Challenges grant No. 101070940.

Appendix

A Task Specific Computation of Π^u for an SMD

This section aims to compute the free energy gradients and the precision on actions (Π^u) for all the three tasks, specific to the SMD system.

Constant State Position Task. Differentiating Eq. 1 with u and substituting $\frac{\partial \dot{x}}{\partial u} = 0$ in it yields $\frac{\partial x}{\partial u} = -(A^{-1}B)^T$. Further substituting it in Eq. 8 yields the free energy gradients (necessary for the meta-TAIC update rule) as:

$$\frac{\partial F}{\partial u} = -(x - x^g)^T P^{x^g}(A^{-1}B) + u^T P^u, \quad \frac{\partial^2 F}{\partial u^2} = (A^{-1}B)^T P^{x^g} A^{-1}B + P^u \quad (18)$$

Substituting A, B, C from Eq. 14 as per Eq. 5, using $\frac{\partial^2(U^g+U^c)}{\partial u^2} = \frac{\partial^2 F}{\partial u^2}$, yields the precision on action for the constant goal position task as:

$$\Pi^u = P^u + \frac{p^{x^g}}{k^2} \tag{19}$$

Constant State Velocity Task. The agent makes an assumption about the consequence of its action on the state evolution as $\frac{\partial x}{\partial u} = \left[\begin{smallmatrix}1\\1\end{smallmatrix}\right]$. Substituting it in Eq. 10 results in:

$$\begin{aligned}\frac{\partial F}{\partial u} &= (Ax + Bu - \dot{x}^g)^T P^{\dot{x}^g}(A\left[\begin{smallmatrix}1\\1\end{smallmatrix}\right] + B) + u^T P^u, \\ \frac{\partial^2 F}{\partial u^2} &= (A\left[\begin{smallmatrix}1\\1\end{smallmatrix}\right] + B)^T P^{\dot{x}^g}(A\left[\begin{smallmatrix}1\\1\end{smallmatrix}\right] + B) + P^u\end{aligned} \tag{20}$$

Substituting A, B, C from Eq. 14, yields the precision on action as:

$$\Pi^u = P^u + p^{\dot{x}^g}\left(\frac{k+b-1}{m}\right)^2 \tag{21}$$

Constant State Acceleration Task. The agent makes the assumptions $\frac{\partial x}{\partial u} = \left[\begin{smallmatrix}1\\0\end{smallmatrix}\right]$, $\frac{\partial \dot{u}}{\partial u} = 0$ and $\dot{u} = \frac{u(t-1)-u(t-2)}{dt}$, resulting in:

$$\begin{aligned}\frac{\partial F}{\partial u} &= (A^2x + ABu + B\dot{u} - \ddot{x}^g)^T P^{\ddot{x}^g}(A^2\left[\begin{smallmatrix}1\\0\end{smallmatrix}\right] + AB) + u^T P^u \\ \frac{\partial^2 F}{\partial u^2} &= (A^2\left[\begin{smallmatrix}1\\0\end{smallmatrix}\right] + AB)^T P^{\ddot{x}^g}(A^2\left[\begin{smallmatrix}1\\0\end{smallmatrix}\right] + AB) + P^u\end{aligned} \tag{22}$$

Substituting A, B, C from Eq. 14, yields the precision on action as:

$$\Pi^u = P^u + p^{\ddot{x}^g}\left(\frac{k-1}{m}\right)^2 \tag{23}$$

B Simulation settings

The parameters of the tools are: i) tool 1 with $k = 0.2N/m$, $m = 0.4kg$, $b = 0.4Ns/m$, ii) tool 2 with $k = 0.3N/m$, $m = 0.3kg$, $b = 0.2Ns/m$ and iii) tool 3 with $k = 0.3N/m$, $m = 0.2kg$, $b = 0.6Ns/m$. The simulation was run for a total time $T0 = 20s$ with a sampling time of $dt = 0.01s$. The prior on u is selected with a mean $\eta^u = 0$ and low precision $P^u = 10^{-5}$. Task 1 has a goal state $x^g = \left[\begin{smallmatrix}0.5\\0\end{smallmatrix}\right]$ with precision $P^{x^g} = \left[\begin{smallmatrix}0.01 & 0\\0 & 0.01\end{smallmatrix}\right]$, task 2 has a goal state velocity $\dot{x}^g = \left[\begin{smallmatrix}0.5\\0\end{smallmatrix}\right]$ with precision $P^{\dot{x}^g} = \left[\begin{smallmatrix}1 & 0\\0 & 1\end{smallmatrix}\right]$, and task 3 has a goal state acceleration $\ddot{x}^g = \left[\begin{smallmatrix}0.5\\0\end{smallmatrix}\right]$ with precision $P^{\ddot{x}^g} = \left[\begin{smallmatrix}10 & 0\\0 & 0\end{smallmatrix}\right]$. A learning rate of $k^l = 1$ was used for task 1, and $k^l = 4$ was used for task 2 and 3.

References

1. Anil Meera, A., Wisse, M.: Dynamic expectation maximization algorithm for estimation of linear systems with colored noise. Entropy **23**(10), 1306 (2021)
2. Baltieri, M., Buckley, C.L.: On kalman-bucy filters, linear quadratic control and active inference. arXiv preprint arXiv:2005.06269 (2020)
3. Fleming, S.M., Daw, N.D.: Self-evaluation of decision-making: a general Bayesian framework for metacognitive computation. Psychol. Rev. **124**(1), 91 (2017)
4. Fleming, S.M., Dolan, R.J.: The neural basis of metacognitive ability. Philos. Trans. R. Soc. B Biol. Sci. **367**(1594), 1338–1349 (2012)
5. Friston, K.: The free-energy principle: a unified brain theory? Nat. Rev. Neurosci. **11**(2), 127–138 (2010)
6. Friston, K.J., Trujillo-Barreto, N., Daunizeau, J.: Dem: a variational treatment of dynamic systems. Neuroimage **41**(3), 849–885 (2008)
7. Hesp, C., Smith, R., Parr, T., Allen, M., Friston, K.J., Ramstead, M.J.: Deeply felt affect: the emergence of valence in deep active inference. Neural Comput. **33**(2), 398–446 (2021)
8. Khalvati, K., Kiani, R., Rao, R.P.: Bayesian inference with incomplete knowledge explains perceptual confidence and its deviations from accuracy. Nat. Commun. **12**(1), 5704 (2021)
9. Lanillos, P., et al.: Active inference in robotics and artificial agents: survey and challenges. arXiv preprint arXiv:2112.01871 (2021)
10. Qin, M., Brawer, J.N., Scassellati, B.: Robot tool use: a survey. Front. Robot. AI **9**, 369 (2022)
11. Sanz, R., López, I., Rodríguez, M., Hernández, C.: Principles for consciousness in integrated cognitive control. Neural Netw. **20**(9), 938–946 (2007)

Understanding Tool Discovery and Tool Innovation Using Active Inference

Poppy Collis[1(✉)], Paul F. Kinghorn[1], and Christopher L. Buckley[1,2]

[1] School of Engineering and Informatics, University of Sussex, Brighton, UK
{pzc20,p.kinghorn,c.l.buckley}@sussex.ac.uk
[2] VERSES AI Research Lab, Los Angeles, CA, USA

Abstract. The ability to invent new tools has been identified as an important facet of our ability as a species to problem solve in dynamic and novel environments [17]. While the use of tools by artificial agents presents a challenging task and has been widely identified as a key goal in the field of autonomous robotics, far less research has tackled the invention of new tools by agents. In this paper, (1) we articulate the distinction between tool discovery and tool innovation by providing a minimal description of the two concepts under the formalism of active inference. We then (2) apply this description to construct a toy model of tool innovation by introducing the notion of tool affordances into the hidden states of the agent's probabilistic generative model. This particular state factorisation facilitates the ability to not just discover tools but invent them through the offline induction of an appropriate tool property. We discuss the implications of these preliminary results and outline future directions of research.

Keywords: active inference · tool innovation · model factorisation · one-shot generalization

1 Introduction

Tool innovation has been identified as a core feature of human cognitive and cultural development, and has provided us with a key adaptive advantage as a species to survive adverse environments [4,17,24]. While both the use and innovation of tools was initially seen as a uniquely human capability, evidence has shown that a phylogenetically widespread variety of non-human animals engage in forms of tool manipulation, innovation and manufacture [19]. A large body of research has approached the topic of tool use in humans, animals and robotic systems [3,7,18]. Here, we define tool use to be "the exertion of control over a freely manipulable external object (the tool) with the goal of altering the physical properties of another object, substance, surface or medium (the target, which may be the tool user or another organism) via a dynamic mechanical

P. Collis and P. F. Kinghorn—These authors contributed equally to this work.

C. L. Buckley et al. (Eds.): IWAI 2023, CCIS 1915, pp. 43–58, 2024.
https://doi.org/10.1007/978-3-031-47958-8_4

interaction" [23]. However, developing the understanding of how to use a given tool is significantly different from the process of inventing a new tool.

Tool innovation refers to the process by which an agent independently constructs novel tools without relying on social demonstration or observation. This requires the ability to envision and conceptualise the appropriate tool for a given problem, while the knowledge of how to physically transform materials during construction is referred to as tool manufacture [2]. The task of tool innovation presents a challenging problem in artificial agents, yet it is one that we as humans are inherently very good at. Indeed, research indicates that we develop innovation skills at a very early age [6]. The animal innovation literature suggests that we can distinguish between two different classes of tool innovation: 1) that which arises as a result of incidental discovery where the animal then simply repeats this action in the same context and 2) that which is the result of intentional action by the animal resulting from some process of causal inference [25]. Herein, we define these two classes of innovation as *tool discovery* and *tool innovation* respectively.

Making such a distinction for both animals and human infants is challenging given the difficulty in determining the intentions driving subjects' proposed solutions to a problem [6]. While human behavioural experiments often explore the putative cognitive abilities required for tool innovation, no attempt is made to model such behaviour [8]. We therefore seek to offer a simple model of the cognitive phenomena underpinning the process of tool innovation. In the interest of a focused inquiry and to maintain conceptual clarity, we limit ourselves to being concerned with the causal reasoning involved in the process of tool innovation, while choosing to omit the challenges associated with the motor skills required to manipulate objects and manufacture tools from physical materials.

In recent years, theories which describe the brain as broadly Bayesian have gained considerable traction in the field of neuroscience. The 'Bayesian brain hypothesis' posits that perception arises as a result of Bayesian model inversion, with incoming sensory data updating these causal models of the world in accordance with Bayes' rule [10]. The theory of active inference (AIF) extends this idea and casts action, perception and learning as being underwritten by the same underlying process of Bayesian inference. Derived from first principles, the theory provides a formal account of behaviour arising as a result of the imperative to minimise of the information-theoretic quantity of surprisal. In other words, an autonomous agent is continually in the act of accumulating Bayesian model evidence ("self-evidencing") and it is from this perspective that we can understand decision-making under uncertainty [21]. AIF offers a rich description of the internal mechanisms of belief-based reasoning and principled account of the natural emergence of curious and insightful adaptive behaviour [11]. It has also recently been proposed as a framework well-suited to robotics [9]. We therefore chose to explore the concept of tool innovation using this framework.

The main contribution of this paper is (1) the articulation of the distinction between tool discovery and tool innovation within the AIF framework and (2) a minimal model of non-trivial tool innovation that requires generalised inferences about the tool structure required to solve a task. First of all, we show that with a perfect generative model, the agent can straightforwardly use tools optimally to solve a task. We then demonstrate that the agent can discover the correct tools and learn to solve the task when it is not provided with this information in its model *a priori*. Finally, we provide evidence that factorising the hidden states of the generative model into the affordances of the tool can enable the agent to conceive offline the appropriate properties of the tool required to solve the task. It is this difference between the generative model and the generative process which is key to facilitating tool invention. This enables the agent to not simply happen upon the appropriate tool during exploration of environmental contingencies, but to invent them through the induction of an appropriate tool property. We discuss the implications of these preliminary results and outline future directions of research.

2 Active Inference in Discrete State Space

In AIF, the minimisation of sensory surprisal is achieved through the minimisation of a tractable quantity called the variational free energy $\mathcal{F}$, known as (negative) evidence lower-bound (ELBO) in the variational inference literature [5]. This minimisation is performed via the maintenance of a probabilistic generative model of the environment. AIF has been widely implemented using discrete-time stochastic control processes known as partially-observable Markov decision processes (POMDPs) [9]. We therefore implement our simulations agent with an AIF framework in discrete state space using the Python package *pymdp* [13]. This specifies a standard POMDP generative model as a joint probability distribution over observations o, hidden states s, policies π and model parameters ϕ. In contrast to much of the reinforcement learning literature, a policy in this case is defined as a fixed sequence of control states u_τ for each timestep τ that together represent a plan of action of length T, $\pi = \{u_1, ..., u_T\}$ [21]. We assume the standard factorisation of the POMDP as a product of conditional (likelihood) distributions and prior distributions over a finite time horizon $[1 : T]$.

The most important distributions when specifying this generative model are the observation likelihood $P(o_\tau \mid s_\tau; \phi)$, the transition likelihood $P(s_\tau \mid$

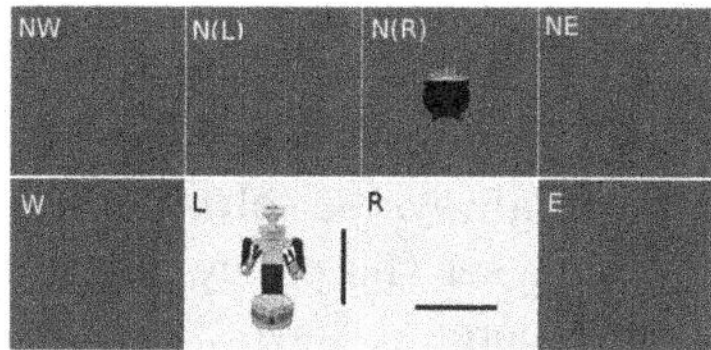

Fig. 1. The task of the agent is to reach the reward by using the tools provided. The simulation environment shows the agent (*robot*) can only move between the left and right rooms (*grey*) and the reward (shown as a pot of gold) can be placed in any of the other rooms (*blue*). A vertical tool (V) can be picked up in the left room, and a horizontal tool (H) in the right room (Color figure online)

$s_{\tau-1}, \pi; \phi)$, and the prior preference over observations $P(o_\tau)$, known in *pymdp* as the A, B and C matrices respectively. We also further factorise our representations of o_τ and s_τ into separate modalities and factors: $o_\tau = \{o_\tau^1, o_\tau^2, ..., o_\tau^M\}$ and $s_\tau = \{s_\tau^1, s_\tau^2, ..., s_\tau^F\}$ in which M is the number of modalities and F the number of hidden state factors such that the likelihood distributions can be written as:

$$P(\mathbf{o}_\tau \mid \mathbf{s}_\tau) = \prod_{m=1}^{M} P(o_\tau^m \mid \mathbf{s}_\tau) \tag{1}$$

$$P(\mathbf{s}_\tau \mid \mathbf{s}_{\tau-1}, \mathbf{u}_{\tau-1}) = \prod_{f=1}^{F} P(s_\tau^f \mid \mathbf{s}_{\tau-1}, \mathbf{u}_{\tau-1}) \tag{2}$$

We allow state factors in the transition likelihoods to depend on themselves and a specified subset of other state factors.[1] Since we are working in discrete space, the probability of states and observations can be described by a categorical probability distribution.

In this work, we consider the simple environment shown in Fig. 1. It consists of a 2×4 grid of locations in which the agent can only move between two rooms: left and right (shown here in grey). The agent is always initialised in the left-hand room. A vertical tool (V) is located in the left-hand room while a horizontal tool (H) is located in the right-hand room. In one of the remaining rooms (shown in blue), a reward is located, and it is the goal of the agent to try and reach this reward using the tools provided. For example, if the reward is in the room directly north of the right-hand room as shown, the agent is required to be in the right-hand room holding tool V in order to reach it. The agent can choose to pick up the tool if it is in the relevant room, while it may drop tools whilst it is in any room (in which case, any of the tools in the agent's possession are dropped and returned to their original rooms). If the agent already possesses a tool and picks up a different tool, this creates a compound tool (HV). The rooms directly north, east and west of the left and right rooms are known as the *adjacent rooms* and these only require the individual tools V or H to solve. The northeastern and northwestern rooms are termed the *corner rooms* and present a greater challenge for the agent as they require the construction of the compound tool (HV) to solve.

For the initial experiments, the hidden states of the environment are factorised into two factors, $s_\tau = \{s_\tau^1, s_\tau^2\}$, which consist of: room state and tool state (see Table 1). A policy length of 4 time-steps is chosen given that the task of retrieving the reward can always be solved optimally within 4 steps (for any reward location). As we have set the policy length to be 4 time-steps and we have 4 possible actions, we therefore have 256 (4^4) possible policies which we must individually evaluate by calculating the expected free energy for every time-step (see Sect. 3). In all experiments, the agent is equipped with a strong prior preference for the observation of reward in the reward modality. In terms

[1] This requires a recent branch of *pymdp* which enables this kind of factorisation. See https://github.com/infer-actively/pymdp/tree/sparse_likelihoods_111.

Table 1. Generative Model Structure

States	Factors	Dimensions	Values
Hidden states	Room	2	Left, Right
	Tool	4	Null, V, H, HV
Observations	Room	2	Left, Right
	Tool	4	Null, V, H, HV
	Reward	2	Null, Reward
Control States		4	Null, Move, Pick-up, Drop

of relative log probabilities, we specify this to be 0 for an observation of null and 50 for an observation of reward. Observations in all other modalities have a flat prior (i.e. no preference given).

3 Policy Inference

In AIF, policy inference is effectively a search procedure in which a free energy functional of expected states and observations under a policy is evaluated for each possible policy. Once we have calculated this quantity (known as the expected free energy, $\mathcal{G}$) for each policy, we can convert this into a probability distribution over the set. Action selection then simply amounts to sampling from this distribution accordingly. Policies which most minimise $\mathcal{G}$ will be assigned a higher probability and are therefore more likely to be chosen. Since the variational posterior factorises over time, we can calculate $\mathcal{G}$ for each time step independently. The expected free energy for a particular future time step under a particular policy is given by:

$$\mathcal{G}_\tau(\pi) = \mathbb{E}_Q[\ln Q(s_\tau|\pi) - \ln \tilde{P}(o_\tau, s_\tau|\pi)] \tag{3}$$

where $\tilde{P}(o_\tau, s_\tau|\pi) = P(s_\tau|o_\tau, \pi)\tilde{P}(o_\tau)$, representing a generative model that is biased to produce preferred observations (for full derivations, see [13]). $\mathcal{G}_\tau(\pi)$ can be rearranged in various ways to give intuition about what it actually represents. One such representation decomposes this free energy functional into an epistemic value (information gain) term and a pragmatic value (utility) term:

$$\mathcal{G}_\tau(\pi) \leq \underbrace{-\mathbb{E}_{Q(o_\tau|\pi)}[D_{KL}\left[Q(s_\tau|o_\tau, \pi) \parallel Q(s_\tau|\pi)\right]]}_{\text{State Information Gain}} - \underbrace{\mathbb{E}_{Q(o_\tau|\pi)}[\ln \tilde{P}(o_\tau)]}_{\text{Utility}} \tag{4}$$

Epistemic value refers to the information gain from the expected outcomes of hidden states. Given a policy, it measures the divergence between the expected states and the expected states conditioned on the observations. This gives rise to curious behaviour in which the agent is compelled to minimise uncertainty about its environment via exploration. On the other hand, the utility term simply measures the extent to which the observations expected under a policy align with the observations the agent wishes to encounter. This promotes the exploitation of knowledge in order to satisfy preference over outcomes. This trade-off between exploration and exploitation therefore naturally arises in AIF; both imperatives are cast as ways in which an agent acts to resolve uncertainty.

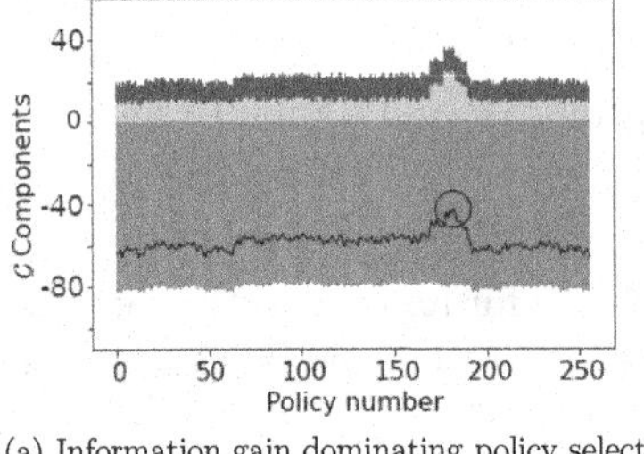

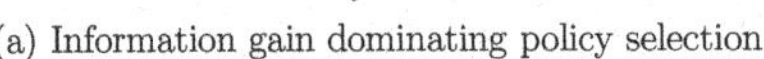

(a) Information gain dominating policy selection

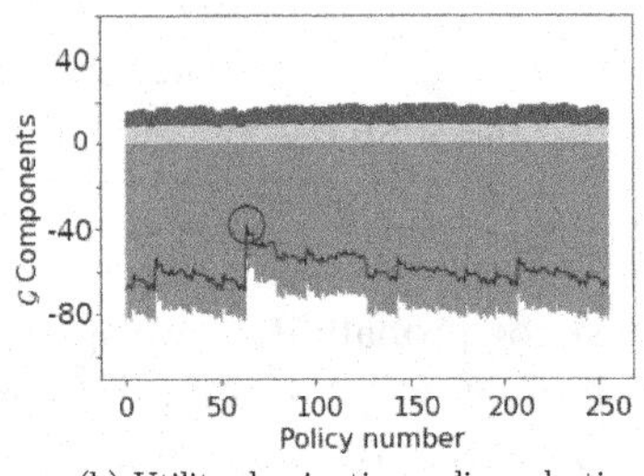

(b) Utility dominating policy selection

Fig. 2. Decomposing expected free energy $\mathcal{G}$ into respective information gain and utility contributions can elucidate the agent's intended consequences of an action. The expected free energy $\mathcal{G}$ (*black line*) is evaluated over a set of 256 policies. The components which contribute to the selection of the best policy (*circled*) are state information gain (*dark green*), parameter information gain (*light green*) and utility (*orange*). Examples shown are instances when the selected policy is a) driven by information gain as there is little variation in utility and b) driven by utility as there is little variation in information gain (Color figure online)

We can visualise this trade-off by plotting the respective utility and information gain components of the total $\mathcal{G}$. Figure 2a shows an example in which each of the policies vary little with respect to their expected utility and the policy selected has been driven by the high information gain component. In contrast, Fig. 2b shows an example of when the dominant driving force in policy selection is the utility component while information gain remains largely invariant across policies. Note that we also include a parameter information gain term which is explained in Sect. 4.

4 Parameter Inference

Learning in AIF is a process of inference over the model parameters, ϕ, which are simply the categorical likelihood distributions. We treat these parameters as something over which the agent maintains and updates beliefs (i.e. as random variables). Consider the example of an A matrix, which encodes the observation likelihood model $P(o|s)$, with the entry $A[i,j]$ representing the probability of seeing observation i given state j. There is therefore a separate categorical distribution for each state (i.e. each column sums to 1). The Dirichlet distribution is a conjugate prior for the categorical distribution, and we therefore model prior beliefs over the categorical as a Dirichlet. It can be shown that, when the agent obtains new empirical information, the Bayesian process of updating this prior is simply a count-based increase of the Dirichlet parameters according to the observation o and inferred state s [13,15]:

$$\alpha_{posterior} = \alpha_{prior} + o \otimes s \tag{5}$$

where α represents the Dirichlet parameters. Now that we are treating model parameters as random variables, we can expand $\mathcal{G}$ to include the expected parameter information gain component:

$$\mathcal{G}_\tau(\pi) \leq \underbrace{-\mathbb{E}_{Q(o_\tau|\pi)}[D_{KL}[Q(s_\tau|o_\tau,\pi) \,\|\, Q(s_\tau|\pi)]]}_{\text{State Information Gain}} \underbrace{-\mathbb{E}_{Q(o_\tau|\pi)}[D_{KL}[Q(\phi|o_\tau,\pi) \,\|\, Q(\phi|\pi)]]}_{\text{Parameter Information Gain}} \underbrace{-\mathbb{E}_{Q(o_\tau|\pi)}[\ln \tilde{P}(o_\tau)]}_{\text{Utility}} \tag{6}$$

This will drive the agent to seek observations which lead to a larger change in the categorical distribution.

5 Experiment 1: Tool Use

In the first set of experiments, the agent has a perfect probabilistic generative model of the world. This means that the correct transition likelihood and observation likelihood distributions are provided and therefore no learning is required. We then show that the agent can straightforwardly infer the optimal actions in order to achieve its goal of reaching the reward. We use this as a simple model of tool use in an autonomous agent, given the definition of tool use defined previously [23]. By this account, our simulated agent conducts tool use by acting to “exert control over” tools V, H or HV in order to “alter the physical properties” of the tool user (by extending the agent's reach) enabling it to successfully retrieve the reward. In this sense, we reduce tool use to an action sequencing problem.

Table 2. Comparing optimal number of steps required to solve each reward location with actions taken. When the generative model is perfectly known, the agent solves the task optimally

Reward Location	Optimal No. of Steps	Actions of Agent
North-left	1	Pick-up
North-right	2	Pick-up, Move
East	2	Move, Pick-up
West	3	Move, Pick-up, Move
Northeast	3	Pick-up, Move, Pick-up
Northwest	4	Pick-up, Move, Pick-up, Move / Move, Pick-up, Move, Pick-up

For each trial, we place the reward in one of the possible reward locations and allow the agent 12 time-steps in which to act in the world and obtain the reward. The agent uses all 12 time-steps, and therefore if it has found the reward, the

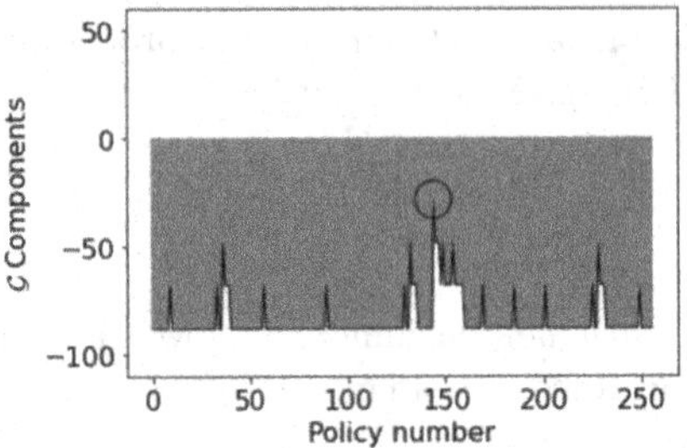

Fig. 3. When the generative model is perfectly known, the selected policy is based solely on the utility component of $\mathcal{G}$. An example of $\mathcal{G}$ (*black line*) evaluated for all 256 policies and the selected policy (circled) which is the one with the highest utility (*orange*). Note that since the generative model is fully known, and the environment is fully observed, all policies have zero information gain component (Color figure online)

optimal behaviour would be to perform an action that will keep it in the same state (i.e. the action "Null").

As expected, the agent solves the task of obtaining the reward optimally for each reward location (Table 2). Given the stochastic nature of policy selection, we note in that the agent solves the northwest room via two different methods, yet both are optimal (i.e. of length 4). Since the generative model is fully known, the agent gains no new information about states or parameters during inference. Indeed, Fig. 3 shows that if we plot the relative utility and information gain contributions to the expected free energy of each policy during action selection, we see that it only comprises of a utility component compared to Fig. 2 (i.e. there is no epistemic value contribution to $\mathcal{G}$).

6 Experiment 2: Tool Discovery

Next, we investigate the ability of the agent to learn how a particular tool solves the task. We present this as a toy example of tool discovery given that knowledge about how to create a tool arises incidentally as a result of environmental exploration. Whilst we again provide the agent with a fully known observation likelihood distribution, for the following experiment we initialise the agent with a uniformly distributed transition likelihood model. This means that the agent initially knows nothing about how states and actions effect future states. It therefore must learn these state transitions rather than being provided with this information from the outset (as in experiment 1). The agent happens upon the correct tool to use for a given reward location, and then repeats this action in the same contexts. This is in line with our previous definition of tool discovery [25].

Figure 4 shows that the number of steps the agent takes to find the reward decreases over the number of runs. In this continual learning task, each time the reward location changes (at runs 0, 10, 20, 30 and 40) it demands the learning of a new tool and we see an initial increase in the number of steps requires to solve the task. This is because the information the agent has about state transitions (i.e. how states and actions give rise to states at the next time-step) is not sufficient to solve the task. The agent therefore explores more of the environment before encountering the correct tool required to satisfy its preference for the reward observation. We then see a sharp drop after the agent has learned about the required state transitions, and the number of steps taken to solve the task quickly plateaus to the optimal number shown in Table 2.

Interestingly, as a result of the ordering in which the reward locations are presented (north-right, west, north-left, east, ...), the agent solves the north-left and east reward locations optimally from the outset. This is due to the fact that the solving of previous adjacent rooms (north-right and west) resulted from the learning of tool V and H respectively. When the agent then encounters the reward in the remaining adjacent rooms, it has already learned about the correct actions to create these tools to solve the task despite never having seen these particular reward locations before. The corner rooms require more steps despite having already learned V and H, as the agent must still discover the new tool HV. Given that the agent is always initialised in the left-hand room, the northwest corner (Fig. 4b) takes more steps to solve that the northeast corner (Fig. 4b) because it involves a more complex action sequence to retrieve the reward (see Table 2).

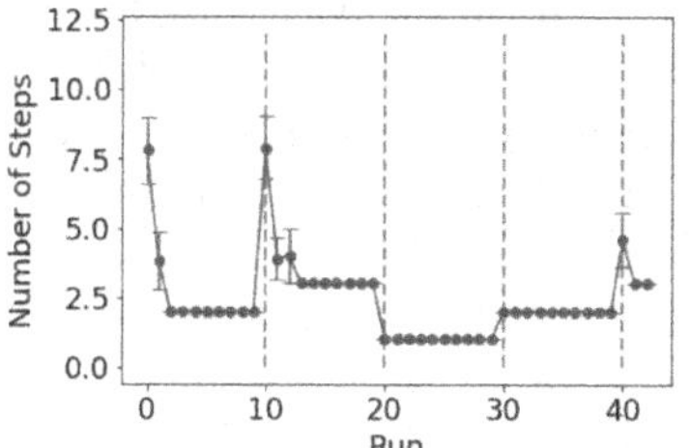

(a) Finishing with reward in northeast room

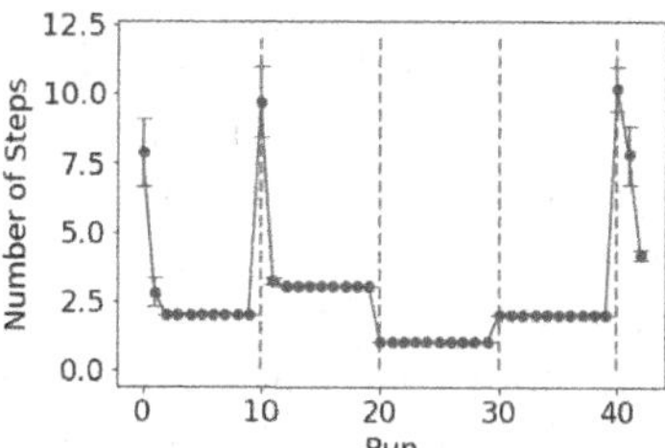

(b) Finishing with reward in northwest room

Fig. 4. The number of steps taken to solve the task for each reward location decreases quickly over runs to the optimal number of steps, reflecting the agent learning via discovery. Graphs show the mean ($\pm$ ste) number of steps to solve reward location averaged over 20 independent trials. The agent is exposed to a different reward locations every 10 runs (*dashed lines*). The reward is first located in the adjacent rooms (in the order north-right, west, north-left, east) before being presented with a) the northeast or b) the northwest room for final 3 runs (40–42). For both cases, despite learning how to create a V and H tool in the earlier runs, the agent still has to learn about the HV tool when the reward is placed in a corner room

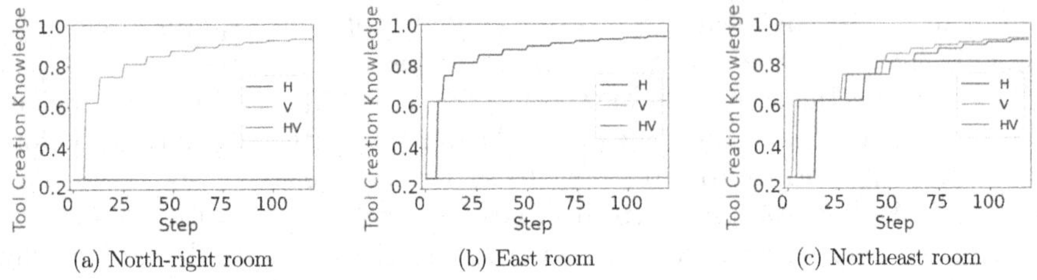

(a) North-right room (b) East room (c) Northeast room

Fig. 5. The agent only learns the tools that it needs to learn in order to solve the task. We provide a measure of how well the agent knows each tool by looking at the posterior probability associated with the correct control state (i.e. action) for creating each tool when solving for rooms a) north-right b) east and c) northeast over 125 steps

Importantly, given that the minimisation of $\mathcal{G}$ naturally incorporates two competing imperatives (utility and information gain), this means that the agent learns only the tools that it needs to learn in order to solve the task, and does not continue exploring its environment if it is able to leverage its current knowledge to effectively realise prior preferences. Figure 5a shows that for the north-right room, the agent only learns the vertical tool (V). This is because the first tool it picked up (V) allowed it to solve the task and therefore the agent did not need to continue exploring the hidden states of the environment as it had all of the knowledge it needed. Figure 5b) shows that for the east room, the agent first tried the vertical tool (V), however this did not lead to the agent observing preferred observations (reward) and therefore it does not infer the action of picking up this tool again. Instead, it pursues policies which yield high information gain (i.e. it explores new states of the environment) and finds that picking up tool H leads to a rewarding observation. By selecting policies which maximise utility, it therefore repeats this action ("pick-up") in the same context, and learns this tool with more confidence while neglecting to explore other options. Finally, Fig. 5c shows that in order to discover the compound tool (HV), the agent first happens upon tools V and H (as these tools are more likely to be stumbled across given they require a less complex sequence of actions in order to learn about them). However, these do not provide it with high utility. Since there are unknown states (such as tool HV) that provide it with high information gain, the agent continues exploring and then finds that creating the compound tool brings about its preferred observations.

A policy is selected on its value of $\mathcal{G}$ which is composed of both expected utility and expected information gain. We can visualise the evolution of this trade-off in driving policy selection during a continual learning trial. For each time-step, we see how the chosen policy ranks in the ordered list of all policies with respect to utility and the ordered list of all policies with respect to information gain. This rank provides us with a measure of the relative contributions of utility and information gain in the selection of a policy. For example, if the chosen policy ranks very highly for utility, and yet ranks very low in the context

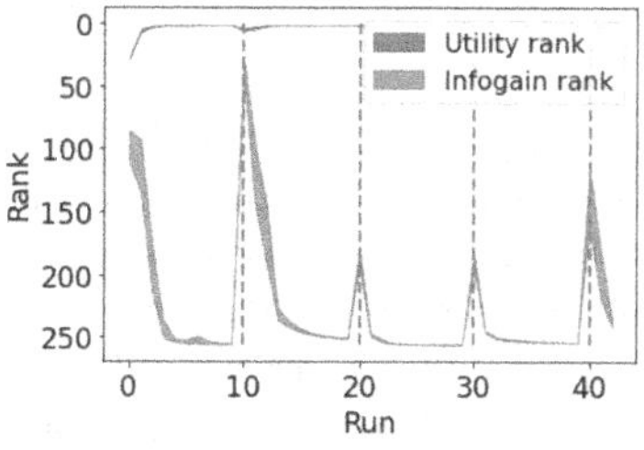

(a) Finishing with reward in northeast room

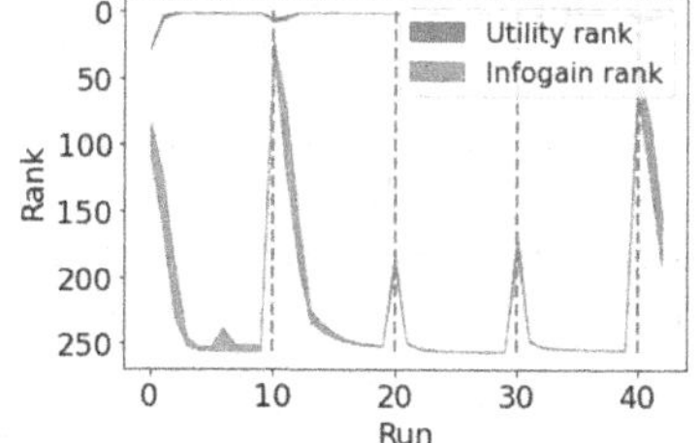

(b) Finishing with reward in northwest room

Fig. 6. Policy selection is initially dominated by information gain, but is then very quickly driven by utility as the agent learns new information. Graph shows how the selected policy ranks in the context of all possible policies in terms of utility and information gain (averaged over 20 independent trials) (best rank is 0, worst is 256). Like Fig. 4, the reward location changes every 10 runs (*dashed lines*) in the order north-right, west, north-left, east. The agent is then presented with a) the northeast or b) northwest room for the final 3 runs

of the best policies for information gain, we know that the policy (and therefore resultant action) has been selected primarily due to its high utility.

As Fig. 6 shows, for each reward location, the information gain component is initially very high and therefore dominates action selection. This is because when the reward location is changed, the state transition information is not adequate to solve the task. The gain in information quickly drops as the agent learns transitions via exploration, while the utility rank of the policy increases as it can leverage this newly learned information to seek the preferred observation of the reward. Note that at runs 20 and 30, this spike in information gain is lower that at 0 and 10. This is because the agent has already learned about creating tool V and H in the north-right and west reward locations respectively. When the agent is then presented with the novel adjacent reward locations (north-left and east), it has the advantage of already having the knowledge of how to pick up the correct tool to use to solve the problem. For the final reward location (northeast for Fig. 6a and northwest for Fig. 6b), we also see a spike in information gain. This is in agreement with Fig. 4 which shows that we do indeed see an increase in the number of steps taken to solve these final rooms. Despite having knowledge about the individual tools H and V, the agent must explore further to 'discover' the compound tool.

We have therefore shown that the agent can leverage the knowledge gained in the incidental discovery of required state transitions to solve the task. This amounts to a simple model of tool discovery behaviour in accordance with our previously defined definition.

7 Experiment 3: Tool Innovation

The following experiment investigates the concept of tool innovation in our AIF agent. In order to achieve this, the agent must be able to analyse the problem and identify the kind of the tool required to solve the task. This entails developing a grounded understanding of the objects in the environment which can then be leveraged to construct a suitable tool through a process of generalisation. For the acquisition of grounded knowledge about the world, we turn to the concept of 'affordances' from ecological psychology [12]. This refers to opportunities for action provided by the environment. In the robotics literature, this is defined as the "relationship between an actor (i.e., robot), an action performed by the actor, an object on which this action is performed, and the observed effect" [1].

We adjust our generative model to incorporate the following tool affordances into the hidden states: the horizontal reach (x-reach) and vertical reach (y-reach) afforded by each tool and the room state $s_\tau = \{s_\tau^1, s_\tau^2, s_\tau^3\}$. Each affordance state can take a binary value. We refer to this as the *Affordance Model* while the previous model which included an unfactorised tool state is referred to as the *Tool State Model.* Importantly, these affordances do not depend on one another, which allows for generalisation of learning in novel situations (i.e. the agent does not need to separately explore the x-reach state in the context of two different y-reach states). This aligns with the concept of *disentangled representations*, characterised as disjoint representations of the underlying transformation properties of the world [14]. That is, transformations that vary a subset of properties of the world state, while leaving all others invariant.

In this sense, the agent can learn solely about the tool V and tool H, and when faced with a new reward location in which it requires both a positive x-reach and y-reach, it should spontaneously produce the compound tool (HV) in an optimal way. This is a simple yet non-trivial notion of innovation in which the agent does not merely just discover a new tool (as in experiment 2). The agent is able to encounter a new situation (reward location), understand the structure of the required solution (both a non-zero x-reach and y-reach) and generate the required solution (tool HV). We can think of this as a simple example of 'one-shot' generalisation to novel stimuli [20,22].

To test this hypothesis, we have the agent learn the entries of the transition likelihood distribution model from scratch (i.e. we initialise it as a uniform distribution as in experiment 2). However, our transition likelihood now includes the new factorised tool states (see Fig. 7). In a continual learning task, we present the agent with the adjacent rooms (which only require the learning about H and V) and then test it on the northeast room (which requires tool HV).

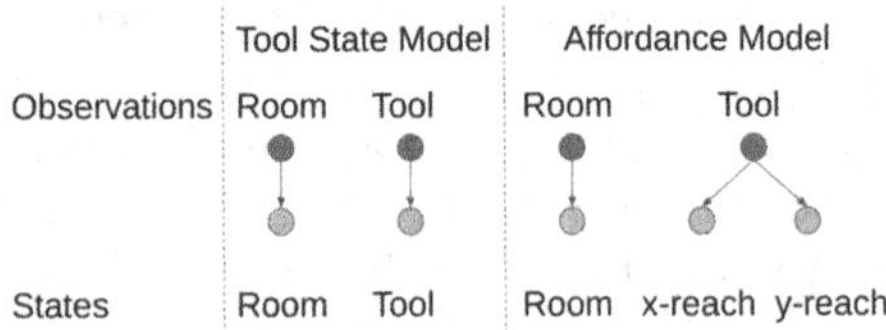

Fig. 7. In the Tool State Model used of experiment 2, there is a one to one mapping between the tools the agent observes, and the internal representations it has for them (None, V, H, HV). In the Affordance Model in experiment 3, the agent separates the lantent tool states into properties of x-reach and y-reach

Figure 8a shows that, indeed, when the Affordance Model agent has only previously learned about tools H and V, it successfully creates tool HV optimally (having never seen this observation before). With the Tool State Model in experiment 2, this task was not solved optimally (as it initially took an average of roughly 5 steps to solve). As Fig. 8d shows, this coincides with a greater information gain component driving action selection, meaning the agent is exploring in order to discover the compound tool. On the other hand the information gain component for the agent with the Affordance Model is much lower. This suggests that the factorisation of hidden states into affordances indeed equips the agent with the ability to leverage its current knowledge in order to compose relevant affordances and spontaneously 'invent' the new tool.

It is worth noting, that when repeating this experimental procedure of exposing the Affordance Model agent to the adjacent rooms and then testing on the northwest (rather than the northeast) room, the agent does not solve this optimally, but 'near-optimally'. As Fig. 8b shows, the Affordance Model agent solves this task marginally quicker than the agent with the Tool State Model, however it does not immediately find the optimal solution of 4 steps. Upon inspection of the learned transition likelihood distributions, it appears that there is a large information gain component of $\mathcal{G}$ that drives the agent to select the action 'drop' (and this is reflected in Fig. 8e). The agent has never explored what this action 'drop' does in the left-hand room with no tools, and therefore it repeats this action until it no longer yields high state information gain. Once it has learned this particular fact, it then goes on to select the optimal policy and solves the task in the next 4 steps.

To confirm that this is indeed what is causing the sub-optimal behaviour, we tailor our policy selection strategy on the critical runs. We repeat the experimental trial, but once the reward location has changed to the final northwest room, we ignore the information gain components of $\mathcal{G}$. The agent therefore selects policies based on utility alone. After this adjustment, the Affordance Model agent then solves the northwest room optimally (see Fig. 8c). Importantly, when information gain is ignored for the Tool State Model, this still does not lead the agent to solve the task optimally. This is because it does not have the required knowledge about the compound tool while the Affordance Model has all of the information it needs in order to solve the task by a process of induction.

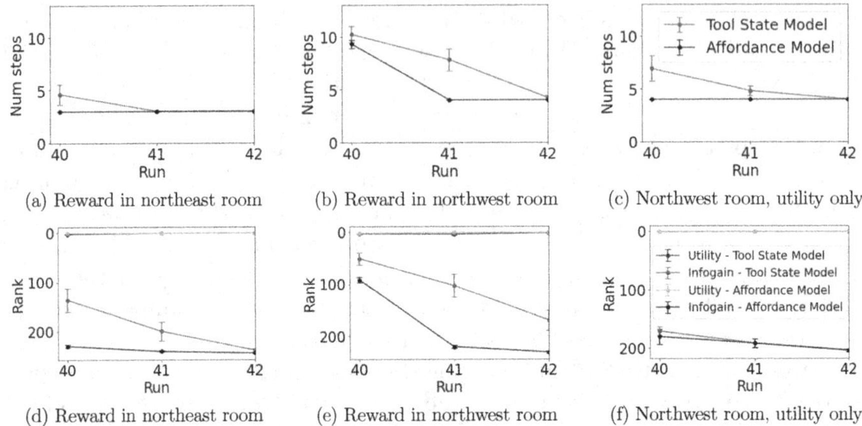

(a) Reward in northeast room (b) Reward in northwest room (c) Northwest room, utility only

(d) Reward in northeast room (e) Reward in northwest room (f) Northwest room, utility only

Fig. 8. Factorising the hidden states into tool affordances enables the agent to perform one-shot generalisation. All graphs are the results of 3 runs following exposure to all adjacent rooms (10 runs per room) and averaged over 10 independent trials. The top panel compares the Tool State Model to the Affordance Model in terms of the mean number of steps (±se) taken to solve a) the northeast room b) northwest room and c) northwest room selecting policies based only on utility. The bottom panel shows the utility and information gain rank of the selected policy for d) the northeast room e) northwest room and f) northwest room selecting policies based only on utility

8 Discussion

We have distinguished between tool use, tool discovery and tool innovation and asked what this might look like using the framework of AIF. We then ground this work with the construction of a simple model in order to take seriously this distinction and see what insights can be drawn. We provide the first evidence for the necessary properties associated with the process of tool innovation: namely that of offline induction of appropriate tool structure through composing relevant affordances.

We have identified that when solving the northwest room, the agent with the Affordance Model is not (sub-optimally) solving the task by having to discover the tool, as is the case with the agent with the Tool State Model. Rather, the agent seeks to investigate a specific state which it has never seen before and when it has sufficiently learned this fact (such that the information gain that it yields is significantly diminished), it subsequently solves the task in the optimal number of steps. Further investigation is required to ask why the utility is not enough to override this high information gain when it already has the knowledge of the correct tool to employ and the state transitions to create this tool.

We acknowledge that in our choice to factorise the hidden state of the agent's generative model into the tool affordances of x-reach and y-reach, we play the role of an intelligent designer. Ideally, we would like to have autonomous systems that choose what to learn from the environment and factorise their model in a way that best explains the latent causes of sensory observations. Smith *et al.* [22] introduce an approach whereby a probabilistic generative model has flexibility in the hidden states. The idea is one of furnishing of extra "slots" in the hidden states, allowing the agent to expand its generative model to incorporate new information when encountering new concepts. A process of Bayesian model reduction then acts to prune the model, ensuring that model complexity is reduced if in fact two concepts can be explained by the same underlying cause. This approach has been further extended to deep hierarchical AIF models, facilitating the formation of flexible and generalisable abstractions during a spatial foraging task [16]. This kind of adaptive structure learning would be useful in the context of tool innovation, allowing us to infer the best affordances to represent a tool. We therefore identify this approach as an interesting avenue for further research in the context of tool innovation in AIF agents.

Finally, we note that our model is limited given our intentional choice to omit the sensorimotor challenges associated with both tool manipulation and tool construction. Given that tool manufacture has been identified by Beck *et al.* [2] as a key component in the process of tool innovation, future work should look towards constructing models which can effectively handle more physically realistic tasks.

9 Conclusion

Overall, we have provided a minimal description of the distinction between tool discovery and tool innovation under the formalism of active inference. We have used this to then explore a simple model of tool innovation in an AIF agent by introducing a factorisation of hidden states of the generative model into affordances. This particular structural choice affords the agent with the ability to generalise what it has learned about state transitions and conceptualise a suitable tool via a process of induction. We have discussed the implications and limitations of our results and outlined directions for further research.

Acknowledgements. This research was funded under the UKRI Horizon Europe Guarantee scheme as part of the METATOOL project led by the Universidad Politécnica De Madrid.

Author contributions. P.F.K. conceived the project and both designed and conducted experiments. P.C. designed and conducted experiments and wrote the manuscript. C.L.B. supervised the project.

References

1. Andries, M., Chavez-Garcia, R.O., Chatila, R., Giusti, A., Gambardella, L.M.: Affordance equivalences in robotics: a formalism. Front. Neurorobot. **12**, 26 (2018)
2. Beck, S.R., Apperly, I.A., Chappell, J., Guthrie, C., Cutting, N.: Making tools isn't child's play. Cognition **119**(2), 301–306 (2011)

3. Bentley-Condit, V.: Smith: animal tool use: current definitions and an updated comprehensive catalog. Behaviour **147**(2), 185-32A (2010)
4. Biro, D., Haslam, M., Rutz, C.: Tool use as adaptation (2013)
5. Blei, D.M., Kucukelbir, A., McAuliffe, J.D.: Variational inference: a review for statisticians. J. Am. Stat. Assoc. **112**(518), 859–877 (2017)
6. Breyel, S., Pauen, S.: The beginnings of tool innovation in human ontogeny: how three-to five-year-olds solve the vertical and horizontal tube task. Cogn. Dev. **58**, 101049 (2021)
7. Cabrera-Álvarez, M.J., Clayton, N.S.: Neural processes underlying tool use in humans, macaques, and corvids. Front. Psychol. **11**, 560669 (2020)
8. Chappell, J., Cutting, N., Apperly, I.A., Beck, S.R.: The development of tool manufacture in humans: what helps young children make innovative tools? Philos. Trans. R. Soc. B Biol. Sci. **368**(1630), 20120409 (2013)
9. Da Costa, L., Lanillos, P., Sajid, N., Friston, K., Khan, S.: How active inference could help revolutionise robotics. Entropy **24**(3), 361 (2022)
10. Friston, K.: The history of the future of the Bayesian brain. Neuroimage **62**(2), 1230–1233 (2012)
11. Friston, K.J., Lin, M., Frith, C.D., Pezzulo, G., Hobson, J.A., Ondobaka, S.: Active inference, curiosity and insight. Neural Comput. **29**(10), 2633–2683 (2017)
12. Gibson, J.J.: The theory of affordances. Hilldale USA **1**(2), 67–82 (1977)
13. Heins, C., et al.: pymdp: a python library for active inference in discrete state spaces. arXiv preprint arXiv:2201.03904 (2022)
14. Higgins, I., et al.: Towards a definition of disentangled representations (2018)
15. Murphy, K.P.: Machine Learning: A Probabilistic Perspective. MIT Press, Cambridge (2012)
16. Neacsu, V., Mirza, M.B., Adams, R.A., Friston, K.J.: Structure learning enhances concept formation in synthetic active inference agents. PLoS ONE **17**(11), 1–34 (2022)
17. O'Brien, M.J., Shennan, S.: Innovation in Cultural Systems: Contributions from Evolutionary Anthropology. MIT Press, Cambridge (2010)
18. Qin, M., Brawer, J.N., Scassellati, B.: Robot tool use: a survey. Front. Robot. AI **9**, 369 (2022)
19. Reader, S.M., Morand-Ferron, J., Flynn, E.: Animal and human innovation: novel problems and novel solutions (2016)
20. Rezende, D.J., Mohamed, S., Danihelka, I., Gregor, K., Wierstra, D.: One-shot generalization in deep generative models (2016)
21. Sajid, N., Ball, P.J., Parr, T., Friston, K.J.: Active inference: demystified and compared. Neural Comput. **33**(3), 674–712 (2021)
22. Smith, R., Schwartenbeck, P., Parr, T., Friston, K.J.: An active inference approach to modeling structure learning: concept learning as an example case. Front. Comput. Neurosci. **14**, 41 (2020)
23. St Amant, R., Horton, T.E.: Revisiting the definition of animal tool use. Anim. Behav. **75**(4), 1199–1208 (2008)
24. Stout, D.: Stone toolmaking and the evolution of human culture and cognition. Philos. Trans. R. Soc. B Biol. Sci. **366**(1567), 1050–1059 (2011)
25. Whiten, A., Van Schaik, C.P.: The evolution of animal 'cultures' and social intelligence. Philos. Trans. R. Soc. B Biol. Sci. **362**(1480), 603–620 (2007)

Efficient Motor Learning Through Action-Perception Cycles in Deep Kinematic Inference

Matteo Priorelli(✉) and Ivilin Peev Stoianov

Institute of Cognitive Sciences and Technologies (ISTC), National Research Council of Italy (CNR), 35100 Padova, Italy
matteo.priorelli@istc.cnr.it, ivilinpeev.stoianov@cnr.it

Abstract. How does the brain adapt to slow changes in the body's kinematic chain? And how can it perform complex operations that need tool use? Here, we consider both processes through the same perspective and propose that the kinematic chain is represented by an Active Inference model encoding, in a hierarchical fashion, intrinsic and extrinsic information separately. However, the several pathways through which prediction errors can be minimized introduce some optimization problems. We show that an agent can rapidly change its kinematic chain online using action-perception cycles, similar to how learning and inference processes are handled in Predictive Coding Networks.

Keywords: Deep kinematic inference · Motor learning · Active Inference · Cortical oscillations · Tool use

1 Introduction

In normal conditions, the kinematic chain of an organism remains constant or only gradually changes on a lifetime scale. But there are situations where it is modified in much faster timescales, e.g., when using a tool to solve a task. It has been demonstrated that when monkeys are trained to use a tool to reach an object, their internal bodily representations in parietal and motor areas change to represent the tool [15]. This finding suggests that the kinematic chain encoded in the motor cortex is not fixed but modifies dynamically, i.e., when an external object is used for a sufficient amount of time. One hypothesis is that this mechanism is the result of an increase in the boundary between the self and the environment which, according to Predictive Coding theories, happens when the agent can predict the consequences of its actions – in this case, the movement of the tool – through a closed loop between motor commands and sensory evidence [11,12]. But a similar behavior can be also seen when patients with lesions to the motor cortex are trained to move, through implanted devices, an

Supported by European Union H2020-EIC-FETPROACT-2019 grant 951910 to IPS and Italian PRIN grant 2017KZNZLN to IPS.

C. L. Buckley et al. (Eds.): IWAI 2023, CCIS 1915, pp. 59–70, 2024.
https://doi.org/10.1007/978-3-031-47958-8_5

external robotic arm, which with extensive training becomes an integral part of the patient. Or, to the other extreme, in patients with an amputated limb, when the cortical region previously devoted to its control shrinks [10].

It is therefore critical (i) to understand how the motor cortex can take into account and predict such slow and rapid changes in the kinematic chain, and (ii) to efficiently simulate the same scenario in robotic experiments. In Optimal Control theories, complicated cost functions usually have to be defined to tackle such dynamic elements [22,23], and while the maturity of the framework has led to interesting results, it seems unlikely that the same mechanisms are at work in biological organisms [6]. In contrast, Predictive Coding based theories such as Active Inference, which tackles the motor control inversion by generating proprioceptive predictions from high-level latent states, provide a simpler and more biologically plausible solution that does not use any cost function [1,17].

In particular, it assumes that agents are endowed with a generative model specifying the dynamics of their hidden states and that desired goals are encoded as priors over the dynamics, which act as attracting states. Goal-directed movements are then realized by first generating predictions from the hidden states and then minimizing the corresponding prediction errors, or the discrepancy between predicted and current sensations. The main difference with respect to Optimal Control is that the mapping between proprioceptive predictions and control signals for the muscles can be implemented easily using reflex arcs in the spinal cord rather than requiring complex inverse dynamics computations [1]. In fact, the inverse model maps from peripheral proprioceptive sensations to movements, not from central hidden states to actions, as in Optimal Control [7].

The advantages of the Active Inference framework are even more evident when using hierarchical models, which are able to construct a richer representation of the environment. Despite such capabilities, the current literature comprises few hierarchical models [3,9,18], with no implementations of deep kinematic structures for realistic settings. As concerns the kinematic inversion, this is usually done through methods borrowed from Optimal Control such as the pseudoinverse [16]. However, these approaches are not biologically plausible since the exteroceptive generative model has to be duplicated into the dynamics of the hidden states. Importantly, no studies today tackle motor learning from an Active Inference perspective, since the forward kinematics typically generates only the end effector position and the agent has no access to all the information inside its kinematic chain, greatly limiting the range of tasks it can solve. Instead, we propose that a deep hierarchical model encoding beliefs over all segments of the kinematic chain [19] is capable of not only inferring the correct kinematic chain during perception but also during action, which may answer why changes in the motor cortex can be recorded after extensive tool use. As will be shown, the simultaneous learning of the joint angles and limb lengths during the movement is made possible through action-perception cycles – with some analogies to the optimization of Predictive Coding Networks [13] – allowing the agent not to get stuck during the free energy minimization process that happens when both phases are not run rhythmically.

2 Deep Kinematic Inference

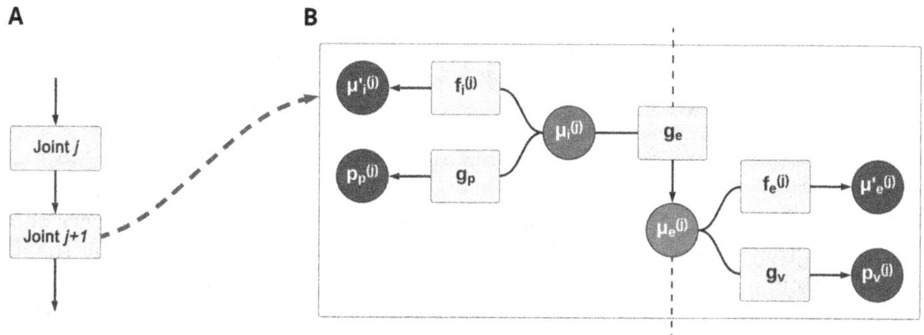

Fig. 1. Generative models for deep kinematic inference. (**A**) An example of a kinematic chain. (**B**) Factor graph of a single level of a hierarchical structure where each block is an IE model. Note that the extrinsic belief acts as a prior for the layer below.

The deep kinematic inference grounds on a simple block called "Intrinsic-Extrinsic (IE) model" [19], shown in Fig. 1B. This model has two different beliefs encoding respectively intrinsic (e.g., joint angles $\boldsymbol{\mu}_\theta$ and limb lengths $\boldsymbol{\mu}_l$) and extrinsic (e.g., absolute position and orientation of a limb $\boldsymbol{\mu}_e$). The two beliefs of a level $j-1$ are used to compute the extrinsic belief of level j through the following kinematic generative model:

$$\boldsymbol{\mu}_e^{(j)} = \boldsymbol{g}_e(\boldsymbol{\mu}_\theta^{(j)}, \boldsymbol{\mu}_l^{(j)}, \boldsymbol{\mu}_e^{(j-1)}) = \begin{bmatrix} x^{(j-1)} + l^{(j)} c_{\theta,\phi}^{(j)} \\ y^{(j-1)} + l^{(j)} s_{\theta,\phi}^{(j)} \\ \phi^{(j-1)} + \theta^{(j)} \end{bmatrix} \tag{1}$$

where we used a more compact notation to indicate the sine and cosine of the angles:

$$\begin{aligned} c_{\theta,\phi} &= \cos(\theta + \phi) \\ s_{\theta,\phi} &= \sin(\theta + \phi) \end{aligned} \tag{2}$$

This block is replicated so as to match the whole agent's kinematic chain; the resulting hierarchical structure allows connecting several nodes to a single layer, thus encoding complex kinematic models with ramifications (e.g., fingers). The extrinsic belief then performs the inference by averaging every contribution of its children through the corresponding precisions π_e and kinematic prediction errors $\boldsymbol{\varepsilon}_e$:

$$\dot{\boldsymbol{\mu}}_e^{(j)} \propto \sum_m^M \pi_e^{(j+1,m)} \boldsymbol{\varepsilon}_e^{(j+1,m)} \tag{3}$$

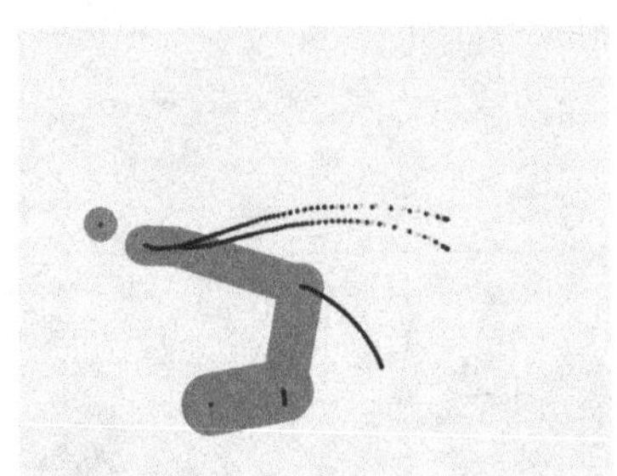

(a) A 4-DoF robotic arm has to reach a static target (represented in red) with its end effector.

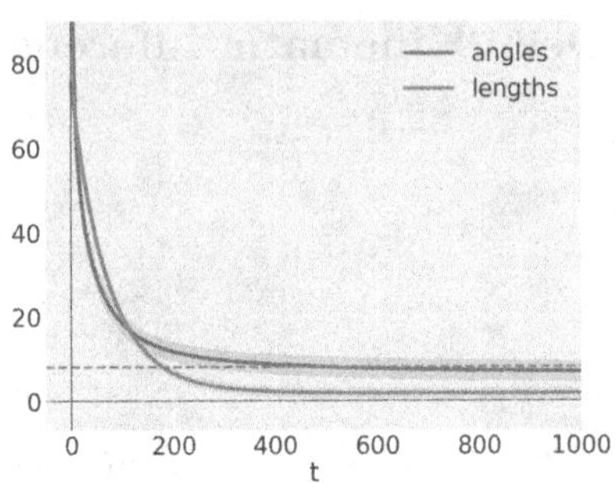

(b) Evolution over time of the difference between true and estimated joint angles (blue line), and between true and estimated limb lengths (red line), aggregated over 1000 trials during inference only.

Fig. 2. (a) A 4-DoF robotic arm has to reach a static target (represented in red) with its end effector.(b) Evolution over time of the difference between true and estimated joint angles (blue line), and between true and estimated limb lengths (red line), aggregated over 1000 trials during inference only. (Color figure online)

Note that the extrinsic kinematic precision $\pi_e^{(j+1)}$ modulates the update dynamics of the length belief $\boldsymbol{\mu}_l^{(j)}$, the angle belief $\boldsymbol{\mu}_\theta^{(j)}$, and the extrinsic belief $\boldsymbol{\mu}_e^{(j)}$ of level j.

Intrinsic and extrinsic beliefs also generate proprioceptive and exteroceptive (e.g., visual) sensations, respectively through the generative models $\boldsymbol{g}_p$ and $\boldsymbol{g}_v$. The kinematic inversion is automatically performed by inference – thus without requiring explicit functions into the dynamics of the hidden states – through the gradients of the kinematic generative model $\partial_\theta \boldsymbol{g}_e$ and $\partial_e \boldsymbol{g}_e$ over joint angles and extrinsic information, respectively. This architecture also allows solving a wide range of tasks through the definition of flexible functions that generate future goals based on the current belief [21], such as obstacle avoidance, trajectory planning in Cartesian space, or maintaining a vertical orientation while reaching a target [19].

For the scope of this study, we only consider a simple 4-DoF robotic arm whose goal is to reach a static target with the end effector, as shown in Fig. 2a. Note that in the following simulations, we assume that visual and proprioceptive observations directly provide the Cartesian position and angles of the limbs, respectively.

3 Perceptual Motor Learning

The model illustrated above allows not only to solve complex tasks that require the simultaneous coordination of several limbs, but also to learn the kinematic chain. In fact, the gradient of the kinematic generative model of Eq. 1 with respect to the length belief:

$$\frac{\partial \boldsymbol{g}_e}{\partial \boldsymbol{\mu}_l^{(j)}} \& = \begin{bmatrix} c_{\theta,\phi}^{(j)} & s_{\theta,\phi}^{(j)} & 0 \end{bmatrix} \tag{4}$$

allows inferring and learning the segment lengths of every level:

$$\dot{\boldsymbol{\mu}}_l^{(j)} = \partial_{\mu_l} \boldsymbol{g}_e^T \pi_e^{(j+1)^T} \boldsymbol{\varepsilon}_e^{(j+1)} \tag{5}$$

This adaptive behavior has several practical applications: for instance, an agent with a tool in its hand could infer the extremity of the tool by extending the length of its end effector. In addition, the hierarchical nature of the model allows specifying different learning dynamics for each segment, so that the belief over the end effector augmented with the new tool could be inferred in a much faster timescale than the rest of the arm. As shown in Fig. 2b, the agent is able to correctly infer – even in single trials – both joint angles and limb lengths randomly initialized each time. Note that in this case we did not use proprioceptive information on purpose, so the agent had to simultaneously infer them through exteroceptive sensations only.

4 Online Motor Learning and Action-Perception Cycles

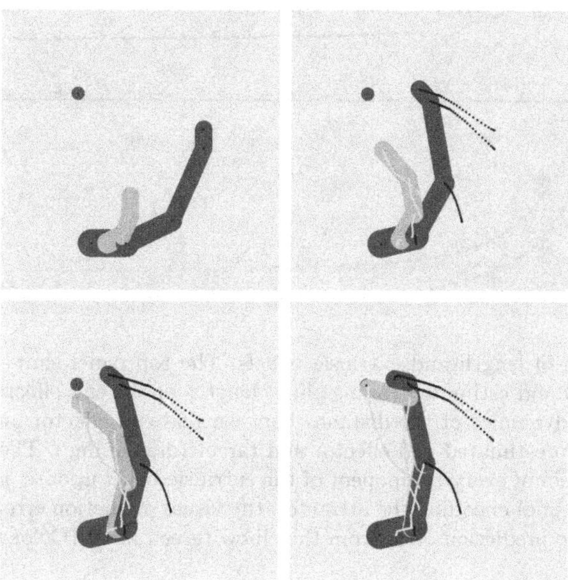

Fig. 3. Sequence of frames for the reaching task with adaptable limb lengths. Real and estimated arms are represented in blue and cyan, respectively. In this simulation, the beliefs over limb lengths are initialized to a wrong value. (Color figure online)

When performing the same kind of inference during goal-directed movements – e.g., target reaching – a few issues arise. In this case, an attractor is embedded into the dynamics of the extrinsic belief of the last layer (corresponding to the end effector), whose update is:

$$\dot{\tilde{\boldsymbol{\mu}}}_e^{(4)} = \begin{bmatrix} \boldsymbol{\mu}_e'^{(4)} - \pi_e^{(4)}\boldsymbol{\varepsilon}_e^{(4)} + \partial \boldsymbol{g}_v^T \pi_v^{(4)}\boldsymbol{\varepsilon}_v^{(4)} + \partial \boldsymbol{f}_e^{(4)^T} \pi_{\mu_e}^{(4)}\boldsymbol{\varepsilon}_{\mu_e}^{(4)} \\ -\pi_{\mu_e}^{(4)}\boldsymbol{\varepsilon}_{\mu_e}^{(4)} \end{bmatrix} \tag{6}$$

The attractor expresses the difference between the current belief and a desired state, multiplied by a gain, i.e., $\boldsymbol{f}_e^{(4)}(\boldsymbol{\mu}_e) = \lambda(\boldsymbol{\mu}_e - \boldsymbol{\mu}^*)$. In brief, the extrinsic belief is subject to an attractive force encoding the target location $\boldsymbol{\varepsilon}_{\mu_e}^{(4)}$, a forward kinematic prediction error coming from the layer below (e.g., the elbow) $\boldsymbol{\varepsilon}_e^{(4)}$, and a visual prediction error $\boldsymbol{\varepsilon}_v^{(4)}$. Thus, the kinematic prediction error acts both on the extrinsic belief in Eq. 6 and on the length belief in Eq. 4.

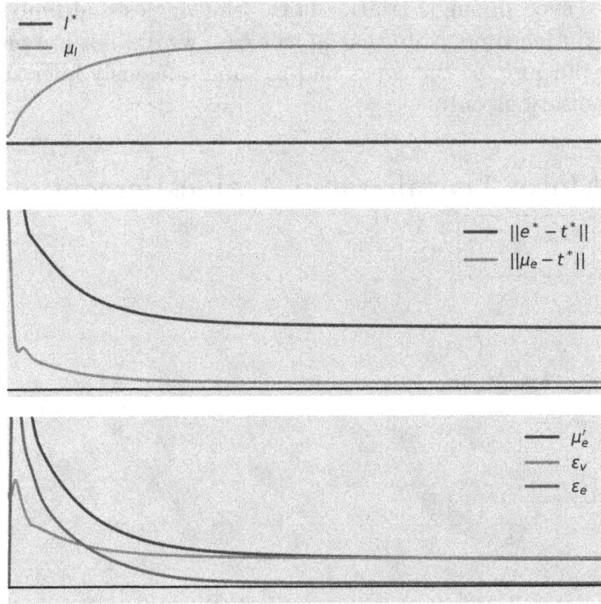

Fig. 4. Evolution of length and extrinsic beliefs. The top panel shows the dynamics of real (blue line) and estimated (orange line) lengths of the end effector. The middle panel shows the dynamics of the distance between real end effector and target (blue line), and between estimated end effector and target (orange line). The bottom panel shows the dynamics of every component of the extrinsic belief update, namely the 1st-order belief (blue line) encoding the attractor, the visual prediction error (orange line), and the kinematic prediction error from the elbow (green line). (Color figure online)

Figure 3 shows a sequence of frames of a simple reaching task when the agent is allowed to infer the length of its limbs, which are initialized to a wrong value.

The joint angles rapidly stabilize, and so do almost all the limb lengths, resulting in the estimated arm gradually growing during the reaching movement, until it matches the real one. However, the agent fails to estimate the length of the last limb – where the attractor is defined – and the end effector stops before reaching the destination. What happens is that the kinematic prediction error ε_e of the end effector affects the minimization of the length belief while the extrinsic belief is pulled toward the desired state.

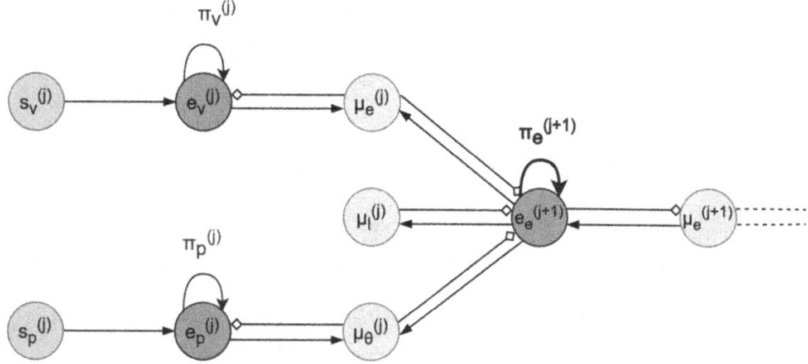

Fig. 5. Neural implementation of a single level of the model. For simplicity, the dynamics functions are not displayed. Here, $\boldsymbol{s}_v^{(j)}$ and $\boldsymbol{s}_p^{(j)}$ indicate visual and proprioceptive observations of level j, respectively.

In order to understand this behavior, let us analyze the evolution of the extrinsic and length beliefs of the end effector, as shown in Fig. 4. In this case, the kinematic prediction error ε_e that results from the attractor of the dynamics function (green line in the bottom panel) climbs up the hierarchy and flows into the angle belief, so that the end effector is gradually pulled toward the target (orange line in the middle panel). However, note that the kinematic prediction error tries to exert a force on the length belief (orange line in the top panel) as well. These different pathways are displayed in Fig. 5, showing a neural implementation for a single level of the hierarchical model. The result is that, as clear in the last frame of Fig. 3, the extrinsic belief settles to the correct value – i.e., the agent thinks that the target has been reached – but the length belief is overestimated. If we focus on the last panel of Fig. 4, we can note that the 1st-order extrinsic belief and the visual prediction error are never really minimized but get stuck pushing in opposite directions.

This behavior is similar to what happens during optimization of a Predictive Coding Network. Since the length belief is not constrained by sensory observations directly but is free to change, we can consider it as a parameter of the network. Changing such parameters – i.e., learning – before the network has settled to a steady configuration where all prediction errors have been minimized leads to some issues in the optimization because the distributions which the predictions are sampled from constantly change. For this reason, whenever a new pair of input and output is presented to the network, learning is allowed after the inference has stabilized [14,24].

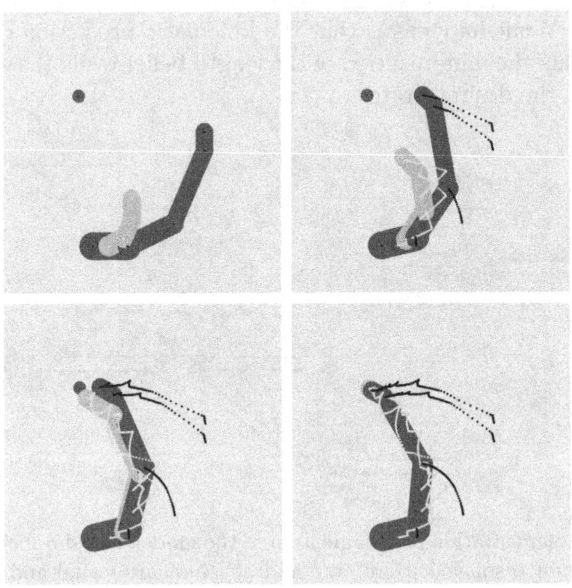

Fig. 6. Sequence of frames for the reaching task with adaptable limb lengths. Real and estimated arms are represented in blue and cyan, respectively. The beliefs over limb lengths are initialized to a wrong value. (Color figure online)

Similarly, the abnormal behavior for online motor learning can be avoided if we alternate between: (i) perceptual phases where, as before, the length likelihood is minimized without imposing any bias over the extrinsic dynamics; and (ii) action phases where the length belief is kept fixed but the extrinsic attractor results in the end effector moving toward the target. As shown in Fig. 6, in this case the agent is able to reach the target while correctly inferring the length of all segments. During the first phase, it tries to match the estimated kinematic chain to the real one; during the second phase, it imposes a false belief in the end effector dynamics, ultimately driving the arm movement. More specifically, Fig. 7, representing the dynamics of the task with action-perception cycles, shows that the 1st-order extrinsic belief, the visual prediction error, and the kinematic prediction error all approach zero. Crucially, after an initial overestimate of the end effector's length, the correct value is gradually found at the end of the trial.

The oscillating behavior of the end effector in Fig. 6 is due to the number of time steps of the action-perception cycles. As shown in Fig. 8, increasing this value results in decreased time needed to reach the target but less stable behavior. To the extreme, a very low frequency has the consequence of splitting the task into a pure perceptual phase and a pure reaching motion, resulting in the

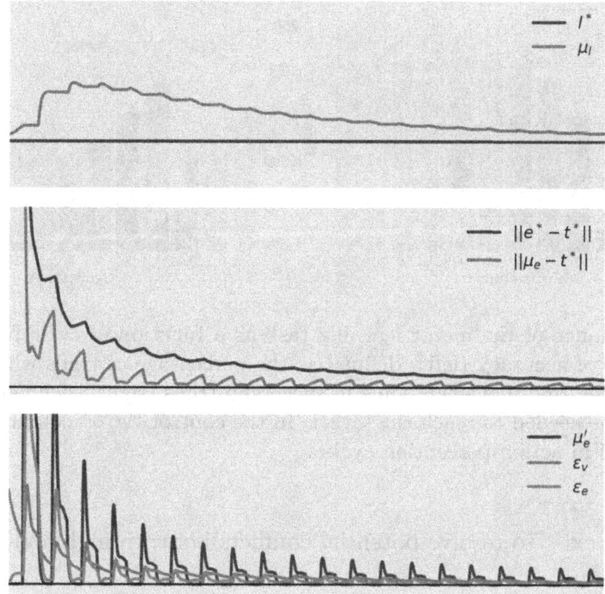

Fig. 7. Evolution of length and extrinsic beliefs with action-perception cycles.

most stable behavior but at the expense of the highest time needed and being unable to react rapidly if environmental changes are introduced in the middle of the trial. On the other hand, if the cycle window decreases the end effector presents less oscillations but at the cost of increased overall time. However, beyond a certain limit, the agent fails again to infer the correct length and hence reach the target. There is thus a tradeoff between stable behavior, time efficiency, and flexibility. For comparison, we also performed a simulation without action-perception cycles: in this case, the agent is not able to reach the target in almost none of the trials. Note however that the oscillating behavior could be avoided by keeping a steady motor command depending on the proprioceptive error of the previous phase.

5 Discussion

Rhythmic oscillations are found throughout all cortical areas. From an Active Inference perspective, action-perception cycles emerge from the modulation of the precisions of prediction errors: specifically, attention has been associated with the estimation of high-level beliefs depending on the evidence accumulated, while salience is related to the uncertainty minimization process that decides what sensory data to sample next [2]. Here, we proposed that this mechanism may be key to the correct estimation and learning of the latent states when goal-directed actions are involved. This finding is in line with the hypothesis that

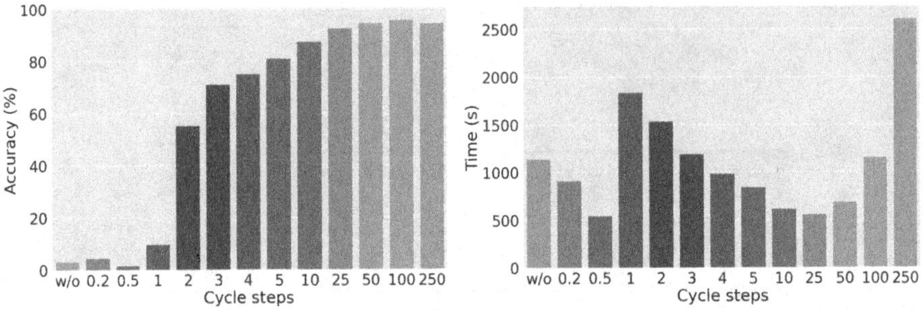

Fig. 8. Performance of the motor learning task as a function of cycle frequency presented in terms of accuracy (left), defined as the percentage of trials where the agent reaches the target, and movement time in successful trials (right), defined as the number of time steps needed to reach the target. In the control "w/o" condition, inference is not separated in action-perception cycles.

theta rhythms exist to resolve potential conflicts between high and low levels of the hierarchy [5].

As explained in the previous section, in a hierarchical model with multiple inputs and multiple outputs prediction errors can concurrently flow into several pathways. When performing state estimation and action at the same time, the agent might get stuck in local minima during the process of free energy minimization whenever particular priors are imposed in the belief dynamics to realize goal-directed movements. In Predictive Coding Networks, phases of inference where the network settles to stable values and prediction errors are totally minimized follows a learning phase where the network's parameters are updated with the new state values – in an analogous way to the optimization steps of an EM algorithm [4]. Similarly, we showed that a correct behavior for an online motor learning task is obtained by splitting it into separate cycles of perception and action. In the most extreme scenario, a pure perceptual phase is followed by a pure reaching phase, leading to the best reaching behavior. However, this condition does not allow the agent to react to environmental changes, e.g., if one has to rapidly modify its kinematic chain in order to grab a tool. Interestingly, cycles with too-high frequencies need even more time to complete the task than the previous case, and good performances that can also account for dynamic flexibility are obtained somewhat in between the two conditions.

The action-perception cycles are performed by modulating high- and low-level precisions. In particular, perceptual phases are realized by increasing high-level precisions and decreasing the ones of the belief dynamics where the attractors are defined - although it has been hypothesized that such phases may arise from switching off the proprioceptive input through sensory attenuation [8]. This has the effect that the network can stabilize to the correct state inferred through the observations, before the beliefs are left free to change by the biased internal dynamics while keeping fixed parameters. This mechanism may generalize to all cases where goal-directed dynamics is embedded into the dynamics function of a hierarchical structure, as we showed with a model that had to concurrently estimate the depth of an object and fixate it through the perspective projections from each eye [20]. In the latter approach, some interesting parallelisms arise with higher-level processes that cycle between saccades and evidence sampling.

We also propose that this online motor learning might be critical not only when considering the most intuitive conditions, i.e., when an organism grows or abruptly loses a limb, but also in voluntary actions where one has to use external tools to solve a specific task. Although not implemented here, the model presented can easily address this scenario since it is possible to extend the length of the last level (i.e., the end effector) without changing the overall structure. The beliefs would be updated according to the new sensory evidence (e.g., visual observations of the tool) through local message passing of prediction errors. A hierarchical model might also explain how patients with implanted devices can adapt their motor cortex so as to represent the new arm attached: in this scenario, a joint would be added to a particular location of the kinematic hierarchy allowing a new Degree of Freedom to the patient. How this is possible through self-modeling of the agent's representation of its kinematic chain would be an interesting direction of research. Finally, future studies will be done to simulate tasks requiring tool use: in particular, an agent might be required to solve a multi-step task involving reaching a tool, grabbing it, and finally reaching an object with its extremity.

References

1. Adams, R.A., Shipp, S., Friston, K.J.: Predictions not commands: active inference in the motor system. Brain Struct. Funct. **218**(3), 611–643 (2013). https://doi.org/10.1007/s00429-012-0475-5
2. Anil Meera, A., Novicky, F., Parr, T., Friston, K., Lanillos, P., Sajid, N.: Reclaiming saliency: rhythmic precision-modulated action and perception. Front. Neurorobot. **16**, 1–23 (2022). https://doi.org/10.3389/fnbot.2022.896229
3. Çatal, O., Verbelen, T., Van de Maele, T., Dhoedt, B., Safron, A.: Robot navigation as hierarchical active inference. Neural Netw. **142**, 192–204 (2021)
4. Dempster, A.P., Laird, N.M., Rubin, D.B.: Maximum likelihood from incomplete data via the EM algorithm. J. Roy. Stat. Soc. Ser. B (Methodol.) **39**(1), 1–22 (1977)
5. Fiebelkorn, I.C., Kastner, S.: A rhythmic theory of attention. Trends Cogn. Sci. **23**, 87–101 (2019). https://doi.org/10.1016/j.tics.2018.11.009
6. Friston, K.: What is optimal about motor control? Neuron **72**(3), 488–498 (2011). https://doi.org/10.1016/j.neuron.2011.10.018
7. Friston, K.J., Daunizeau, J., Kilner, J., Kiebel, S.J.: Action and behavior: a free-energy formulation. Biol. Cybern. **102**(3), 227–260 (2010). https://doi.org/10.1007/s00422-010-0364-z
8. Friston, K.J., Mattout, J., Kilner, J.: Action understanding and active inference. Biol. Cybern. **104**(1–2), 137–60 (2011). https://doi.org/10.1007/s00422-011-0424-z
9. Friston, K.J., Rosch, R., Parr, T., Price, C., Bowman, H.: Deep temporal models and active inference. Neurosci. Biobehav. Rev. **77**, 388–402 (2017). https://doi.org/10.1016/j.neubiorev.2017.04.009

10. Fuhr, P., et al.: Physiological analysis of motor reorganization following lower limb amputation. Electroencephalogr. Clin. Neurophysiol./Evoked Potentials **85**(1), 53–60 (1992). https://doi.org/10.1016/0168-5597(92)90102-H
11. Hohwy, J.: The Predictive Mind. Oxford University Press, Oxford (2013)
12. Lanillos, P., Pages, J., Cheng, G.: Robot self/other distinction: active inference meets neural networks learning in a mirror (ECAI) (2020). https://arxiv.org/abs/2004.05473
13. Millidge, B., Osanlouy, M., Bogacz, R.: Predictive Coding Networks for Temporal Prediction, pp. 1–59 (2023)
14. Millidge, B., Tschantz, A., Buckley, C.L.: Predictive coding approximates backprop along arbitrary computation graphs. Neural Comput. **34**(6), 1329–1368 (2022). https://doi.org/10.1162/neco_a_01497
15. Obayashi, S., et al.: Functional brain mapping of monkey tool use. Neuroimage **14**(4), 853–861 (2001). https://doi.org/10.1006/nimg.2001.0878
16. Oliver, G., Lanillos, P., Cheng, G.: An empirical study of active inference on a humanoid robot. IEEE Trans. Cogn. Dev. Syst. **8920**(c), 1–10 (2021). https://doi.org/10.1109/TCDS.2021.3049907
17. Parr, T., Pezzulo, G., Friston, K.J.: Active inference: the free energy principle in mind, brain, and behavior (2022)
18. Pezzulo, G., Rigoli, F., Friston, K.J.: Hierarchical active inference: a theory of motivated control. Trends Cogn. Sci. **22**(4), 294–306 (2018)
19. Priorelli, M., Pezzulo, G., Stoianov, I.P.: Deep kinematic inference affords efficient and scalable control of bodily movements. bioRxiv (2023). https://doi.org/10.1101/2023.05.04.539409. https://www.biorxiv.org/content/early/2023/05/05/2023.05.04.539409
20. Priorelli, M., Stoianov, I.P.: Intention modulation for multi-step tasks in continuous time active inference. In: Active Inference, Third International Workshop, IWAI 2022, Grenoble, France, 19 September 2022 (2022). https://link.springer.com/book/9783031287206
21. Priorelli, M., Stoianov, I.P.: Flexible intentions: an active inference theory. Front. Comput. Neurosci. (2023). https://doi.org/10.3389/fncom.2023.1128694
22. Stengel, R.F.: Optimal control and estimation (1994)
23. Todorov, E.: Optimality principles in sensorimotor control. Nat. Neurosci. **7**, 907–915 (2004). https://doi.org/10.1038/nn1309
24. Whittington, J.C., Bogacz, R.: Theories of error back-propagation in the brain. Trends Cogn. Sci. **23**(3), 235–250 (2019)

Active Inference and Psychology

Towards Understanding Persons and Their Personalities with Cybernetic Big 5 Theory and the Free Energy Principle and Active Inference (FEP-AI) Framework

Adam Safron[1,2,3,4(✉)] and Zahra Sheikhbahaee[5]

[1] Center for Psychedelic and Consciousness Research, Johns Hopkins University School of Medicine, Baltimore, MD, USA
asafron@gmail.com
[2] Institute for Advanced Consciousness Studies, Santa Monica, CA, USA
[3] Cognitive Science Program, Indiana University, Bloomington, IN, USA
[4] Kinsey Institute, Indiana University, Bloomington, IN, USA
[5] University of Montreal, Montreal, QC, Canada

Abstract. Here we review recent work attempting to combine the first principles formalism of the Free Energy Principle and Active Inference (FEP-AI) framework with a recently proposed integrative model that attempts to ground personality as control variables for goal-seeking systems: Cybernetic Big 5 Theory (CB5T). First we summarize core aspects of this synthesis, then introduce some novel (and speculative) hypotheses, and then finally consider future implications for personality modeling with FEP-AI and CB5T.

Keywords: Personality · Cybernetic Big 5 Theory (CB5T) · Free Energy Principle and Active Inference (FEP-AI) Framework

1 Introduction

As AI models continue to gain sophistication, we find ourselves with both new opportunities and challenges. In terms of benefits, we have the potential for AIs to act as tools for understanding (e.g. computational psychiatry), helpers (e.g. industrial applications and labor augmentation), and perhaps even companions (e.g. elder care). With respect to risks, we have the possibility of these systems to learn unexpected behavior patterns that could have potentially undesirable consequences. In what follows, we briefly review some recent work on personality modeling [1], which we believe could have far reaching consequences for our abilities to realize positive outcomes with respect to a future where AI becomes an increasingly central part of our lives.

Personality can be thought of as a "phenomenological" description of the most relevant features for explaining overall behavior and cognition. In dynamic systems terms, we may think of this as a "normal form" description, that attempts to capture the maximal amount of detail of a particular system with minimal description lengths [2]. In the

C. L. Buckley et al. (Eds.): IWAI 2023, CCIS 1915, pp. 73–90, 2024.
https://doi.org/10.1007/978-3-031-47958-8_6

realm of psychology, personality can be considered as the 'essence' of individuality, in terms of describing more enduring features that are stable across circumstances. With Cybernetic Big 5 Theory (CB5T) [3], DeYoung proposes that personality is constituted by both "characteristic adaptations" (i.e., policies people learn for responding to different classes of situations) as well as the well-known (evolutionarily selected) traits such as Openness, Extraversion, Agreeableness, Conscientiousness, and Neuroticism. CB5T further argues that these traits are best understood in the context of modeling individuals in cybernetic terms, or as goal-seeking systems that are governed by various forms of feedback processes. This kind of functional understanding of persons and other complex adaptive systems suggests potentially fruitful intersections with computational frameworks, which is the issue we turn towards next.

2 FEP-AI

The Free Energy Principle (FEP) understands all persisting systems as entailing predictive (generative) models of the conditions under which they maintain their particular forms through intelligent actions. The processes enacted by generative models of persons come in many varieties, ranging from unconscious habits, to emotionally charged reactive dispositions, to declarative knowledge and self-organization via autobiographical narratives [4, 5]. To the extent that persons have identifiable traits and characteristic adaptations (i.e., personalities), these would represent enduring parameter values for the generative models governing dynamics, where this stability could be due to being genetically specified, epigenetically canalized [6], or as stable (to degrees) emergent equilibria. Different personality configurations would correspond to different models by which persons attempt to achieve their goals, including the primary goal of preserving essential features at the core of identity.

It is worth emphasizing the extremely broad scope of the FEP, which is as far-reaching as the purview of generalized Darwinism, with which it may be fully isomorphic [7]. Not only do nervous systems entail predictive models, but so do entire populations of organisms and their extended phenotypes as (previously selected, teleonomical) 'predictions' with respect to evolutionary fitness. By this account, nervous systems are merely a (very) special case of generative modelling, where not only is it the case that such *systems are models* in their very existence, but where such *systems also have models* that function as cybernetic controllers [8–10]. In these ways, active inference provides a formalism in which all persisting dynamical systems can be understood as (self-)generative models, grounded in first principles regarding the necessary preconditions for continued existence in a world governed by the 2nd law.

This universal Bayesian/Darwinian account extends all the way down to neuronal oscillations [11], to habitual reactions [12, 13], and all the way up to narrative selves as stories that achieve degrees of truth with the telling-doing-enacting [5, 14, 15], including with respect to shared narratives by which we more effectively collaborate with each other in pursuing valued goals [16, 17]. Within active inference, all characteristics of persons are selected—in the sense of both generalized Darwinism [18] and Bayesian model selection [19]—according to their relative abilities to minimize their respective free energies, which is suggested to be equivalent to maximizing self-model-evidence.

Specifically, each characteristic of the person represents its own replicative dynamic that teleonomically 'attempts' to maximize model evidence for itself [20, 21]. From this perspective, personalities represent relatively stable evolutionary game theoretic equilibria among competing and cooperating quasi-species [22–25].

Work within the FEP paradigm has yielded a normative model of behavior in *Active Inference (AI)* to describe the processes by which free energy is minimized [26]. Further, advances in deep reinforcement learning appear to be converging on the kinds of solutions that are predicted to be necessary for (bounded) optimality in *the FEP and Active Inference (FEP-AI) framework* [27, 28]. The notion of active inference rests on the insight that perception takes place within the context of adaptively shaping actions, which alter patterns of likely perceptions. Rather than being the result of passive sensations, perception is an active process of foraging for information and resolving model uncertainty [29–31], often driven by discrete actions as a kind of hypothesis testing [32–35]. Both perception and action are understood as kinds of inferences in the FEP-AI framework, in that they both represent means by which systems can engage in comparing predictions against sensations. One way systems can reduce prediction-error is by updating internal models, thus changing predictions; in this way, perception is understood as a kind of best-guess inference as to the causes of sensations. However, another way systems can reduce prediction-error is by updating the world through action, and thus making its predictions more accurate by changing likely perceptions; in this way, active inference represents a means by which not just perception, but also adaptive goal-oriented behavior can be realized via prediction-error minimization. The degree to which dynamics are governed by these two strategies—of updating of states either internal or external to the system—is determined (by gradient descent) according to whichever combination is expected to minimize overall free energy (i.e., cumulative precision-weighted prediction error) (Fig. 1). As we will describe in greater detail below, this foundational (intertwined and synergistic) duality between perceiving and acting may also have implications for understanding fundamental aspects of personality as well (Fig. 2).

Within the FEP-AI framework, all cybernetic systems necessarily minimize free energy for their generative models. However, in order to effectively achieve this objective, adaptive goal-seeking systems (such as organisms) select actions anticipated to result in free energy minimizing consequences in the future. Under this regime of *expected free energy*, model accuracy becomes expected utility, or opportunities for realizing the extrinsic value of ensuring preferred outcomes. Further, model complexity becomes the ambiguity and risk associated with pursuing particular courses of action, or opportunities for realizing the intrinsic value of reducing uncertainty via learning. Optimizing for *extrinsic value* involves minimizing discrepancies between preferred system-world configurations and observations, which entails pragmatically exploiting particular *policies* (i.e., sets of state-action mappings for goal realization, broadly construed to include the covert behavior of cognition). Optimizing for *intrinsic value* involves model refinement through seeking out sources of uncertainty as opportunities for maximizing information gain, so allowing for epistemic exploration of hypothesis spaces regarding adaptive actions. These two sources of value relate to exploitation/exploration tradeoffs, which in this case are navigated [36] by selecting policies based on whatever combination of actions is estimated to most effectively minimize overall expected free energy. If these

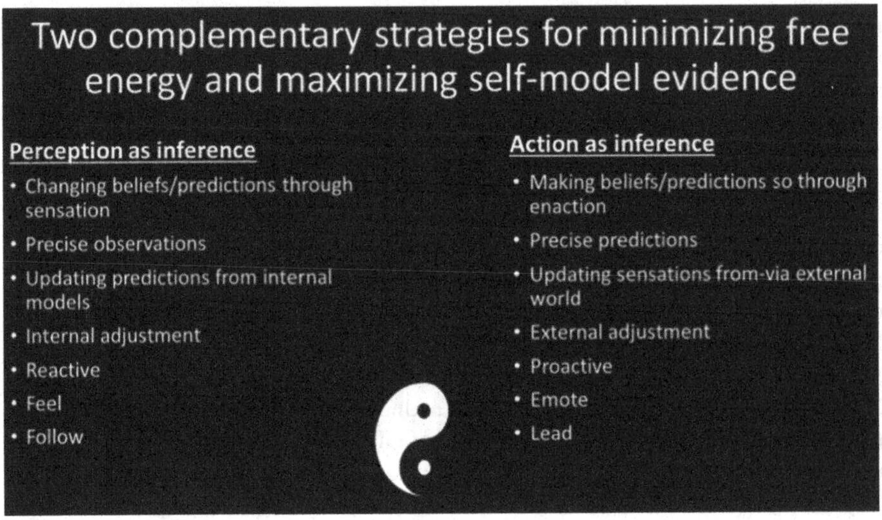

Two complementary strategies for minimizing free energy and maximizing self-model evidence

Perception as inference
- Changing beliefs/predictions through sensation
- Precise observations
- Updating predictions from internal models
- Internal adjustment
- Reactive
- Feel
- Follow

Action as inference
- Making beliefs/predictions so through enaction
- Precise predictions
- Updating sensations from-via external world
- External adjustment
- Proactive
- Emote
- Lead

Fig. 1. Perception and action as active inference.

Fig. 2. Internal attunement and external attunement as active inference.

actions occur in the context of a novel task environment about which little is known, then policy selection in FEP-AI will tend to primarily involve the exploration of optimizing for information gain, followed by a shift to more exploitative behavior as the task structure becomes sufficiently clear to afford informed actions. However, if actions fail to be as successful as anticipated, then this will tend to result in shifting back to exploratory behavior until a better "grip" on the situation can be acquired [37, 38]. In this way, given well-calibrated prior expectations, agents governed by FEP-AI will tend to exhibit flexibly balanced levels of curiosity as they engage in goal-oriented behavior.

These dual strategies for minimizing expected free energy mean that cybernetic systems governed by FEP-AI may decide to forego pursuing a goal in favor of model refinement, if the latter will lead to a bigger reduction in prediction error [39]. This may be a surprising implication of these models, as certainly evolutionary fitness depends more on achieving goals that on enhancing the accuracy of perceptual maps. Refining internal models will be preferred over achieving valued goals through action only to the extent that this choice serves the more fundamental goal of promoting continued existence and effective goal pursuit over time [20]. Indeed, evolved cybernetic systems will tend to prioritize stable goal pursuit and homeostasis as a pre-requisite for existence, with "interoceptive inference" providing a potentially useful case example, in terms of those modeling efforts tending to center on the avoidance of excessive risk with respect to the preconditions for basic life management [40]. FEP-AI tries to address this prioritization in that we should expect "adaptive priors" to make it such that all predictions are ultimately chained to evolutionary fitness [41], such that we would expect from systems selected to minimize free energy with respect to maintaining the preconditions for their existence.

3 CB5T and Cybernetic Control Variables; Focus on Stability and Plasticity

Parallels between FEP-AI and models of personality have recently been explored by Safron and DeYoung [1] with respect to Cybernetic Big 5 Theory (CB5T) [3]. In brief, CB5T contextualizes the Big 5 trait hierarchy as reflecting evolved control parameters for systems that attempt to minimize entropy with respect to the goals by which they preserve themselves. This is highly compatible with FEP-AI. Intriguingly, above the Big 5 in the trait hierarchy, two higher-order factors have been identified as Stability (shared variance of Conscientiousness, Agreeableness, and (inverse) Neuroticism) and Plasticity (shared variance of Openness and Extraversion). One may be tempted to map onto FEP-AI's dual optimization for extrinsic/pragmatic and intrinsic/epistemic value, respectively. However, this potential functional mapping should not be overstated, since a cybernetic interpretation of Extraversion as indicating "reward sensitivity" could be applied to situations where either pragmatic or epistemic value are at play, depending on what a particular individual finds rewarding (e.g. persons with high or low trait Openness as appreciating intrinsic value to respectively greater or lesser degrees). The dynamic tension between Stability and Plasticity has been observed in countless systems, with intriguing recent work suggesting that a major dimension of cultural variation may exist in the degree of "tight" or "loose" attitudes with respect to norms (i.e., relative degrees of precision on prediction-errors that would release policies for realizing deontic value) [42, 43]. An integration (and convergent support) between active inference and a predominant model of personality has important implications for computational psychiatry [2], and variations in the Big 5 trait hierarchy have recently been shown to converge with principled taxonomies of psychopathology (e.g. HiTOP) [44] (Figs. 3 and 4).

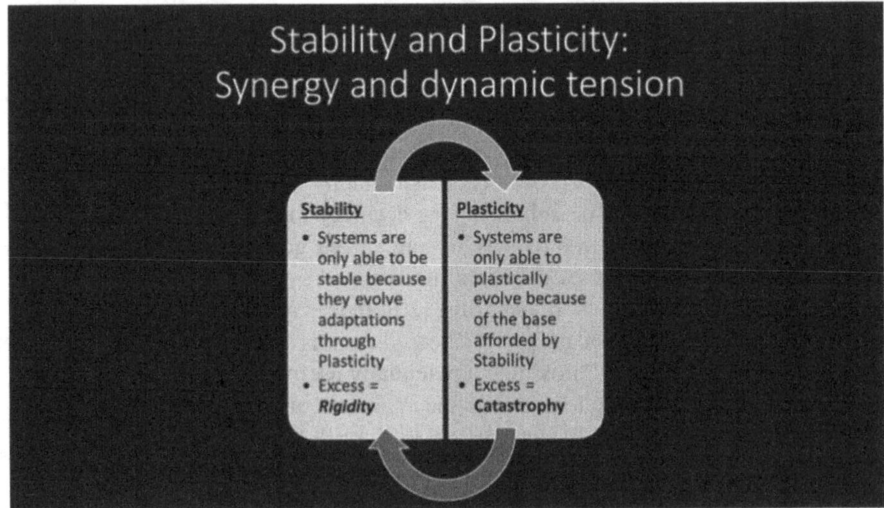

Fig. 3. Stability and Plasticity in personality theory.

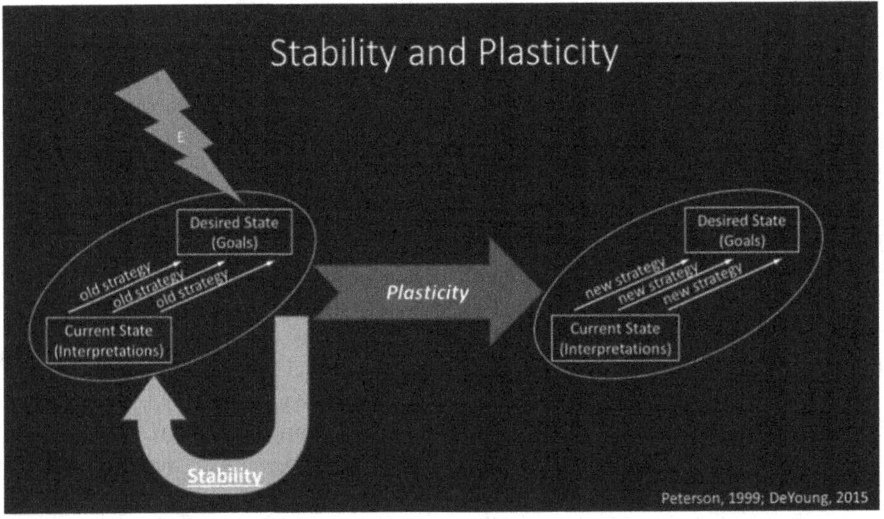

Fig. 4. Stability and Plasticity as the respective protection and updating of policies for enaction.

Interpretations of biophysical processes in terms of parameter settings for FEP-AI may help to provide substantial convergent support for CB5T, as well as an additional means of interpreting similar phenomena. For example, differing levels of dopaminergic function appear to have major impacts on personality with respect to Extraversion and probably also Openness/Intellect [45], and has been interpreted as indicating tendencies towards exploration (or Plasticity) in CB5T. In FEP-AI, tonic dopaminergic function is associated with precision (an inverse temperature parameter) over policies, indicating certainty (subjectively, confidence) and more deterministic action selection; phasic dopamine, however, indicates changing estimates with respect to expected free energy and updating of likelihoods for selecting different policies for enaction, as in reward prediction errors [46–49], including overt behavior as well as cognitive 'acts' [50]. That is, nervous systems tuned to exhibit high dopaminergic signaling may more readily deploy mental acts such as discrete attentional shifts and more extended simulated plans, potentially contributing to more exploratory and flexible cognitive styles. While utilizing distinct conceptual frames, this account of dopamine in FEP-AI has strong correspondences with CB5T's interpretation of dopamine as a "neuromodulator of exploration" (DeYoung, 2013) and contributor to personality Plasticity. Further, FEP-AI's interpretation of changes in phasic dopamine as updating likelihoods for policy deployment has clear parallels with CB5Ts interpretations of Plasticity as a capacity for updating strategies for goal attainment when confronted with obstacles and associated psychological entropy [51, 52].

While CB5T and FEP-AI both associate dopamine with potentially more exploratory behavioral and cognitive styles—and possibly more extraverted and open personalities—they also emphasize the context-sensitivity of functional consequences from varying patterns of neuromodulation. In both frameworks, overly simple exploration/exploitation distinctions are problematized based on the fact that action selection is governed by control hierarchies with multiple (potentially nested) goals unfolding over varying timescales (DeYoung, 2015; Pezullo, Rigoli, Friston, 2018). For example, a person with highly confident beliefs regarding goal realization may choose to exploit a particular opportunity, or they may venture out into the unknown and explore new territories if they predict that course of action could realize even greater value.

The relative positioning of goals and related representations within overall hierarchies in many ways speaks to the core of what we tend to mean by 'personality.' That is, we might expect hierarchically higher (or deeper) representations to be somewhat shielded from disruption by particular events on account of their being more opportunities (or multiple realizability) for minimizing prediction error via hierarchically lower patterns. This would be consistent with the ways in which systems distal from primary modalities have both close connections with neuromodulatory value signals, which tend to be most responsive to overall surprisal from relatively abstract action-outcome associations (e.g. attainment of particular goals from specific patterns of enaction). If personality is understood as a way of summarizing the most impactful and enduring features of a goal-seeking system, then perhaps we should not be too surprised to observe that personalities are often most powerfully impacted by disruption to hierarchically higher (or deeper) systems such as the ventral prefrontal cortex (c.f. pseudopsychopathy phenomena and the case of Phineas Gage (who became "no longer Gage" after a steel rod blasted through his head during a railroad construction accident) [53].

The hierarchical organization of goals is crucially important for multiple reasons. Firstly, the world as a whole tends to have hierarchical structure with smaller (more quickly evolving) things tending to be nested within larger (more slowly evolving) things. As such, there could be advantages for reactive dispositions to entities/events at these different scales having a similar kind of organization (cf. optimization via local gradients, locality-sensitive hashing, etc.). The attainment of complex goals via extended sequences necessarily requires larger (and likely more slowly evolving) zones of integration, with coherent orchestration among the various sub-goals required to achieve the broader aims to which they might contribute (or hinder) [54]. Conscientiousness is the personality trait most associated with the coherent management of goal hierarchies, and seems to only be reliably identifiable in animals with more complex nervous systems [55]. Speculatively, the nature of Conscientiousness (as a feature of cybernetic systems) may provide conceptual linkage between consciousness as knowledge and the character virtue of conscience as wise/integrated self-knowing and self-governance [56–58]. That is, the common etymological roots of these words may also point to their overlapping functions and potential inter-dependencies during the processes by which personhood is bootstrapped in sophisticated cognitive systems, like us [59].

While the functional significances of the brain's serotonin systems have not been thoroughly explored within FEP-AI, compelling parallels can nonetheless be identified. Within CB5T, moderate levels of serotonergic signaling would tend to correspond to Stability, or the protection of pre-existing strategies from disruption. These functions may have been conserved throughout evolution, as can be observed both in the locomotory modes of *C. elegans*, and even the foraging consequences of single celled organisms reducing directed motion upon encountering and consuming nutrient-rich meal (tryptophan → serotonin) [60, 61]. In this way, serotonin's functionality as a satiety—and potentially safety and successful sociality—promoting signal would provide a countervailing force to dopaminergic disinhibition of action, consistent with the opponent-process dynamics that have been observed on multiple levels of organization ranging from hypothalamic nuclei to frontal lobe attractor dynamics [62–65]. Within FEP-AI, physiological levels of serotonin (potentially resulting in greater occupancy of 5-HT1a relative to 5-HT2a receptors) have been associated with greater precision over interoceptive states, whose (allostatic) connections to the internal milieu and life management would be consistent with an association with Stability in CB5T. Notably with respect to computational psychiatry (and also ethology, these are the neurotransmitter systems agonized by SSRIs for depression and anxiety (and also dominance within social hierarchies). However, extreme levels of serotonergic signaling have been associated with the "relaxation" of beliefs and greater exploration in FEP-AI [66–68]. While CB5T has previously emphasized serotonin's potential role in modulating Stability, potential roles of this system for enhancing Plasticity is a promising direction for future research.

There is increasing interest in the effects of stimulating serotonergic 5-HT2a receptors for providing means for "how to change your mind" [69], and for achieving "altered traits via altered states" [70]. Compounds that act on these pathways have been described with various forms of suggestive terminology including psychedelics ("mind manifesting" and "higher states of consciousness"), hallucinogens (perception as inference),, entheogens (self-actualization and transpersonal psychology and spirituality), and even "entactogens" for non-classic psychedelics (healing from trauma and repairing excessively jagged/ruptured—and so non-navigable—free energy landscapes) [71]. Openness is the trait most commonly associated with potential change under psychedelic psychotherapy [72–74], but little work has examined the Big 5 aspects (where substantial effect sizes are most likely to be observed) [75, 76], nor what kinds of personality change could be possible with targeted interventions. With respect to specific mechanisms, it is notable that 5-HT2a receptors appear to be particularly concentrated on deep association cortices [77], which would be consistent with our suggestion that hierarchically-higher representations may be particularly relevant for stabilizing personality. The possibility for changing persons by changing the function of upper levels of cortical hierarchies has been compellingly described in terms of "hub collapse" and disrupted personhood with psychedelics and meditative states [78–82], the cybernetic properties of bowtie architectures allowing for an "allostatic overload" mechanisms for flexibly adjusting functional depth [54], as well as in terms of predictive processing and other machine learning principles [66, 67, 83, 84].

4 A Hypothesis on Personality Aspects and Social Power

The discovery of two (and only two) aspects underneath each of the Big 5 trait domains could potentially be partially explainable in terms of a fundamental axis of variation in active inference: that is, the degree to which prediction-error is minimized via either perception (i.e., updating internal models) or action (i.e., updating system-world states). This may lead to new testable hypotheses regarding the bio-computational processes contributing to personality variation. Could personalities be influenced by general tendencies with respect to adjusting the relative gain on predictions (including actions) or prediction-errors (i.e., sensory observations) in different inferential control hierarchies?

For example, increased gain on interoceptive precision has been associated with social power [85], potentially corresponding to more opportunities for inwardly focused attention (due to not having to constantly attend outwards in order to monitor environmental contingencies). If prediction errors are allowed to ascend to hierarchically higher (anterior) levels of the insula, then somatic information may be more likely to be accompanied by conscious access, and potentially the disinhibition of action (via coupling with frontoparietal control hierarchies over the predictive enactment of sequences of proprioceptive poses). Any neuromodulator or hormonal factor that agonizes the mesolimbic dopamine system may further contribute to the likelihood of connecting interoceptive percepts with actions, both via lowering disinhibition thresholds in the striatum and increasing coupling between relevant networks [86]. In terms of both phenomenology and functional significances, this may correspond to the experience of willing and empowerment through the exercising of agency [15, 87, 88].

Given the fact that sex/gender roles have evolved (on phylogenetic, cultural, and ontogenetic levels) in conjunction with power (or dominance) differentials, this could potentially help to explain the especially large effects observed with respect to male-female differences at the aspect level [76]. Similar differences could potentially be observed with respect to other social power differentials, such as race or class, and across all cases, may help to explain differences in either internalizing or externalizing in psychopathology, with the former being more likely to result in autonomic dysregulation as observed with respect to cardiac and gastrointestinal dysfunction [89, 90], but with the latter being more likely to result in accident proneness—or daring/improbable, but potentially great and heroic accomplishments [91, 92]. However, given the potentially rapid pace of cultural evolution, we might not expect empirical correlations between aspect-level personality traits and gender/race/class to be constant over time either across or within individuals [93–95].

5 Psychological Integration, Mindfulness, and Wellbeing

FEP-AI and CB5T both emphasize the importance of hierarchical organization for achieving complex goals. However, the stable pursuit of complex/distal objectives may depend on the integrity of hierarchical active inference [96]. This may further require the ability to down-regulate (predominantly interoceptive) prediction errors (i.e., emotionally self-regulate), so allowing for flexibility in prioritizing policies and not have the integrity of goal hierarchies be disrupted by proximal setbacks [55]. This flexible balancing of priorities could have transdiagnostic relevance [97, 98], theoretically promoting the formation of a kind of reflective equilibrium where goal-hierarchies become more elegant/realizable. Over time, this balanced adjustment and personal evolution could even help contribute to the kinds of integration and individuation discussed by self-actualization psychologists [3]. While the precise nature of mindfulness remains unclear in personality psychology [99], this state (and possibly a trait) may correspond to one of the most effective means of engaging in the kind of emotional self-regulation required for inter-temporally coherent active inference [24, 100]. Without this skill at internal navigation, we might expect excessive responses to prediction-errors, which may result in elevated Neuroticism when considered at the level of personality traits.

In theory, the difficulty of cultivating the kinds of cognitive (and affective) flexibility associated with mindfulness could potentially account for the failure to consistently observe a general factor of personality (GFP) [101–104]. For example, part of the reason that a GFP may not be reliably observed could be due to opposing relations between Stability and Plasticity. However, similarly to the relationship between epistemic and pragmatic value in active inference, synergy between these objectives ought to be possible (and is normatively required). Mindfulness may allow people to more readily (and flexibly achieve this dynamic balance, potentially even allowing individuals to more robustly occupy an "edge of chaos" regime where Stability and Plasticity positively interact/correlate. Even more, mindfulness could promote greater control through meta-awareness [105], so allowing one to be more effective in occupying inter-regimes without falling into one attractor or another. From this cybernetic view, mindfulness could be thought of as adaptively/flexibly stretching the region of phase space where adaptively balanced dynamics are possible [106].

Theoretically, systems that succeed in approaching this optimality frontier may exhibit the hallmarks of self-organized criticality [107]. The dynamic balance between order and chaos is characteristic of governance by critical-point attractors, an essential property for adaptive systems [108], and a promising avenue for research. Relationships may potentially be observable between personality variables and putative metrics of criticality in neural measures such as network flexibility [109], critical slowing [107], fractal dimension [110], and power-law distributions [111]. These criticality measures may exhibit positive correlations with Plasticity as reflecting a general potential for adaptive reorganization. [Relationships between criticality and Stability may be best accounted for by an inverted-U function (after controlling for Plasticity).] Correlations between these criticality measures, personality, and effectively minimizing free energy—as indexed by learning ability or positive affect [112–115]—could provide compelling evidence that "edge of chaos" dynamics may allow for optimality to be achieved across multiple regimes, including persons [98, 116] (Figs. 5 and 6).

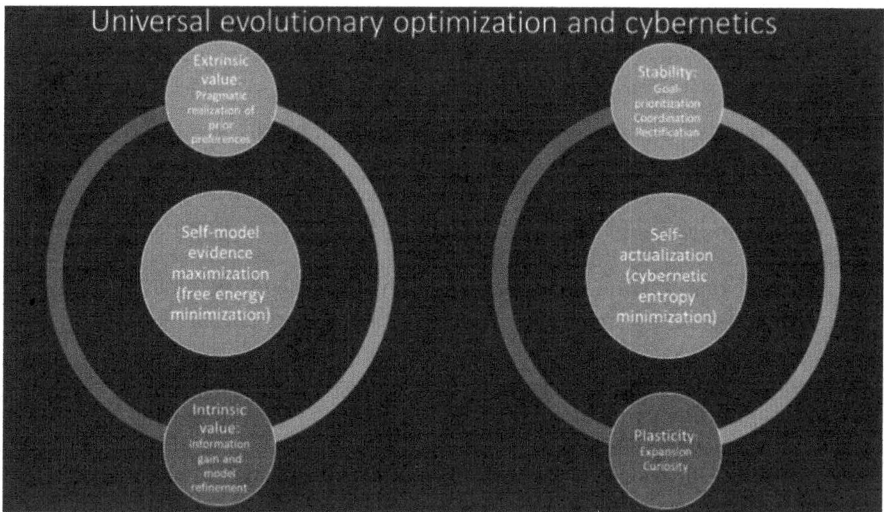

Fig. 5. Dynamic balance of Stability and Plasticity as cybernetic universality class.

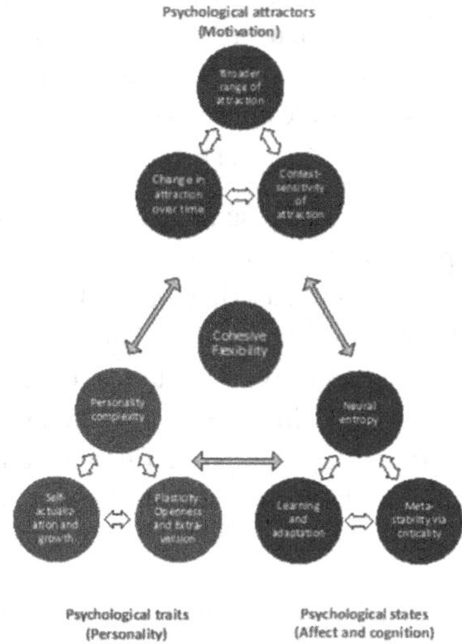

Fig. 6. Dynamic balancing of Stability and Plasticity and enhanced adaptivity via (cohesive) psychological flexibility.

6 Conclusions

In reviewing points of intersection between FEP-AI and CB5T, our intention was to provide a general sense for what might be possible for personality modeling. For more details, we refer interested readers to a recent book chapter on this topic [1]. The potential intersections of personality science with FEP-AI is a complex and deep topic, and as such we consider the preceding discussion to be more of a suggestion of potentially helpful directions for future work, rather than a definitive statement of canonical cross-mappings. More specifically, we believe it could be useful to investigate the following hypotheses:

- The meta-traits of Stability and Personality can be fruitfully applied to cybernetic systems at all scales.
- These meta-traits may have useful functional mappings with neuromodulators (understood as kinds of machine learning parameters) such as dopamine and serotonin.
- The identification of two aspects beneath each of the big 5 trait domains suggests potentially functionally significant opponent processes (or opposing modes of policy selection with respect to coherent organismic/agentic life histories).
- There may be fruitful correspondences between personality aspects and tendencies towards minimizing free energy via updating either internal models or external world states via respective perception or overt enaction.
- Understanding self-actualization may help to illuminate the question of what tends to be 'predicted' overall by systems such as us, with implications for both (multi-level) evolutionary theory and the psychology of wellbeing.
- Computational frameworks such as FEP-AI may be useful for explaining some of the cybernetic significances of major personality traits and their neurophysiological and behavioral correlates; e.g. neuroticism as sensitivity to overall cybernetic entropy, with centralized control structures such as the anterior cingulate and amygdala potentially acting as expected free energy integrators, and also potential sources of intervention for clinical condition (cf. loci for deep brain stimulation in depression).

We chose to discuss this work here because we are in the early stages of implementing a major research program inspired by these theories, where we will be demonstrating how stable prosocial personalities (as preferences) can be made to emerge in AI agents as the iteratively select policies and update of priors [117]. This project will initially be focused on developing architectures and integrative simulation environments for the FEP-AI community and personality modelers. However, we will expand this research program over the coming years to develop increasingly powerful (and human(ely)-aligned) agents. As such being able to precisely model the personalities of these agents may eventually be a matter of more than academic importance.

References

1. Safron, A., DeYoung, C.G.: Chapter 18 - Integrating cybernetic big five theory with the free energy principle: a new strategy for modeling personalities as complex systems. In: Wood, D., Read, S.J., Harms, P.D., Slaughter, A. (eds.) Measuring and Modeling Persons and Situations, pp. 617–649. Academic Press (2021). https://doi.org/10.1016/B978-0-12-819200-9.00010-7
2. Friston, K.J., Redish, A.D., Gordon, J.A.: Computational nosology and precision psychiatry. Comput. Psychiatry **1**, 2–23 (2017). https://doi.org/10.1162/CPSY_a_00001
3. DeYoung, C.G.: Cybernetic big five theory. J. Res. Pers. **56**, 33–58 (2015). https://doi.org/10.1016/j.jrp.2014.07.004
4. Damasio, A.: Self Comes to Mind: Constructing the Conscious Brain, Reprint Vintage, New York (2012)

5. Hirsh, J.B., Mar, R.A., Peterson, J.B.: Personal narratives as the highest level of cognitive integration. Behav. Brain Sci. **36**(3), 216–217 (2013). https://doi.org/10.1017/S0140525X12002269
6. Waddington, C.H.: Canalization of development and the inheritance of acquired characters. Nature **150**(3811), 150563a0 (1942). https://doi.org/10.1038/150563a0
7. Campbell, J.O.: Universal darwinism as a process of Bayesian inference. Front. Syst. Neurosci. **10**, 49 (2016). https://doi.org/10.3389/fnsys.2016.00049
8. Ramstead, M.J.D., Kirchhoff, M.D., Friston, K.J.: A tale of two densities: active inference is enactive inference (2019). http://philsci-archive.pitt.edu/16167/. Accessed 12 Dec 2019
9. Stepp, N., Turvey, M.T.: On strong anticipation. Cogn. Syst. Res. **11**(2), 148–164 (2010). https://doi.org/10.1016/j.cogsys.2009.03.003
10. Safron, A.: Integrated World Modeling Theory (IWMT) Expanded: Implications for Theories of Consciousness and Artificial Intelligence. PsyArXiv (2021). https://doi.org/10.31234/osf.io/rm5b2
11. Palacios, E.R., Isomura, T., Parr, T., Friston, K.J.: The emergence of synchrony in networks of mutually inferring neurons. Sci. Rep. **9**(1), 6412 (2019). https://doi.org/10.1038/s41598-019-42821-7
12. FitzGerald, T.H.B., Dolan, R.J., Friston, K.J.: Model averaging, optimal inference, and habit formation. Front. Hum. Neurosci. **8**, 457 (2014). https://doi.org/10.3389/fnhum.2014.00457
13. Hesp, C., Tschantz, A., Millidge, B., Ramstead, M., Friston, K., Smith, R.: Sophisticated affective inference: simulating anticipatory affective dynamics of imagining future events. In: Verbelen, T., Lanillos, P., Buckley, C.L., De Boom, C. (eds.) Active Inference. CCIS, pp. 179–186. Springer, Cham (2020). https://doi.org/10.1007/978-3-030-64919-7_18
14. Dennet, D.: The self as a center of narrative gravity. In: Kessel, F.S., Cole, P.M., Johnson, D.L. (eds.) Self and Consciousness: Multiple Perspectives. Lawrence Erlbaum (1992)
15. Safron, A.: The radically embodied conscious cybernetic Bayesian brain: from free energy to free will and back again. Entropy **23**(6), Art. no. 6 (2021). https://doi.org/10.3390/e23060783
16. Damasio, A.R.: The Strange Order of Things: Life, Feeling, and the Making of Cultures. Pantheon Books, New York (2018)
17. Friston, K.J., Frith, C.D.: Active inference, communication and hermeneutics. Cortex **68**, 129–143 (2015). https://doi.org/10.1016/j.cortex.2015.03.025
18. Edelman, G., Mountcastle, V.B.: The Mindful Brain: Cortical Organization and the Group-Selective Theory of Higher Brain Function, 1st edn. MIT Press, Cambridge (1978)
19. Seth, A.: Being You: A New Science of Consciousness. Dutton, New York (2021)
20. Deacon, T.W.: Incomplete Nature: How Mind Emerged from Matter, 1st edn. W. W. Norton & Company, New York (2011)
21. Dennett, D.: From Bacteria to Bach and Back: The Evolution of Minds, 1st edn. W. W. Norton & Company, New York (2017)
22. Ainslie, G.: Précis of breakdown of will. Behav. Brain Sci. **28**(5), 635–650; discussion 650–673 (2005). https://doi.org/10.1017/S0140525X05000117
23. Ainslie, G.: Selfish goals must compete for the common currency of reward. Behav. Brain Sci. **37**(2), 135–136 (2014). https://doi.org/10.1017/S0140525X13001933
24. Fujita, K., Carnevale, J.J., Trope, Y.: Understanding self-control as a whole vs. part dynamic. Neuroethics **11**(3), 283–296 (2018). https://doi.org/10.1007/s12152-016-9250-2
25. Minsky, M.: Society of Mind. Simon and Schuster, New York (1988)
26. Friston, K.J., FitzGerald, T., Rigoli, F., Schwartenbeck, P., Pezzulo, G.: Active inference: a process theory. Neural Comput. **29**(1), 1–49 (2017). https://doi.org/10.1162/NECO_a_00912
27. Bengio, Y., Deleu, T., Hu, E.J., Lahlou, S., Tiwari, M., Bengio, E.: GFlowNet Foundations. arXiv (2022). https://doi.org/10.48550/arXiv.2111.09266

28. Safron, A., Çatal, O., Verbelen, T.: Generalized Simultaneous Localization and Mapping (G-SLAM) as unification framework for natural and artificial intelligences: towards reverse engineering the hippocampal/entorhinal system and principles of high-level cognition. PsyArXiv (2021). https://doi.org/10.31234/osf.io/tdw82
29. Parr, T., Friston, K.J.: Uncertainty, epistemics and active inference. J. R. Soc. Interface **14**(136) (2017). https://doi.org/10.1098/rsif.2017.0376
30. Wade, S., Kidd, C.: The role of prior knowledge and curiosity in learning. Psychon. Bull. Rev. (2019). https://doi.org/10.3758/s13423-019-01598-6
31. Pezzulo, G., Friston, K.J.: The value of uncertainty: an active inference perspective. Behav. Brain Sci. **42** (2019). https://doi.org/10.1017/S0140525X18002066
32. Parr, T., Friston, K.J.: The discrete and continuous brain: from decisions to movement-and back again. Neural Comput. **30**(9), 2319–2347 (2018). https://doi.org/10.1162/neco_a_01102
33. Gopnik, A.: The Philosophical Baby: What Children's Minds Tell Us About Truth, Love, and the Meaning of Life. Macmillan, New York (2009)
34. Gopnik, A., et al.: Changes in cognitive flexibility and hypothesis search across human life history from childhood to adolescence to adulthood. PNAS **114**(30), 7892–7899 (2017). https://doi.org/10.1073/pnas.1700811114
35. Gopnik, A., Griffiths, T.L., Lucas, C.G.: When younger learners can be better (or at least more open-minded) than older ones. Curr. Dir. Psychol. Sci. **24**(2), 87–92 (2015). https://doi.org/10.1177/0963721414556653
36. Kaplan, R., Friston, K.J.: Planning and navigation as active inference. Biol. Cybern. **112**(4), 323–343 (2018). https://doi.org/10.1007/s00422-018-0753-2
37. Bruineberg, J., Rietveld, E.: Self-organization, free energy minimization, and optimal grip on a field of affordances. Front. Hum. Neurosci. **8**, 599 (2014). https://doi.org/10.3389/fnhum.2014.00599
38. Kiverstein, J., Miller, M., Rietveld, E.: The feeling of grip: novelty, error dynamics, and the predictive brain. Synthese **196**(7), 2847–2869 (2019). https://doi.org/10.1007/s11229-017-1583-9
39. Parr, T., Friston, K.J.: Working memory, attention, and salience in active inference. Sci. Rep. **7**(1), 14678 (2017). https://doi.org/10.1038/s41598-017-15249-0
40. Seth, A.K., Friston, K.J.: Active interoceptive inference and the emotional brain. Phil. Trans. R. Soc. B **371**(1708), 20160007 (2016). https://doi.org/10.1098/rstb.2016.0007
41. Badcock, P.B., Friston, K.J., Ramstead, M.J.D.: The hierarchically mechanistic mind: a free-energy formulation of the human psyche. Phys. Life Rev. (2019). https://doi.org/10.1016/j.plrev.2018.10.002
42. Constant, A., Ramstead, M.J.D., Veissière, S.P.L., Friston, K.J.: Regimes of expectations: an active inference model of social conformity and human decision making. Front. Psychol. **10**, 679 (2019). https://doi.org/10.3389/fpsyg.2019.00679
43. Gelfand, M.: Rule Makers, Rule Breakers: How Tight and Loose Cultures Wire Our World. Simon and Schuster, New York (2018)
44. Latzman, R.D., DeYoung, C.G.: Using empirically-derived dimensional phenotypes to accelerate clinical neuroscience: the Hierarchical Taxonomy of Psychopathology (HiTOP) framework. Neuropsychopharmacology **45**, 1–4 (2020). https://doi.org/10.1038/s41386-020-0639-6
45. DeYoung, C.G.: The neuromodulator of exploration: a unifying theory of the role of dopamine in personality. Front. Hum. Neurosci. **7** (2013). https://doi.org/10.3389/fnhum.2013.00762
46. Adams, R.A., et al.: Variability in action selection relates to striatal dopamine 2/3 receptor availability in humans: a PET neuroimaging study using reinforcement learning and active inference models. Cereb Cortex (2020). https://doi.org/10.1093/cercor/bhz327

47. FitzGerald, T.H.B., Dolan, R.J., Friston, K.J.: Dopamine, reward learning, and active inference. Front. Comput. Neurosci. **9** (2015). https://doi.org/10.3389/fncom.2015.00136
48. Friston, K.J., et al.: Dopamine, affordance and active inference. PLoS Comput. Biol. **8**(1), e1002327 (2012). https://doi.org/10.1371/journal.pcbi.1002327
49. Friston, K.J., Schwartenbeck, P., FitzGerald, T., Moutoussis, M., Behrens, T., Dolan, R.J.: The anatomy of choice: dopamine and decision-making. Philos. Trans. R. Soc. Lond. B Biol. Sci. **369**(1655) (2014). https://doi.org/10.1098/rstb.2013.0481
50. Parr, T., Corcoran, A.W., Friston, K.J., Hohwy, J.: Perceptual awareness and active inference. Neurosci. Conscious **2019**(1) (2019). https://doi.org/10.1093/nc/niz012
51. Dalege, J., Borsboom, D., van Harreveld, F., van der Maas, H.L.J.: The attitudinal entropy (AE) framework as a general theory of individual attitudes. Psychol. Inq. **29**(4), 175–193 (2018). https://doi.org/10.1080/1047840X.2018.1537246
52. Hirsh, J.B., Mar, R.A., Peterson, J.B.: Psychological entropy: a framework for understanding uncertainty-related anxiety. Psychol. Rev. **119**(2), 304–320 (2012). https://doi.org/10.1037/a0026767
53. Damasio, A.R.: Descartes' Error: Emotion, Reason, and the Human Brain, 1st edn. Harper Perennial, New York (1995)
54. Goekoop, R., de Kleijn, R.: How higher goals are constructed and collapse under stress: a hierarchical Bayesian control systems perspective, arXiv:2004.09426 (2020). http://arxiv.org/abs/2004.09426. Accessed 25 Apr 2020
55. Rueter, A.R., Abram, S.V., MacDonald, A.W., Rustichini, A., DeYoung, C.G.: The goal priority network as a neural substrate of Conscientiousness. Hum. Brain Mapp. **39**(9), 3574–3585 (2018). https://doi.org/10.1002/hbm.24195
56. Metzinger, T.: The Ego Tunnel: The Science of the Mind and the Myth of the Self, 1st edn. Basic Books, New York (2009)
57. Lewis, C.S.: Studies in Words, 2 edn. Cambridge University Press, Cambridge; New York (2013)
58. Frith, C., Metzinger, T.: What's the use of consciousness? (2016). https://doi.org/10.7551/mitpress/9780262034326.003.0012
59. Tomasello, M.: A Natural History of Human Thinking. Harvard University Press, Cambridge (2014)
60. Azmitia, E.C.: Modern views on an ancient chemical: serotonin effects on cell proliferation, maturation, and apoptosis. Brain Res. Bull. **56**(5), 413–424 (2001). https://doi.org/10.1016/S0361-9230(01)00614-1
61. Chase, D.: Biogenic amine neurotransmitters in C. elegans. WormBook (2007). https://doi.org/10.1895/wormbook.1.132.1
62. Chudasama, Y., Robbins, T.W.: Functions of frontostriatal systems in cognition: comparative neuropsychopharmacological studies in rats, monkeys and humans. Biol. Psychol. **73**(1), 19–38 (2006). https://doi.org/10.1016/j.biopsycho.2006.01.005
63. Di Pietro, N.C., Seamans, J.K.: Dopamine and serotonin interactions in the prefrontal cortex: insights on antipsychotic drugs and their mechanism of action. Pharmacopsychiatry **40**(Suppl. 1), S27-33 (2007). https://doi.org/10.1055/s-2007-992133
64. Koster, R., et al.: Big-loop recurrence within the hippocampal system supports integration of information across episodes. Neuron **99**(6), 1342-1354.e6 (2018). https://doi.org/10.1016/j.neuron.2018.08.009
65. Olvera-Cortés, M.E., Anguiano-Rodríguez, P., López-Vázquez, M.A., Alfaro, J.M.C.: Serotonin/dopamine interaction in learning. Prog. Brain Res. **172**, 567–602 (2008). https://doi.org/10.1016/S0079-6123(08)00927-8
66. Carhart-Harris, R.L., Friston, K.J.: REBUS and the anarchic brain: toward a unified model of the brain action of psychedelics. Pharmacol. Rev. **71**(3), 316–344 (2019). https://doi.org/10.1124/pr.118.017160

67. Safron, A.: On the Varieties of Conscious Experiences: Altered Beliefs Under Psychedelics (ALBUS). PsyArXiv (2020). https://doi.org/10.31234/osf.io/zqh4b
68. Safron, A., Sheikhbahaee, Z.: Dream to explore: 5-HT2a as adaptive temperature parameter for sophisticated affective inference. In: Kamp, M., et al. (eds.) Machine Learning and Principles and Practice of Knowledge Discovery in Databases, pp. 799–809. Springer, Cham (2021). https://doi.org/10.1007/978-3-030-93736-2_56
69. Pollan, M.: How to Change Your Mind: The New Science of Psychedelics. Penguin Books Limited, London (2018)
70. Goleman, D., Davidson, R.J.: Altered Traits: Science Reveals How Meditation Changes Your Mind, Brain, and Body, Illustrated Avery, New York (2017)
71. Safron, A., Johnson, M.: Classic psychedelics: past uses, present trends, future possibilities. PsyArXiv (2022). https://doi.org/10.31234/osf.io/eys2j
72. Blain, S.D., Longenecker, J.M., Grazioplene, R.G., Klimes-Dougan, B., DeYoung, C.G.: Apophenia as the disposition to false positives: a unifying framework for openness and psychoticism. J. Abnorm. Psychol. **129**(3), 279–292 (2020). https://doi.org/10.1037/abn0000504
73. Erritzoe, D., Smith, J., Fisher, P.M., Carhart-Harris, R., Frokjaer, V.G., Knudsen, G.M.: Recreational use of psychedelics is associated with elevated personality trait openness: exploration of associations with brain serotonin markers. J. Psychopharmacol. 269881119827891 (2019). https://doi.org/10.1177/0269881119827891
74. MacLean, K.A., Johnson, M.W., Griffiths, R.R.: Mystical experiences occasioned by the hallucinogen psilocybin lead to increases in the personality domain of openness. J. Psychopharmacol. **25**(11), 1453–1461 (2011). https://doi.org/10.1177/0269881111420188
75. DeYoung, C.G., Carey, B.E., Krueger, R.F., Ross, S.R.: 10 aspects of the big five in the personality inventory for DSM-5. Personal Disord. **7**(2), 113–123 (2016). https://doi.org/10.1037/per0000170
76. Weisberg, Y.J., DeYoung, C.G., Hirsh, J.B.: Gender differences in personality across the ten aspects of the big five. Front. Psychol. **2**, 178 (2011). https://doi.org/10.3389/fpsyg.2011.00178
77. Kringelbach, M.L., et al.: Dynamic coupling of whole-brain neuronal and neurotransmitter systems. PNAS (2020). https://doi.org/10.1073/pnas.1921475117
78. Letheby, C.: Philosophy of Psychedelics. Oxford University Press, Oxford (2021)
79. Letheby, C., Gerrans, P.: Self unbound: ego dissolution in psychedelic experience (2017). https://philarchive.org/rec/LETSUE. Accessed 31 Aug 2023
80. Brouwer, A., Carhart-Harris, R.L.: Pivotal mental states. J. Psychopharmacol. 269881120959637 (2020). https://doi.org/10.1177/0269881120959637
81. Hinton, D.E., Kirmayer, L.J.: The flexibility hypothesis of healing. Cult. Med. Psychiatry **41**(1), 3–34 (2017). https://doi.org/10.1007/s11013-016-9493-8
82. Deane, G.: Dissolving the self: active inference, psychedelics, and ego-dissolution. PhiMiSci **1**(1), 1–27 (2020). https://doi.org/10.33735/phimisci.2020.I.39
83. Carhart-Harris, R.L., et al.: Canalization and plasticity in psychopathology. Neuropharmacology **226**, 109398 (2023). https://doi.org/10.1016/j.neuropharm.2022.109398
84. Juliani, A., Safron, A., Kanai, R.: Deep CANALs: A Deep Learning Approach to Refining the Canalization Theory of Psychopathology. PsyArXiv (2023). https://doi.org/10.31234/osf.io/uxmz6
85. Moeini-Jazani, M., Knoeferle, K., de Molière, L., Gatti, E., Warlop, L.: Social power increases interoceptive accuracy. Front. Psychol. **8**, 1322 (2017). https://doi.org/10.3389/fpsyg.2017.01322
86. Dang, L.C., O'Neil, J.P., Jagust, W.J.: Dopamine supports coupling of attention-related networks. J. Neurosci. **32**(28), 9582–9587 (2012). https://doi.org/10.1523/JNEUROSCI.0909-12.2012

87. de Abril, I.M., Kanai, R.: A unified strategy for implementing curiosity and empowerment driven reinforcement learning, arXiv:1806.06505 (2018). http://arxiv.org/abs/1806.06505. Accessed 15 Dec 2018
88. Safron, A.: On the degrees of freedom worth having: psychedelics as means of understanding and expanding free will. PsyArXiv (2021). https://doi.org/10.31234/osf.io/m2p6g
89. van der Kolk, B.: The Body Keeps the Score: Brain, Mind, and Body in the Healing of Trauma, Reprint Penguin Books, New York (2015)
90. Porges, S.W.: The polyvagal theory: New insights into adaptive reactions of the autonomic nervous system. Cleve. Clin. J. Med. **76**(Suppl. 2), S86–S90 (2009). https://doi.org/10.3949/ccjm.76.s2.17
91. Peterson, J.B.: Maps of Meaning: The Architecture of Belief. Psychology Press (1999)
92. Trofimova, I.: Do psychological sex differences reflect evolutionary bisexual partitioning? Am. J. Psychol. **128**(4), 485–514 (2015). https://doi.org/10.5406/amerjpsyc.128.4.0485
93. Hernandez, L., Gonzalez, L., Murzi, E., Páez, X., Gottberg, E., Baptista, T.: Testosterone modulates mesolimbic dopaminergic activity in male rats. Neurosci. Lett. **171**(1–2), 172–174 (1994)
94. Höfer, P., Lanzenberger, R., Kasper, S.: Testosterone in the brain: Neuroimaging findings and the potential role for neuropsychopharmacology. Eur. Neuropsychopharmacol. (2012). https://doi.org/10.1016/j.euroneuro.2012.04.013
95. Travison, T.G., Araujo, A.B., O'Donnell, A.B., Kupelian, V., McKinlay, J.B.: A population-level decline in serum testosterone levels in American men. J. Clin. Endocrinol. Metab. **92**(1), 196–202 (2007). https://doi.org/10.1210/jc.2006-1375
96. Pezzulo, G., Rigoli, F., Friston, K.J.: Hierarchical active inference: a theory of motivated control. Trends Cogn. Sci. **22**(4), 294–306 (2018). https://doi.org/10.1016/j.tics.2018.01.009
97. DeYoung, C.G., Krueger, R.F.: A cybernetic theory of psychopathology. Psychol. Inq. **29**(3), 117–138 (2018). https://doi.org/10.1080/1047840X.2018.1513680
98. Hayes, S.C.: A Liberated Mind: How to Pivot Toward What Matters. Penguin (2019)
99. Hanley, A.W., Garland, E.L.: The mindful personality: a meta-analysis from a cybernetic perspective. Mindfulness **8**(6), 1456–1470 (2017). https://doi.org/10.1007/s12671-017-0736-8
100. Ainslie, G.: Picoeconomics: The Strategic Interaction of Successive Motivational States within the Person, Reissue Cambridge University Press, Cambridge (2010)
101. Caspi, A., et al.: The p factor: one general psychopathology factor in the structure of psychiatric disorders? Clin. Psychol. Sci. **2**(2), 119–137 (2014). https://doi.org/10.1177/2167702613497473
102. Hopwood, C.J., Wright, A.G.C., Donnellan, M.B.: Evaluating the evidence for the general factor of personality across multiple inventories. J. Res. Pers. **45**(5), 468–478 (2011). https://doi.org/10.1016/j.jrp.2011.06.002
103. Martel, M.M., et al.: A general psychopathology factor (P factor) in children: structural model analysis and external validation through familial risk and child global executive function. J. Abnorm. Psychol. **126**(1), 137–148 (2017). https://doi.org/10.1037/abn0000205
104. Veselka, L., Schermer, J.A., Petrides, K.V., Cherkas, L.F., Spector, T.D., Vernon, P.A.: A general factor of personality: evidence from the HEXACO model and a measure of trait emotional intelligence. Twin Res. Hum. Genet. **12**(5), 420–424 (2009). https://doi.org/10.1375/twin.12.5.420
105. Sandved Smith, L., Hesp, C., Lutz, A., Mattout, J., Friston, K., Ramstead, M.: Towards a formal neurophenomenology of metacognition: modelling meta-awareness, mental action, and attentional control with deep active inference, PsyArXiv preprint (2020). https://doi.org/10.31234/osf.io/5jh3c
106. Moretti, P., Muñoz, M.A.: Griffiths phases and the stretching of criticality in brain networks. Nat. Commun. **4**, 2521 (2013). https://doi.org/10.1038/ncomms3521

107. Friston, K., Breakspear, M., Deco, G.: Perception and self-organized instability. Front. Comput. Neurosci. **6**, 44 (2012). https://doi.org/10.3389/fncom.2012.00044
108. Hoffmann, H., Payton, D.W.: Optimization by self-organized criticality. Sci. Rep. **8**(1), 2358 (2018). https://doi.org/10.1038/s41598-018-20275-7
109. Bassett, D.S., Wymbs, N.F., Porter, M.A., Mucha, P.J., Carlson, J.M., Grafton, S.T.: Dynamic reconfiguration of human brain networks during learning. PNAS **108**(18), 7641–7646 (2011). https://doi.org/10.1073/pnas.1018985108
110. Tagliazucchi, E., Balenzuela, P., Fraiman, D., Chialvo, D.R.: Criticality in large-scale brain fMRI dynamics unveiled by a novel point process analysis. Front. Physiol. **3**, 15 (2012). https://doi.org/10.3389/fphys.2012.00015
111. Atasoy, S., Roseman, L., Kaelen, M., Kringelbach, M.L., Deco, G., Carhart-Harris, R.L.: Connectome-harmonic decomposition of human brain activity reveals dynamical repertoire re-organization under LSD. Sci. Rep. **7**(1), 17661 (2017). https://doi.org/10.1038/s41598-017-17546-0
112. Betzel, R.F., Satterthwaite, T.D., Gold, J.I., Bassett, D.S.: Positive affect, surprise, and fatigue are correlates of network flexibility. Sci. Rep. **7**(1), 520 (2017). https://doi.org/10.1038/s41598-017-00425-z
113. Hesp, C., Smith, R., Allen, M., Friston, K., Ramstead, M.: Deeply Felt Affect: The Emergence of Valence in Deep Active Inference, PsyArXiv, preprint (2019). https://doi.org/10.31234/osf.io/62pfd
114. Joffily, M., Coricelli, G.: Emotional valence and the free-energy principle. PLoS Comput. Biol. **9**(6), e1003094 (2013). https://doi.org/10.1371/journal.pcbi.1003094
115. Reddy, P.G., et al.: Brain state flexibility accompanies motor-skill acquisition. Neuroimage **171**, 135–147 (2018). https://doi.org/10.1016/j.neuroimage.2017.12.093
116. Jennings, J.R., Allen, B., Gianaros, P.J., Thayer, J.F., Manuck, S.B.: Focusing neurovisceral integration: cognition, heart rate variability, and cerebral blood flow. Psychophysiology **52**(2), 214–224 (2015). https://doi.org/10.1111/psyp.12319
117. Safron, A., Sheikhbahaee, Z., Hay, N., Orchard, J., Hoey, J.: Value cores for inner and outer alignment: simulating personality formation via iterated policy selection and preference learning with self-world modeling active inference agents. In: Buckley, C.L., et al. (eds.) Active Inference. CCIS, vol. 1721, pp. 343–354. Springer, Cham (2023). https://doi.org/10.1007/978-3-031-28719-0_24

On Embedded Normativity an Active Inference Account of Agency Beyond Flesh

Avel Guénin–Carlut[1,4,5(✉)] and Mahault Albarracin[2,3]

[1] Department of Engineering and Informatics, University of Sussex, Falmer, UK
avel@disroot.org
[2] VERSES Lab, Los Angeles, CA, USA
[3] Departement d'informatique, Universite du Quebec a Montreal, Montreal, Canada
[4] Active Inference Institute, Davis, USA
[5] Kairos Research, Chavín de Huántar, Peru
https://kairos-research.org

Abstract. We introduce and motivate the concept of *embedded normativity* to account for the externalization of social norms in the material environment through human social activity. We ground this notion in the Active Inference framework, and more specifically through the derived Skilled Intentionality framework of ecological perception and action. This framework considers that skilled agents experience their world as a landscape of affordances, or opportunities for action. This landscape is inherently normative, as its experience is tied to the agent's anticipations over its own behaviour (and therefore, indirectly, to its motivations). We emphasize that given this framework, normativity does not exist inside or outside the agent's boundaries, but is brought about by its engagement with the world. We discuss the dynamics of *internalization* and *externalization* by which agents come to project normativity onto elements of their environment, and experience this normativity as a simple attraction toward favoured states. Given this account, we revisit earlier descriptions of the shared material and sociocultural niche enable the broadcasting and integration of norms. Finally, we discuss how embedded normativity can be brought into existence by the perception of humans, and relate our discussion to the ontological stance of participatory realism. We hope that our argument contributes to a variety of debates ranging from social ontology to epistemology, but most notably those regarding the relation between cognitive and material culture and the localization of cognition.

Keywords: Embodied cognitive science · Skilled intentionality · Active Inference · Cognitive niche construction · Participatory realism

1 Introduction

Cognitive science has traditionally attempted to explain human behaviour by calling onto computational processes happening in the brain. By construction, this approach abstracts the role of the material and sociocultural environment in

C. L. Buckley et al. (Eds.): IWAI 2023, CCIS 1915, pp. 91–105, 2024.
https://doi.org/10.1007/978-3-031-47958-8_7

the construction of human behaviour. Yet, interindividual coordination entails participation in the set of (implicit or explicit) norms and rules which underlie social activity. Basic examples could be the simple rules of the game of catch, the use of a specific language, or the distribution of speech in a conversation. Failure to respect those norms typically means failure to coordinate. However, the way we understand those norms is inherently reconstructive. The correct actions in a given context are not inherently known to us; instead, we construct them through trial and error throughout our development. In Wittgensteinian terms, we could claim that human social behaviour is influenced by the participation in a collective "form of life" [53], an expression which (while never properly defined by the philosopher) emphatically highlights the constructive aspect of the structure underlying social activity.

We propose to account for the co-construction between encultured cognition and the environment in which it is embedded through the way social normativity becomes integrated and externalized within a given social and material niche, for which we coin the term *embedded normativity*. This term describes forms of *normativity*, *i.e.* the property of judging certain actions or outcomes as desirable or not through specific evaluative norms and values, which are *embedded* within an agent's ecological and cognitive niche - rather than following from purely internal metabolic and cognitive processes. It is meant to highlight that all forms of normativity ultimately constitute embedded normativity, assuming that normativity is produced by a process of attunement in which organisms change their structure (through learning and perception) and the structure of the world (through action) so as to maximise coherence between their embodied expectations about the world and their actual sensorimotor flow. In so doing, the model of the world implicitly encoded within an agent's activity becomes entangled with the statistical structure of the environment from whence they emerged (i.e., the generative process) [7], in a way that prevents the distinction between an *internal* realm of decision making and an *external* realm which to perceive and act upon.

The concept of embedded normativity was recently mobilized in cognitive archaeology to defend the possibility of inferring the structure of social organizations from the material traces of past societies [33,34]. The argument goes as follows: if normativity is indeed embedded in the material and social environment one experiences, rather than in the architecture of the brain, maybe we can reconstruct from the archaeological landscapes we investigate the norms and values of the agents that experienced (and shaped) that landscape. We intend to hereby present this notion in a more focused and systematic manner, exposing its formal grounding as well as the conceptual questions it raises. In particular, we argue that embedded normativity is not present *outside* the perspective of a situated agent, but precisely emerges from the multiscale integration of social and cognitive constraints occurring *through* engagement with the world. Our argument aims to inform an existing debate between artefact-first [41] to cognition-first [50] accounts of the relation between material culture and human cognition, by dissolving the dichotomy between the two.

The account we propose is based on the Active Inference framework (ActInf), a neurocomputational theory positing that cognition works through the systematic prediction of expected sensorimotor states and the minimization of prediction error [37,42]. Active Inference affords a rich conceptual model of how humans integrate and enact cultural norms and values, as developed for example in [38,39,44,52]. Perhaps most importantly, it emerges as a special case of the Free Energy Principle (FEP), a mathematical framework describing how biological systems resist disorder and maintain their existence through minimizing free energy by constructing cognitive meaning (formally speaking, Bayesian belief distributions) from dynamical self-organisation [14,46]. More precisely, the FEP accounts for cognition as a process of dynamical individuation as a well-defined system ongoing simultaneously at many nested scales of analysis [35,36,43,45] - and not only within the brain, but wherever we can identify well-defined systems with measurable statistical regularities.

Under the FEP, the organism's internal states do indeed garner and encode exploitable, action-guiding dynamics about environmental states. However, they are established and maintained through active inference, that is, through patterns of adaptive action. This suggests that cognition is not a passive process of receiving information from the environment, but an active process of engaging with and shaping the environment. If we take this picture seriously, meaning emerges through the process of engagement with the world, while affording the multi-scale integration of nested boundaries of cognitive individuation [43]. In this light, we clarify that embedded normativity does not exist *inside* or *outside* the brain, but rather in the statistical properties of engagement with the world by cultural agents. This accounts provide a novel argument within the existing debate over the localization of cognition, as well as the role of materiality in human minds. Not only do we claim that the norms governing cognition exist at the interface between an agent and their niche, but we also claim that those norms are constructed through the (constrained) activity of the agent.

2 Encultured Cognition as Active Inference

The Free Energy Principle (FEP) is a theoretical framework that first emerged as a description of the mechanics of the human brain [23], and was then extended to describe the behavior of living systems [24,25] and more generally self-organization in dynamical systems [26]. At its core, the FEP posits that all cognitive agents strive to minimize their Variational Free Energy (VFE), a measure of surprise or uncertainty about their sensory inputs and motor outputs[1]. This is achieved by improving their internal model of the world to better predict sensory inputs, and by acting on the world to bring about sensory inputs that conform to their predictions. Critically, the minimization of VFE is a mathematical consequence of the existence of a Markov Blanket, *i.e.* a collection of states

[1] Formally, Variational Free Energy constitutes an upper bound over the surprise. However, we do not need to focus on the specifics of the formalism in the present article.

which mediate the interaction between an agent and its environment[2]. This process of minimizing free energy is thought to underlie perception, learning, and action [5,37], but also the biological processes of development and evolution [40]. The FEP therefore provides a mathematical formulation of the tendency of living systems to maintain themselves in a restricted set of states while embedded in a fluctuating, partially observed environment, and this at multiple nested scales of analysis [35,45]. Although the references discussed here may look esoteric, and that their accessibility is comprised by the shift in the meaning of key concepts over time, we may redirect the reader toward accessible discussion of the underlying theory in its latest articulation [42].

In the context of the human brain, the FEP motivates Active Inference [27], a mechanical theory of the dynamics of the mind drawing heavily from predictive processing. The core idea of this line of research is that the brain produces predictions (or, more minimally, anticipation) of the upcoming sensory states, and "perceives" this world of imagination. This design principle allows a straighforward explanation of the uncanny computational power of the mind, its ability to function with very limited data, the impression we have to experience a continuous environment even though our sensations are architecturally limited in scope and features, and the prevalence of top-down neuronal connections even in regions of the brain which are associated with sensory processing [10]. Active Inference is particular in its two principal features: first, it postulates that the brain optimizes its generative model of the world (*i.e.* its posterior belief over the causes of its sensations) specifically by minimizing variational free energy (which means performing approximately Bayes-optimal inference over sensory and motor states); second, it considers that motor commands themselves constitute predictions/anticipations of motor activity. In other words, action is modeled as a self-fulfilling prophecy, where agents predict their own actions and then generate evidence for these predictions through their actions. This enables the implementation of very rich patterns of regulation in behavior agents interact with their environment, gather information, and update their beliefs based on new data in a nearly optimal way. This is to be contrasted with the more classical picture that the course of action is computed after perception as an explicit series of motor command which is then enacted, which could only enable online adaptive regulation at a prohibitive computational cost [37]. Given those considerations, we consider that Active Inference is essentially correct in identifying and formalizing the general design principles that power human cognitive architecture.

[2] Technically, the mutual information between the attracting distribution of the internal states of the agent (A) and the attracting distribution of the external states of its environment (E) become zero when conditioned on the attracting distribution of the boundary (B), *i.e.* $I(A;E|B) = 0$. The distinction with our initial statement is that it still allows direct causation from the environment to the agent or vice-versa, assuming that this causation does not translate into the dependance of the attracting distributions. We have used the shortcut above as it is largely used, more immediately intuitive, and the distinction appears nowhere in our argument. However, we refer the reader to [1], would they want to deepen their understanding of the issue.

In the general case, Active Inference may be considered as a process of attunement between an agent and the environment it inhabits, as both (or more precisely, their statistics) become predictive of each other. This affords directly an interpretation in terms of niche construction [16], as the agent *de facto* recruits its environment in producing the statistical regularities which underlie its existence. For example, the traces the mammal leaves when foraging may leave a path they will consequently use to assist (and to some extent perform in their stead) the function of spatial navigation. In Active Inference terms, the tunnel "predicts" states that enable the continued existence of the mammal, and conversely it provides a regular niche for the mammal to predict. A more compelling example may be the act of writing to remember things, or to compute complex calculations. Just like regular niche construction can be understood as an extension of the phenotype, cognitive niche construction can be understood as an extension of the mind [15]. The idea that the relevant cognitive process may rely on external states is entirely unproblematic, as (from the perspective of Active Inference) cognition is performed by the dynamical flow of the coupled agent-environment system, and not by any of its subparts [43].

These considerations provide a straightforward account of encultured cognition, acknowledging that the human niche includes other humans and their cultural practices. Indeed, Active Inference suggest that if humans belong to each other's environments, the process of trying to predict each other's actions (and what they would do in a situation such as ours) lead to the active construction of shared goals and narratives in interaction with others through a process known as "Thinking Through Other Minds" (TTOM). It affords the transmission of cultural representations, such as the meanings of specific symbols or the content of social norms, but also the integration of those representation in our expectations and our actions, as well the participative construction of their meaning. Critically, the content of representations or norms is not information that an agent may rationally decide to acknowledge or ignore. The content of representations is integrated in the very flow of the expectations by which we understand and act in the world, and therefore in the opportunity for actions we experience [17,44,52]. What we actually do is determined not by reason or passion, but by the flow of expectations which generates our perceptions, actions and cognitive attitudes [11]. Of course, this is strongly coupled to (and constrained by) our social identity [31], as well as (implicit or explicit) social norms [2]. For example, an agent that would perceive themselves as a parent and believe that parents care for their children would not experience a choice in whether or not to care for their children. They would care for their children simply because they expect themselves to do (or they would need to revise their identity as a parent, or the belief they have about what parents do). In that context, the relevant scale to explain the behaviour of the agent would be the shared cultural niche which led to the development of its identity and the norms it embeds, rather than the individual attitudes of the agent.

3 The Skilled Intentionality Framework and Embedded Normativity

The Skilled Intentionality Framework (SIF) is a theoretical approach that aims to integrate the Active Inference Framework presented above with the earlier approach of ecological psychology. Ecological psychology, unlike the early program of cognitive science, focused on the process of direct engagement with the world, independently from representation and higher-order "computation". A core notion of ecological psychology is the notion of *affordance*, which refers to the possibilities for action that a given environment *affords* to a given agent [30]. For example, a chair *affords* sitting while a lamp does not, and a lamp *affords* lighting while a chair does not. Critically, while an affordance constitutes a direct relationship between the structure of an organism and this of an object, an affordance as an object of perception entails a direct invitation (or, to use the contextually relevant terminology, *solicitation*) for action. Given those consideration, the notion of *skilled intentionality* aims to capture the way agents systematically maintains themselves in a metastable zone[3] where they are able to recognize the structure of their environment, and act selectively on the affordances which enable them to continue this process [8]. The authors argue that this ecological perspective is compatible with the Active Inference paradigm, as the tendency for skilled agents to strive toward an 'optimal grip' on their environment can be described as a process of minimizing surprise or prediction error, thereby maximizing predictability. In the words of the authors: "a skilled climber is anticipating the affordances ahead; she does not just get a grip on the next hold in climbing, say, but also anticipates that she needs to be able to move on after that. [...] One can see again that in such a metastable state, one is flexibly able to switch between different movement regimes and better fit to adapt to the specific details of the environmental aspects". Most importantly, perhaps, Active Inference provides an explanation of how the perception of the world as a landscape of affordances emerges: it is through a process of predictive processing, where the brain generates and updates predictions about the causes of sensory inputs, that the agent anticipates and engages in the perception-action cycle [51].

This account entails a profound duality between the regimes of normativity enacted by the agent and the very structure of its experience. Yet again borrowing from [8]: "For an expert boxer the zone of optimal metastable distance will solicit moving toward, because this zone offers a wide range of action opportunities and the possibility to flexibly switch between them in line with what the dynamically changing environment demands or solicits". In other words, the

[3] Metastability is a concept in physics and dynamical systems theory describing states which, while robust with regard to infinitesimal perturbations, are absorbed within a nearby attractors states with relatively small perturbations. In neuroscience, the term is used somewhat abusively to describe the persistence of oscillations away from equilibrium, which is widely understood to be a critical condition for cognition (see *e.g.* [47] for a recent discussion of the issue).

skilled agent described by the SIF does not experience the states of the world as a given set of perceptual states, on which they compute then enact plans of action to reach their goals. Rather, the skilled agent experiences its world as a stream of invitations for action, on which they can act seamlessly. This can be related to the phenomenology of the state of *flow* described by positive psychology [19], or to the *wei wu wei* ("action without action", or "action without effort") of ancient Chinese philosophy [49].[4] For the skilled agent, intentionality emerges as an attuned synergy between the flow of action and sensory signal, rather than as the intervention of an overarching driver such as the Self or the Will. In consequence, the norms and values that guide behaviour reduce to the experience of affordances and sollications made by the agent, and of their intrinsic valence. For the most part, normativity is experienced as coming from *outside*, in the realm of perceptions which elicit anticipations and actions.

To be clear, there are many nuances to be included in that statement. The norms and value underlying behaviour are not in fact imposed from the outside. Rather, they arise from the agent's own predictive models, which are shaped by both their individual history, by the cultural and social norms through which they understand the world [15], and the concrete features of their sensory environment. Additionally, the sensations that guide behaviour are not limited to so-called "exteroceptive" sensation which inform us of the states of the outside world. They also include interoceptive and proprioceptive sensations, all of which contribute to the generation of anticipations. The normativity embedded in proprioception would feel like it comes from *inside*, given that the agent is capable to discriminate the boundary of its self, and this process indeed grounds the agent's self-attribution of emotions [48]. Additionally, an agent may formulate an explicit plan of action and then strive to follow it, corresponding to what is classically understood as an act of will. However, words may as well be considered as an anchor for cognition which enables the agent to "write" their intentions in language, or in other as an artificial space of sensations which the agent constructs to extend its ability for recalling memories and past decisions [9]. Therefore, they constitute a simple extension of the domain of agency as we described it here, rather than the basis a parallel system of decision making. Our central point remains: to the skilled agent, intentionality feels like attunement between actions and sensations. We hereby propose, in line with [33,34] to account for that phenomenon through the concept of *embedded normativity.*

The role that embedded normativity plays in the argument of [34] is two fold. First, it constitutes an ontological statement that the processes underlying agency cannot be quite pinned down to inside or outside. Indeed, the picture that the active inference framework paints is that normativity arises not just from the internal structure of an organism but also from the interaction between the organism and its environment (see for example [3] for a thermodynamical view,

[4] However, these states of effortless action are not always the norm or the goal, and conscious effort in planning and decision-making processes" [27]. Rather, they represent particular modes of engagement that can emerge when an agent is highly skilled and the conditions are right [10].

or [6] for the classical enactive treatment of the question). Here, the domain of normativity is extended to represent how the structure of the environment of the organism shapes the norms enacted by the agent[5]. This is in line with the classical Active Inference picture of individuation, cognitive niche construction, and enculturation, as was presented in part 2. Second, it entails an epistemological shift from a focus on an agent's structure and cognitive attitudes to a focus on the constraints applying on the agent's behaviour. Normativity is considered not just as a property of the landscapes that agents experience but also as a property of the agents themselves, arising from their ability to set and pursue goals. The authors insist that human societies may participatively construct and enact regimes of embedded normativity which "extrinsically regulating the intrinsic metabolic and modulatory normativity of an autonomous agent, and organizing the way human agents acquire norms". Taken together, those notions paint a radically counter-intuitive picture. Most people would admit that social norms indeed exist, and that to exist they must somehow be inferred from one's environment. But classically, one would account for normativity as a supervenient phenomenon, and ultimately reduce it to individual behaviour and attitudes. A question remains: in what sense can normativity exist in a shared social and material niche, if it can only manifest through the activity of our individual minds?

4 Externalization and Internalization: Bringing About Our World

The answer to that puzzle is deceptively simple. There is no problem with the idea that normativity can act from *outside* while existing *inside* if we don't admit that *outside* versus *inside* constitutes a meaningful dichotomy to begin with. In the context of Active Inference, the boundary between an agent and its environment unambiguously constitutes the Markov Blanket. However, the FEP describes cognition as a process of attunement through the dynamical flow of the coupled agent-environment system, where the flow of agent's internal dynamics is (by construction) dual to a system of beliefs over the environment [21]. There is no reason why the localization of a given object inside or outside an agent's Markov Blanket should determine whether or not it is a driver of cognition. As discussed in [15], this notion echoes the parity principle formulated by the seminal article for the Extended Mind program: "If [...] a part of the world functions as a process which, were it done in the head, we would have no hesitation in recognizing as part of the cognitive process, then that part of the world is [...] part of the cognitive process" [13]. In other words, if we accept that cognitive agents are enactive and predictive systems, then any system outside that is recruited by

[5] Note that enactive theorists would typically agree that enculturation plays a role in the generation of normativity (see e.g. [20]. However, norms are still conceived as a property of the organism, which may or may not be influenced from out there. The concept of embedded normativity, on the contrary, suggests that there is normativity acting from out there.

a cognitive agent to guide its behaviour can be considered part of the cognitive process. Indeed, acting on the environment is a normal way to reduce prediction error under Active Inference, and this include the construction of an adequate ecological niche which enacts biological functions for the agent [16]. Perhaps more problematically for the proponents of a strong role of localization in deciding whether a given system produces cognition, we should outline that there exist no definitive separation between well-individuated domains of "inside" and "outside". Indeed defining a partition between agent and environment only requires that the statistics of both subsystem be independent of each other, given those of the boundary between the two (the Markov Blanket). This allows for the definition of agent boundaries at many nested scales of analysis [43]. This pattern of multiscale integration is arguably the key driver enabling complex information processing in cognitive agents, and it is not compatible with the view that there exist a definitive boundary between "agent" and "non-agent".

Rather than projecting an excessive meaning to the dichotomy between "inside" and "outside" a given agent, we would like to focus on the processes underlying the integration of given objects within its cognitive architecture. We believe most researchers would agree that there is a relevant difference between using an object to understand or manipulate its world, or try to manipulate and understand it. In the case of tool use, this difference is straightforward: someone may look through binoculars or look at them, and they may use a hoe to dig or they may try to change its handle. The difference between both cases is not whether the object is inside or outside *the* Markov Blanket, it is whether it is inside or outside the relevant Markov Blanket *for the specific interaction at play*. We should outline that this distinction directly maps onto the priors the agent (understood here at the exclusion of the tool) is equipped with, as manifested within the regimes of attention enacted in each case. If the agent has the adequate priors to understand the binoculars as a means to see and the hoe as a means to dig, they will experience the related action-oriented affordances and allocate their attention accordingly. If they don't, they can only experience the tools as an affordance for exploratory behaviour, which may (or may not) cause them to discover the aforementioned priors. Critically, the possibility of an account of cognitive integration in terms of priors enables a treatment in terms of information theory. To borrow the terminology of [32], that cognitive agent learn about the world by "asking questions" about the states which are of interest to it, where the nature of the "questions" asked is dual to its Bayesian priors (as understood under Active Inference) [22]. Therefore, the agent's perception of the world is constrained in its outcome to a specific space of possible "answers" conditioned by the intrinsic priors defining the "question", and the associated regime of attention. Followingly, learning how to use a new object so as to gather information (*i.e.* integrating it in the structure of priors underlying "questions") means constructing a new space of possible observed states. This leads to a counter-intuitive state of affairs, where exploratory behaviour led by the imperative of Variational Free Energy minimization leads to the construction of a novel semiotic interface, or a novel "world" to be enacted by the augmented agent.

Given those considerations, we may simply consider than we experience as *outside* what we ask questions about, while we experience as *inside* what we ask question with. In other words, we perceive as an external reality those aspects of the world which we interrogate, while those aspect of the world which enable our inquiry by integrating our embodied engagement with the world become as transparent as are our eyes and ears. [29] describes the process by which agents construct the boundaries of what they experience as an external reality. As agents engage with their environment and progressively master their tools, the means through which they perceive reality become transparent. They stop being experienced as external objects which serves to gather information, and simply become the new mode by which we interface with the external world. In the same process, the outcome of our perception stops being experienced as the product of our own activity, and become objective features of the outside world. In the words of the author, "during these successful perceptual processes, the quality of the agent-environment interaction is transformed; the agent's grip on a specific aspect of the world is improved, and this allows that aspect to be grasped as a distal object. From this point of view, the local activity is still a necessary part of the process, but its role is fundamentally different: it no longer serves as the input for the internal construction of a putative object, but rather becomes part of the coupling through which the object is disclosed to experience". We may call this process *externalization*, as the author does in his discussion of previous work on sensory substitution [4]. This word captures well how the target of perception are objectivized and experienced as external objects. However, we wish to outline that this process is dual to the *internalization* of specific priors (or regimes of attention) which enable perception, and constrain their semiotics.

The discussion of the semiotic role of cognitive *externalization* and *internalization* becomes critical when we export this framework to the perception of normativity in sociocultural landscapes. When an agent learns to navigate a given society, they internalize the proper priors to communicate with other agents through embodied synchronization, language and material symbolism. Mature members of the communities therefore come to fully externalize the associated normative load, and experience it as an objective (naturalized) feature of the world. An early exemple of how this process enables the writing of normative cues in the material niche is the Sierra de Barbanza described in [18]. There, small rock edifices mark segments of the optimal travel path for travellers crossing the Sierra. To an unaware traveller, those edifices would be a feature of the environment among many, which (if they even notice those) they may choose to investigate or not. But to the experienced travellers, those edifices simply constitute direct instructions regarding the path they should follow. While the edifices were most likely built with the explicit intent to mark the way, the fact their efficacy as norms depends on the agent's skill should highlight a core aspect of our treatment. Embedded normativity does not exist inside or outside the agent, it's brought about by specific environmental cues which the agent experiences as carrying normative significance due to their integrated priors. In other words, the internalization of the shared priors underlying engagement with a given cultural niche is dual to the enaction of a specific collection of normative constraints over behaviour, which constitute the locus of social normatvity.

5 Conclusion

This paper has explored the concept of *embedded normativity*, a regime of normativity where the norms and value underlying an agent's actions are somehow embedded into its environment. We account for this phenomenon through the Active Inference framework, and more specifically through the derivative framework of Skilled Intentionality. In this picture, agents experience their environment as a landscape of affordances, or in another words as a space of opportunities for action. Because the perception of affordances elicits the expectation of action, and therefore (under Active Inference) motivate it, the perception of affordances constitute an inherently normative phenomenon. We highlight that, following from this treatment, the constraints of the agent's environment are as meaningfully carriers of normativity as the constraint's of the agent's organism.

We address the main apparent limitation of the concept of embedded normativity, *i.e.* the apparent suggestion that norms may exist outside the domain of cognition. In our account, agents may fluidly internalize (integrate in their own organization) or externalize (instrument as an element of the world) specific systems. The intrinsic normativity of cognitive agents may be externalized by creating traces in the material landscape around them, for example a path marking their most frequent itineraries. Consequently, the same agents may internalize the normative load of those traces by "forgetting" that they are states of the external world which happen result from their own actions as they learn to treat them as direct instruction to modulate their own behaviour. The construction, the integration, and the enaction of embedded normativity all require the active participation of the agent in perceiving and acting on their material niche, as structured by a given generative model of their environment. Assuming that other agents share common constraints over their own generative model, which may for example follow from past engagement with and internalization of similar patterns, action and perception of the shared ecological niche emerges as a locus of shared regimes of normativity.

The concept of embedded normativity becomes especially interesting when we consider how the construction of affordances is, at least in humans, a social activity influenced by (and constitutive of) cultural norms and values. The shared material, social and cultural niche serves as a way to broadcast normativity through material symbols such as word and road signs, as well as through the production of direct constraints over perception and action. In turn, the shared niche serve to scaffold the development of complex cognition and coordination in mature humans, and (to some extent) to externalize the cognitive load it entails. This accounts highlights the mutually constructive relationship between human cognition and material culture, as outlined *e.g.* in [12]. Our argument suggests that the ambiguity intrinsic in the process of cognitive externalization enables the construction of new, previously unaccessible and unconceivable, possibilities for cognition and social organization. In other words, the dynamics of embedded normativity underwrite how human participatively construct the social constraints which constitutes the structure of their societies.

In this perspective, the drive toward prediction error minimization of cultural agents counter-intuitively leads to their active participation in the construction of a shared reality. This produce a novel problem for embedded normativity, namely the difficulty of account for objects or property that do not exist inside or outside the boundaries of the agent but in the terms of the interaction itself. Normativity constitutes a very straightforward and unconvestroversial example of this regime of existence, which may be used to specify and formalize further the framework. In the general case, the conception of material engagement as being enabled by the internalization of regimes of embedded normativity entails the much more radical view that the outcome of any observation is relative to specific biologically grounded and enculturated modes of navigating the world. As a means to alleviate the tensions which emerge from this view, we point toward the more general epistemological and ontological position of participatory realism which entails that cognitive agents (or, in other contexts, physical observers) construct reality by their very activity of engaging with the world in order to understand it [29,32].

More generally, we argue embedded normativity provides a rich landscape of possibilities for future research. We could explore how this concept may be integrated with preexisting theories of cognition and perception to provide a more comprehensive understanding of human behavior and experience. For instance, how do material landscapes constrains the development of human cognition? Does embedded normativity enables the other cognitive processes such as memory, attention, and decision-making? Since those process are embedded in complex social interactions and cultural phenomena, can we really attribute them to single agents? Perhaps most importantly, this line of research admits a straightforward application to artificial intelligence and robotics, in the treatement of explanability through the lenses of externalization-interalization dynamics. By integrating the scales of analysis which underlie the construction, integration, and enaction of cultural norms, embedded normativity provide a natural operational concept for the study of social and ecological robotics across scales of activity - in particular through the "ecosystem of intelligence" approach outlined by the recent work of VERSES Lab [28].

Acknowledgements. We would like to thank Andy Clark for his feedback on the previous publication in which we introduce the notion of embedded normativity, which motivated the present article. We also thank Maxwell Ramstead and the anonymous reviewers for their constructive comments. This work was financed by the XSCAPE project (Synergy Grant ERC-2020-SyG 951631).

References

1. Aguilera, M., Millidge, B., Tschantz, A., Buckley, C.L.: How particular is the physics of the Free Energy Principle? arXiv:2105.11203 [q-bio], May 2021
2. Albarracin, M., Constant, A., Friston, K., Ramstead, M.J.: A variational approach to scripts, June 2020. https://doi.org/10.31234/osf.io/67zy4
3. Allen, J.W., Bickhard, M.H.: Normativity: a crucial kind of emergence: commentary on Witherington. Hum. Dev. **54**(2), 106–112 (2011)
4. Bach-Y-Rita, P., Collins, C.C., Saunders, F.A., White, B., Scadden, L.: Vision Substitution by tactile image projection. Nature **221**(5184), 963–964 (1969). https://doi.org/10.1038/221963a0
5. Badcock, P.B., Friston, K., Ramstead, M.J.: The hierarchically mechanistic mind: a free-energy formulation of the human psyche. Phys. Life Rev. **31**, 104–121 (2019). https://doi.org/10.1016/j.plrev.2018.10.002
6. Barandiaran, X., Di Paolo, E., Rohde, M.: Defining agency: individuality, normativity, asymmetry, and spatio-temporality in action. Adapt. Behav. **17**(5), 367–386 (2009). https://doi.org/10.1177/1059712309343819
7. Bruineberg, J.: Active Inference and the Primacy of the ?I Can? In: Metzinger, T., Wiese, W. (eds.) Philosophy and Predictive Processing, p. 5 (2017)
8. Bruineberg, J., Rietveld, E.: Self-organization, free energy minimization, and optimal grip on a field of affordances. Front. Hum. Neurosci. **8** (2014). https://doi.org/10.3389/fnhum.2014.00599
9. Clark, A.: Language, embodiment, and the cognitive niche. Trends Cogn. Sci. **10**(8), 370–374 (2006). https://doi.org/10.1016/j.tics.2006.06.012
10. Clark, A.: Surfing Uncertainty: Prediction, Action, and the Embodied Mind. Oxford University Press, October 2015
11. Clark, A.: Beyond desire? Agency, choice, and the predictive mind. Australas. J. Philos. **98**(1), 1–15 (2020). https://doi.org/10.1080/00048402.2019.1602661
12. Clark, A.: Mind unlimited? April 2023
13. Clark, A., Chalmers, D.: The extended mind. Analysis **58**(1), 7–19 (1998)
14. Constant, A., Clark, A., Friston, K.: Representation wars: enacting an armistice through active inference. Front. Psychol. **11**, 3798 (2021). https://doi.org/10.3389/fpsyg.2020.598733
15. Constant, A., Clark, A., Kirchhoff, M., Friston, K.: Extended active inference: constructing predictive cognition beyond skulls. Mind Lang. n/a(n/a) (2019). https://doi.org/10.1111/mila.12330
16. Constant, A., Ramstead, M.J., Veissière, S.P.L., Campbell, J.O., Friston, K.: A variational approach to niche construction. J. R. Soc. Interface **15**(141), 20170685 (2018). https://doi.org/10.1098/rsif.2017.0685
17. Constant, A., Ramstead, M.J., Veissière, S.P.L., Friston, K.: Regimes of expectations: an active inference model of social conformity and human decision making. Front. Psychol. **10** (2019). https://doi.org/10.3389/fpsyg.2019.00679
18. Criado-Boado, F., Villoch Vázquez, V.: La monumentalización del paisaje: Percepción actual y sentido original en el Megalitismo de la Sierra de Barbanza (Galicia) (1998)
19. Csikszentmihalyi, M.: The Flow Experience and its Significance for Human Psychology. In: Optimal Experience: Psychological Studies of Flow in Consciousness, pp. 15–35. Cambridge University Press, New York, NY, US (1988)
20. Cuffari, E.C., Di Paolo, E., De Jaegher, H.: From participatory sense-making to language: there and back again. Phenomenol. Cogn. Sci. **14**(4), 1089–1125 (2014). https://doi.org/10.1007/s11097-014-9404-9

21. Da Costa, L., Friston, K., Heins, C., Pavliotis, G.A.: Bayesian mechanics for stationary processes. Proc. R. Soc. A Math. Phys. Eng. Sci. **477**(2256), 20210518 (2021). https://doi.org/10.1098/rspa.2021.0518
22. Fields, C., Levin, M.: How do living systems create meaning? Philosophies **5**(4), 36 (2020). https://doi.org/10.3390/philosophies5040036
23. Friston, K.: The free-energy principle: a unified brain theory? Nat. Rev. Neurosci. **11**(2), 127–138 (2010). https://doi.org/10.1038/nrn2787
24. Friston, K.: A free energy principle for biological systems. Entropy **14**(11), 2100–2121 (2012). https://doi.org/10.3390/e14112100
25. Friston, K.: Life as we know it. J. R. Soc. Interface **10**(86), 20130475 (2013). https://doi.org/10.1098/rsif.2013.0475
26. Friston, K.: A free energy principle for a particular physics. arXiv:1906.10184 [q-bio], June 2019
27. Friston, K., FitzGerald, T., Rigoli, F., Schwartenbeck, P., Pezzulo, G.: Active inference: a process theory. Neural Comput. **29**(1), 1–49 (2016). https://doi.org/10.1162/NECO_a_00912
28. Friston, K.J., et al.: Designing ecosystems of intelligence from first principles, December 2022. https://doi.org/10.48550/arXiv.2212.01354
29. Froese, T.: Scientific Observation is socio-materially augmented perception: toward a participatory realism. Philosophies **7**(2), 37 (2022). https://doi.org/10.3390/philosophies7020037
30. Gibson, J.J.: The Ecological Approach to Visual Perception. Houghton Mifflin (1979)
31. Guénin-Carlut, A.: The clothes of the empire : an active inference account of identity capture. Kairos J. (2022)
32. Fuchs, C.A.: On participatory realism. In: Durham, I.T., Rickles, D. (eds.) Information and Interaction. TFC, pp. 113–134. Springer, Cham (2017). https://doi.org/10.1007/978-3-319-43760-6_7
33. Guénin-Carlut, A.: Creating material and cultural landscapes - a constraints ontology for multiscale socio-historical dynamics, March 2023
34. Guénin-Carlut, A., White, B., Sganzerla, L.: The cognitive archeology of sociocultural lifeforms, March 2023. https://doi.org/10.31219/osf.io/qxszh
35. Hesp, C., Ramstead, M.J., Constant, A., Badcock, P., Kirchhoff, M., Friston, K.: A Multi-scale View of the Emergent Complexity of Life: A Free-Energy Proposal. In: Georgiev, G.Y., Smart, J.M., Flores Martinez, C.L., Price, M.E. (eds.) Evolution, Development and Complexity, pp. 195–227. Springer, Cham (2019). https://doi.org/10.1007/978-3-030-00075-2_7
36. Hipólito, I.: A simple theory of every 'thing'. Phys. Life Rev. **31**, 79–85 (2019). https://doi.org/10.1016/j.plrev.2019.10.006
37. Hipólito, I., Baltieri, M., Friston, K., Ramstead, M.J.: Embodied skillful performance: Where the action is. Synthese (2021). https://doi.org/10.1007/s11229-020-02986-5
38. Hipólito, I., Hutto, D.D., Ilundain-Agurruza, J.: Culture in Mind -An Enactivist Account: Not Cognitive Penetration But Cultural Permeation (Apr 2021)
39. Hipólito, I., van Es, T.: Enactive-Dynamic Social Cognition and Active Inference. Front. Psychol. **13**, 855074 (2022). https://doi.org/10.3389/fpsyg.2022.855074
40. Kuchling, F., Friston, K., Georgiev, G., Levin, M.: Morphogenesis as Bayesian inference: a variational approach to pattern formation and control in complex biological systems. Phys. Life Rev. **33**, 88–108 (2020). https://doi.org/10.1016/j.plrev.2019.06.001

41. Malafouris, L.: Material engagement and the embodied mind. Cognitive Models Palaeolithic Archaeol. 69–82 (2016)
42. Parr, T., Pezzulo, G., Friston, K.: Active Inference: The Free Energy Principle in Mind, Brain, and Behavior. MIT Press, Cambridge (2022)
43. Ramstead, M.J.D., Kirchhoff, M.D., Constant, A., Friston, K.J.: Multiscale integration: beyond internalism and externalism. Synthese **198**(1), 41–70 (2019). https://doi.org/10.1007/s11229-019-02115-x
44. Ramstead, M., Veissière, S.P., Kirmayer, L.J.: Cultural affordances: scaffolding local worlds through shared intentionality and regimes of attention. Front. Psychol. **7**, 1090 (2016)
45. Ramstead, M.J., Constant, A., Badcock, P.B., Friston, K.: Variational ecology and the physics of sentient systems. Phys. Life Rev. **31**, 188–205 (2019). https://doi.org/10.1016/j.plrev.2018.12.002
46. Ramstead, M.J., Friston, K., Hipólito, I.: Is the free-energy principle a formal theory of semantics? from variational density dynamics to neural and phenotypic representations. Entropy **22**(8), 889 (2020). https://doi.org/10.3390/e22080889
47. Safron, A., Klimaj, V., Hipólito, I.: On the importance of being flexible: dynamic brain networks and their potential functional significances. Frontiers Syst. Neurosci. **15** (2022)
48. Seth, A.K.: Interoceptive inference, emotion, and the embodied self. Trends Cogn. Sci. **17**(11), 565–573 (2013). https://doi.org/10.1016/j.tics.2013.09.007
49. Slingerland, E.: Trying Not to Try: The Ancient Art of Effortlessness and the Surprising Power of Spontaneity. Canongate Books, April 2014
50. Stout, D.: The cognitive science of technology. Trends Cogn. Sci. **25**(11), 964–977 (2021). https://doi.org/10.1016/j.tics.2021.07.005
51. Tison, R.: The fanciest sort of intentionality: active inference, mindshaping and linguistic content. Philos. Psychol. 1–41 (2022). https://doi.org/10.1080/09515089.2022.2062315
52. Veissière, S.P.L., Constant, A., Ramstead, M.J., Friston, K., Kirmayer, L.J.: Thinking through other minds: a variational approach to cognition and culture. Behav. Brain Sci. **43** (2020/ed). https://doi.org/10.1017/S0140525X19001213
53. Wittgenstein, L.: Philosophical Investigations. Macmillan, New York (1953)

A Model of Agential Learning Using Active Inference

Riddhi J. Pitliya[1,2](✉) and Robin A. Murphy[2]

[1] VERSES Research Lab, Los Angeles, CA 90016, USA
riddhi.jain@psy.ox.ac.uk
[2] Department of Experimental Psychology, University of Oxford, Oxford, UK
robin.murphy@psy.ox.ac.uk

Abstract. Agential learning refers to the process of forming beliefs regarding one's degree of control over actions and outcomes in their environment. We first provide an overview and evaluation of associative, statistical, and Bayesian models of agential learning. We then argue that the existing models have limitations in explaining the process of agential learning. Finally, we introduce an active inference account of agential learning, and present results from simulations. We propose that the active inference framework may provide a comprehensive model of agential learning describing three fundamental processes: (i) perception, (ii) learning, and (iii) action.

Keywords: Agency · Agential Learning · Active Inference · Computational Psychology

1 Introduction

An agent is some*one* or some*thing* that acts to control their actions and events in the environment. Agency, then, refers to having control over one's own actions, and leveraging that sense to control themselves or events in the environment [1,2]. Agential learning is the process of tracking and forming relevant beliefs [3] regarding one's degree of agency. Having an ongoing registration of the degree of control agents (self and others) have over the states in their environment facilitates individual- and group-level goal-directed behaviours [4].

Rather than a binary concept, degree of agency refers to the amount of control the agent has to generate or prevent the occurrence of the event. When based purely on objective experience, agency can be formalised as a statistical relationship, or contingency, between the action produced by an agent and its consequence/outcome (discrete variables), each with a dichotomous state of being present or absent. Contingencies, and correlations, vary on a scale from -1 to +1: a positive contingency is when an action predicts the outcome (e.g., pressing a button and the light being turned on), negative contingency is when an action signals the absence of the outcome (e.g., pressing a button and the light being turned off), and zero contingency is when the action has no relation

C. L. Buckley et al. (Eds.): IWAI 2023, CCIS 1915, pp. 106–120, 2024.
https://doi.org/10.1007/978-3-031-47958-8_8

to the presence or absence of the outcome (e.g. the pressing of a button does not have an impact on the light).

One experimental task widely used to assess action-outcome contingency learning involves an action that the participant can freely perform (a so-called free-operant procedure [5]), such as pressing a button, and depending on the objective contingency set by the experimenter, an outcome is present or absent, such as a light being on or off. Subsequently, participants report the degree of control they perceive they have on a visual analog or numeric rating scale varying from -1 to +1. It has been well-demonstrated that perceived contingency as reported on the rating scale is aligned with the action-outcome objective contingency [6–12]. In this paper, we examine agential learning in the context of a simple scenario involving a single action and single outcome, though more complex versions could be entertained.

2 Previous Models of Agential Learning

Philosophers and then psychologists have been challenged to explain how agents learn that one event predicts (or causes) the presence or absence of another event [13–17]. Models that were originally used to explain cue-outcome contingency learning in non-human animals have been employed to explain human performance with some success. Several models based on associative learning theory, statistical accounts, inferential reasoning, and Bayesian learning have been proposed. However, none account for the complexity of learning [18–20] as they fail to capture the relations between perception, learning, and action in informing a sense of agency.

2.1 Associative Models

Associative models [21–25] adopt a bottom-up approach and are process-driven. One model first applied to Pavlovian learning and then extended to explain instrumental learning is the Rescorla-Wagner model. Based on reinforcement principles and successfully applied to human statistical learning, the learning rule is as follows:

$$\Delta V_n = \alpha\beta(\lambda_n - V_{total}) \tag{1}$$

ΔV represents the change in associative strength of an action in that trial (n). The learning rate parameters, α and β, represent the associability of the action and outcome respectively, representing how fast a particular action can be learnt. The subtraction in the parenthesis represents the prediction error, which is the discrepancy between the expected and actual occurrence of the outcome given an action. λ represents the absolute value of the outcome on a trial (n). V_{total} is the total current associative prediction of all stimuli presented at that trial, therefore it comprises $V_1 + V_2 + ... + V_n$. In sum, the Rescorla-Wagner model proposes that learning involves forming associations between all actions present in the environment, and those associations compete with one another as there is a limit to the amount of associative strength the outcome can support. The agent's knowledge regarding associations is represented as a single weight value on each action.

Associative learning explanations of action-outcome contingency learning suggest that agents integrate information online as each action's associative strength gets updated, requiring few cognitive resources and being computationally cheap. It is often described as a form of model-free associative learning. However, because the values get updated with each trial, explaining phenomena such as retrospective revaluation, which has been demonstrated in humans [26–28], require additional assumptions.

The statistical account of action-outcome contingency learning [13,29] suggests that the perceived contingency by the agent is related to an estimate of the difference between the probability of outcome occurring given an action and probability of outcome occurring given a lack of action. Models based on such statistical metrics alone, however, are unable to account for learning curves because probabilities are not affected by the amount of evidence on which they are based [30].

The associative learning and statistical models explain agential learning as the perception of a punctate value reflecting the action-outcome contingency, overlooking other processes that may be involved. For example, the selection of actions by the agent are not accounted for, except in the obvious cases where an experimenter impels or instructs action. An agent's action produces data for the agent, which they use to form beliefs about their agency [31]. Indeed, a link between probability of acting and objective contingency has been established in free-operant tasks [32–34]. Moreover, agency may emerge from a form of inferential reasoning [18,35,36], wherein agents not only rely on direct sensory input and statistical metrics, but also engage in processes that involve learning about the dynamics and causes of the latent states of the world. Bayesian models of contingency learning provide an alternate account and address some of the limitations of previous models [37,38].

2.2 Bayesian Associative Models

Inferring a Causal Structure

Researchers have proposed that agents may conduct Bayesian inference to infer the causal structure of the environment [39,40]. The agent may do this by using bottom-up sensory information (observations) to infer the causal hidden states of their environment using an internal model of the world: a generative model that captures the agent's beliefs about how (potentially dynamic) latent states of the world relate to observable sensory data.

A generative model of agential learning would comprise: (i) a prior probability distribution which represents the agent's current beliefs about the hidden states, and (ii) a likelihood probability distribution which captures the agent's knowledge of how observations (the action and outcome) are generated from hidden states by encoding the likelihood of observations given states. Using Bayes' rule, one can compute a posterior probability distribution over hidden states, given observations. This can be interpreted as the agent's beliefs regarding which hidden states best explain its sensory data, i.e., beliefs regarding their degree of agency. In the context of Bayesian cognitive neuroscience, this updating of beliefs via Bayesian inference has been analogised to perception [41].

The discrepancy between the agent's predictions (from the priors) and beliefs about hidden states after receiving observations (posterior) is quantified by Bayesian surprise, a similar metric to prediction error as in the Rescorla-Wagner model. This is a measure of the degree to which the internal model and posterior beliefs get updated to reduce future surprise, which would ensure an internal model of the causal structure of the world to be as close to the real causal structure of the world as possible. Bayesian inference can therefore be framed as an alternative problem of maximising marginal (log) likelihood, or, in other words, minimising surprise.

In traditional models of contingency learning, punctate values represent all of the agent's knowledge. Bayesian approaches assume a different knowledge representation in the generative model as the agent entertains a probabilistic representation of its world, allowing a spectrum of alternative hypotheses to be represented via their posterior beliefs. The probability distributions allow the agent to express uncertainty, where the more spread out the beliefs are (represented by a flatter probability distribution), the greater the uncertainty. Such a representation of knowledge allows the model to keep track of multiple combinations of hypothetical beliefs, making the perceived causal structure malleable. Therefore, when belief regarding an association is highly uncertain, observational data has a rapid influence on changing that belief. These properties of a Bayesian approach account for how an agent perceives an action-outcome contingency [37].

Explaining Actions

The models described so far consider the agent as a passive observer, and predict action based on the action that is strongest associated with the outcome to produce the most favoured outcomes [7]. However, in reality, when agents are learning the degree of agency they have, the agent has the opportunity to explore or manipulate the world in order to extract information. In other words, the agent actively samples the environment, creating observations for itself to infer and perceive (a degree of) agency and test its beliefs in order to attain the preferred outcome state.

The representation of uncertainty in Bayesian models used to explain observational learning can be leveraged to guide active learning. Here, the agent's actions are explained as the agent actively engaging with the environment to maximise expected information gain based on the generative model to reduce uncertainty [42]. While this explains exploratory behaviour, exploitation is explained by a separate function, based on Bayesian decision theory (or expected utility theory), wherein a value function of states is computed, which represents how rewarding the state is for the agent to be in. The value of the states depend on the agent's learning history of state-action pairs, i.e., tracking how many times the agent attains the outcome by conducting that action from that state. The agent would thereby select the action that yields the outcomes it values.

However, while exploitation and exploratory behaviour can be explained by different functions, the balance between the two often must be adjusted by introducing trade-off parameters, and different strategies have been employed according to task constraints [43]. This calls for a universal model of active learning instead of selecting a model from a class of models to optimally conduct the trade-off between exploration and exploitation dependent on the context. In the next section, we introduce an active inference model of agential learning, where a trade-off between information gain and rewards inherently arises as perception and action are not treated as processes optimising two different functions but rather a single function. It is argued that the active inference framework provides a comprehensive model of agential learning.

3 Active Inference

3.1 Perception, Action, and Learning in Active Inference

Active inference is a process theory, based on the free energy principle [44], that provides a unified account of perception, action, and learning in agents. Active inference extends the (variational) Bayesian inferential process described earlier for perception to action, stemming from the notion that the agent minimises surprise. In active inference, however, a proxy for Bayesian surprise, (variational) free energy, is minimised [31,45]. It is argued that while Bayesian frameworks consider surprise to be dependent on the agent's generative model, surprise is also dependent on observations [45]. Active inference leverages this dependence to predict actions, wherein the agent infers the consequences of its own actions and the hidden states of the world, to exhibit behaviour that attains its preferences and actively reduces uncertainty in the agent's world model [46,47].

Under active inference, action selection is not only a function of past and present observations (as in Bayesian accounts), but also a function of prospective forms of inference based on anticipated future observations. The agent infers the best action sequence (policies) on the basis of future observations the actions would engender, which is based on beliefs about likelihoods of observations given the anticipated states in addition to the transitions of states across time as a function of the policy. This formulation of action selection in active inference casts action trajectories as a functional of beliefs (i.e., beliefs of beliefs, with probability distributions) inevitably encompassing the notions of uncertainty and preferences.

According to active inference, action selection occurs by expected free energy (EFE) being calculated for each policy and a policy is selected according to its negative EFE as policies that afford the lowest EFE are the most likely. EFE can be seen as the combination of (i) the anticipated information gain afforded by expected observations under a policy (exploration) and (ii) how well expected observations align with preferences (exploitation). Maximising the exploration term is equivalent to maximising the expected divergence between the expected posterior distribution, with and without observations expected under a policy - maximising this leads to behaviour that actively seeks out observations that resolve the most (posterior) uncertainty. Maximising the exploitation term is equivalent to changing policies to produce those observations that best match the agent's prior beliefs about observations (i.e., its preferences), which is specified in the agent's generative model. Hence, active inference balances exploration and exploitation, ensuring that an optimal agent pursues both. Often, in situations where the agent is uncertain about hidden states that are relevant to preferred observations, active inference agents will first perform more epistemically driven actions to resolve uncertainty, before opting for a more pragmatic action that maximises utility, i.e., exploit the resolved structure of the environment.

Learning occurs in active inference by updating model parameters, such as the likelihood distributions and state transition beliefs. In the discrete state-space models commonly used in active inference, these likelihood and transition distributions are described as categorical distributions with matrices of parameters. These distributions are often equipped with conjugate Dirichlet priors [48], whose parameters take the form of *pseudocounts* or positive real numbers that parameterise prior beliefs about the corresponding categorical parameters. The values of these Dirichlet hyper parameters can be interpreted as *pseudocounts* that are proportional to the prior probability of seeing particular state-outcome contingencies or coincidences between states and actions over time. Learning is thus cast as posterior inference over these Dirichlet hyperparameters [48]. Hence, when a new observation is received by the agent, a posterior distribution over the model parameter is acquired to be used as the prior distribution in the next time step, equipping the agent to sequentially update beliefs about the model parameters. A learning rate parameter can also be specified to control how much the values in the Dirichlet distribution change after each time step, representing how quickly the agent can get stuck in its ways during learning [48].

To summarise, when there is a mismatch between the agent's predictions and sensory inputs, the agent (i) updates its internal model to reduce future surprise by updating its beliefs about the states that caused the observation, and/or (ii) updating its beliefs about the dynamics of the world (updating model parameters), and/or (iii) actively engages with the environment to generate and maximise model evidence, thereby reducing future surprise. These processes of minimising surprise respectively map onto three fundamental processes: (i) perception, (ii) learning, and (iii) action.

3.2 Generative Model of Agential Learning

The Agential Learning Task

In this section, a discrete-time generative model of the classic free-operant agential learning task is presented as in Fig. 1), along with a set of simulations presented in Figs. 3 and 4). In the learning task, the agent produces an action by pressing a button or not, and according to the objective contingency, an outcome is present or absent. In some of our experimental conditions, the agent has 100% or 80% (positive or negative) control over the outcome, referred to as the deterministic and probabilistic conditions, respectively. In other conditions, the agent has no control, i.e., the outcome is produced at random, independent of the agent's actions.

Generative Model of Agential Learning Task
The generative model an active inference agent is equipped with is in the form of a Partially Observable Markov Decision Process (POMDP; [45]). POMDPs express the generative model with a sequence of hidden states (s) that evolve over time. The hidden states inferred by the agent in this agential learning task are the objective contingency (positive, negative, or zero) between the self-produced action and outcome, which is the context (or experimental condition) the agent is in ($s_t^{context}$).

At each time step (t), the current state is conditionally dependent on the state at the previous time step and on the actions (u; aka control states) currently being executed. The actions are dependent on the policy (π; aka action sequence) currently being executed. Each time step is associated with an observation (o) that depends only on the state at that time. The observations the agent receives are of the outcome ($o_t^{outcome}$), wherein the outcome can be present or absent, and the observations of the action the agent conducted (o_t^{action}), wherein the agent observes that it pressed the button or not.

The hidden and control states are classified into state factors, and observations are classified into observation modalities. This means that at any given time, observations will be evinced from each modality, and hidden states will be inferred from each state factor, and an action (control state) is selected accordingly. The s, u, and o are discrete random variables, so all model parameters are categorical distributions too.

The agent's generative model is equipped with model parameters denoted as **A**, **B**, **C**, and **D** tensors that allow the agent to perform active inference. The likelihood tensor (**A**), represents the beliefs of probability of some observation given the states in the agent's environment, $P(o_t|s_t)$. The top-left matrix in **A** tensor panel in Fig. 2 illustrates that the agent believes the probability it will observe the outcome being present given it is in the positive control state and has pressed the button is 0.6. This value is not 1.0 as the agent cannot have already learned the precise likelihood mappings as it does not know the objective contingency in all of the possible contexts in which it could be operating, so it conducts learning by updating the likelihood tensor regarding outcomes via Dirichlet counts ($Dir(A^{\text{outcome}}),$). The degree of control the agent perceives is indicated by the posterior probability of the state.

The state transition tensor (**B**), represents the beliefs of the dynamics of the environment as how hidden states and actions determine subsequent hidden states, $P(s_t|s_{t-1}, u_t)$. The objective contingency does not change over a block of trials, and we assume the agent knows this fact veridically, and thus their generative model has an identity matrix in the left matrix presented in the **B** tensor panel in Fig. 2.

Sampling the environment occurs as a function of preferring each observation, represented in the preference tensor (**C**) in Fig. 2, reducing uncertainty. There is a slight preference for not producing an action as producing an action costs resource. To introduce evidence variance, periods of sub-optimal action would be intentionally conducted by the agent to create variation in the observations and assess the agent's generative model. The **D** tensor represents the agent's beliefs of the prior probability of being in each state, which is a flat distribution to reflect the agent's lack of a bias towards being in a positive, negative, or zero-contingency state.

Simulation Results

All simulations described in this paper were conducted using the sparse_likelihoods_111 branch of pymdp, a freely available Python package for performing active inference in discrete state spaces [49]. The code used for the simulations described in this paper can be found here: https://github.com/riddhipits/iwai_agency_oneagent.

Figures 3 and 4 illustrate the results of simulations of an agent conducting agential learning in the deterministic and probabilistic learning task across three experimental conditions. Panels correspond to each experimental condition: positive control, negative control, and zero control. The three sub-panels in each panel illustrate the agent's beliefs over time (x-axis) regarding the experimental condition (or context), the actions it took, and the outcomes it observed. The strength of the belief is reflected in the grayscale cells, with black cells indicating a value of 0.0 and white cells indicating a value of 1.0. The agent had 50 trials to learn about the degree of control.

In the deterministic (100%) agency simulation (Fig. 3), in the positive and negative control condition, the agent quickly learned that it had full positive and negative control, respectively; this is illustrated by the gradual transition from black to white on the top-most sub-panels. In the first few trials, the agent tracks (via Dirichlet counts) the outcome observation given the states it observes (actions) and infers (context) and reflects its learning of the environment being deterministic by updating the likelihood tensors in its generative model. Accordingly, the agent then infers, with certainty that it is in the positive or negative condition. As predicted, the agent introduces evidence variance by occasionally acting sub-optimally to increase certainty regarding its beliefs. In the probabilistic (80%) agency simulation (Fig. 4), the agent learns similarly, albeit less quickly and with more uncertainty as illustrated with more grey cells.

In the zero-control condition, the agents in both simulations (Fig. 3 and Fig. 4) takes longer learn that its actions have no control over the outcome. To elaborate, in Fig. 3, during the first few trials (Box A), the agent's actions of pressing the button were coincidentally paired with the outcome being present, which is why it had a higher belief of being in the negative control state. And in the middle of the block of trials (Box B), the agents actions aligned with what it would predict to perceive in a positive control condition, which is why its beliefs shift towards the positive control condition until it receives evidence against that belief. Finally, the agent's beliefs increase for the zero-control condition. Throughout the block of trials, the agent tests its hypotheses by variably pressing the button or not.

4 Discussion and Concluding Remarks

These simulations reveal that the active inference framework has potential to provide a comprehensive model of agential learning tying perception, actions, and learning processes, resulting from the minimisation of a single metric: free energy. Previous models have treated these processes as optimising disparate functions.

Compared to Bayesian agents, active inference agents possess a deeper representation of the causal structure and dynamics of the environment as an active inference agent's generative model is equipped with beliefs about state transitions across time. This is leveraged by the active inference agent as it allows the agent to consider future states and observations based on future actions to optimally select an action. The actions maximise evidence for the agent's generative model of their environment by exploring the environment when uncertainty is high and then exploiting the environment to attain preferred observations/outcomes, and introduce evidence variance to continually assess the agent's generative model.

The active inference model of agential learning may allow us to explain individual differences in agential learning. For example, agents experiencing learned helplessness (a key symptom of depression) may have a higher learning rate for the zero-control state due to generalisation from trauma, resulting in them having a bias and getting stuck when the belief of being in a zero-control state is higher. Over time, this may result in them developing a habit of not producing an action (due to deep temporal active inference models; see [50]), resulting in reduced variance in sampling the environment.

The simulation results in this paper emphasise that observations of different action-outcome combinations make a big difference to the perceived contingency in a zero-control condition. This predicts that agents who produce actions would experience more (but accidental) action-outcome-present observations and thereby perceive an illusion of (positive) control, whereas agents who withhold actions would perceive more no-action-outcome-present observations, resulting in perceiving zero control. The predictions are in line with data from humans as experimenters showed that non-depressed individuals produced more actions in the zero-control condition, perceiving an illusion of (positive) control, and individuals experiencing depression withheld actions, perceiving a lack a control, potentially explaining their lack of sense of agency [51].

Nonetheless, further examination of the active inference formulation of agential learning is warranted. In future research studies, we intend to: (i) conduct statistical model comparisons between the different accounts of agential learning via model fitting to human behavioural data, (ii) examine if active inference

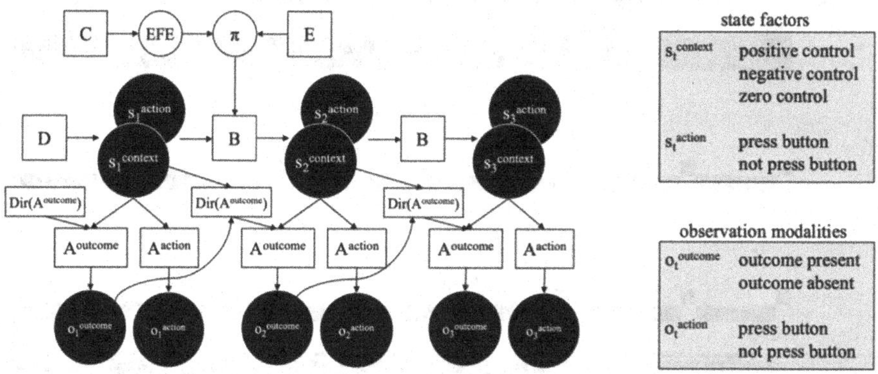

Fig. 1. A graphical representation [52] of the active inference based generative model of the agential learning task. The variables of the model are illustrated as circles and model parameters as squares and rectangles. The arrows indicate the direction of influence. Please see the main text for a description of the variables and parameters.

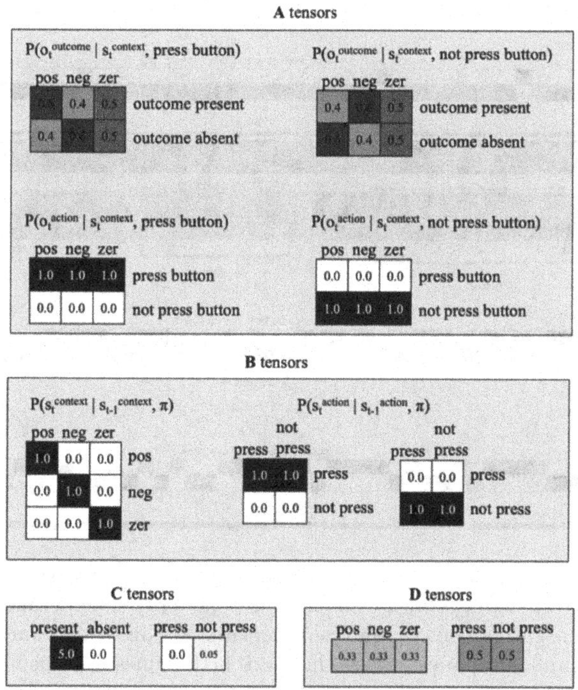

Fig. 2. The details of the model parameters of the generative model of the agential learning task.

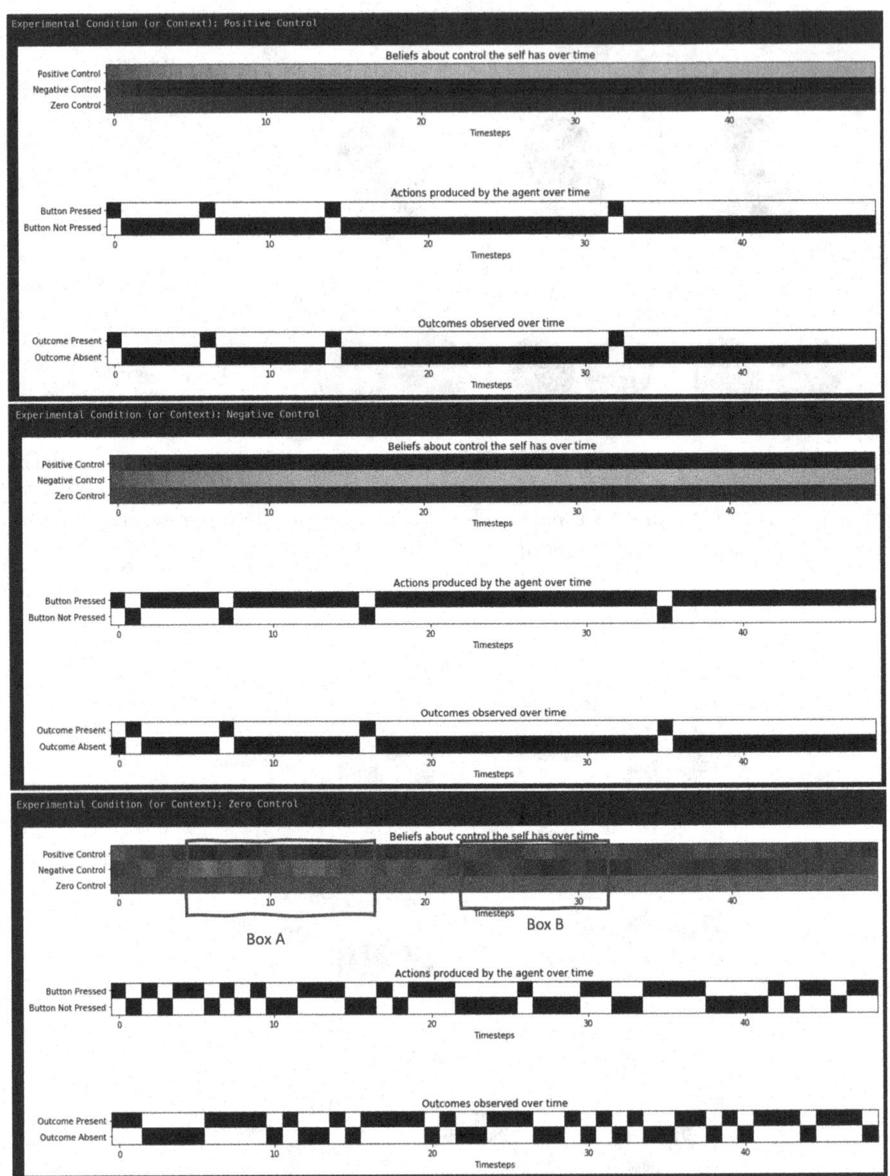

Fig. 3. Simulation results for deterministic (100% control) agential learning task. The three panels illustrate three separate simulations, one for each experimental condition: positive control, negative control, and zero control. Within each panel of simulation result, there are three sub-panels, where the x axis is the timestep. The black cells represent the value of 0.0 and white cells represent the value of 1.0, so the grayscale cells are values within that range. The top sub-panel illustrates the beliefs the agent has regarding the context states, the middle sub-panel illustrates the actions the agent selected over time, and the bottom sub-panel illustrates the outcomes the agent observed over time.

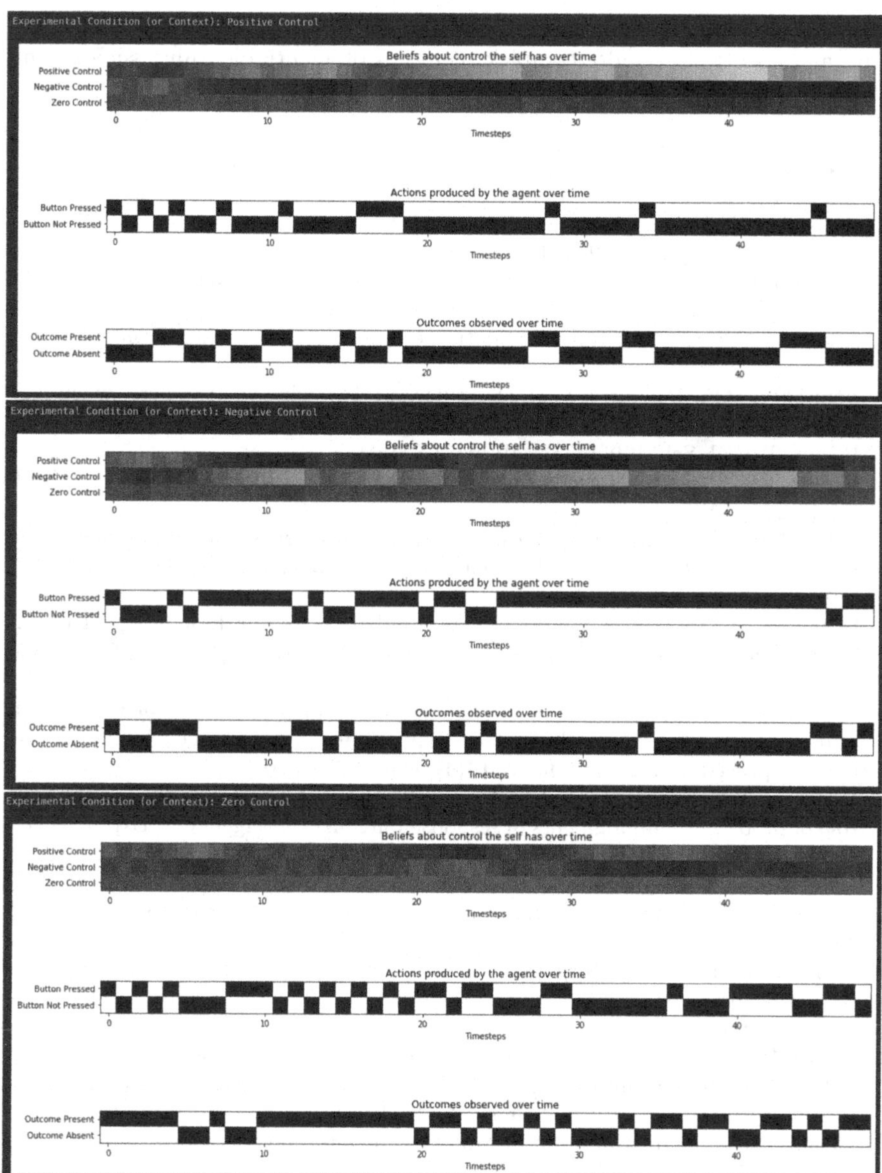

Fig. 4. Simulation results for probabilistic (80% control) agential learning task. The three panels illustrate three separate simulations, one for each experimental condition: positive control, negative control, and zero control. Within each panel of simulation result, there are three sub-panels, where the x axis is the timestep. The black cells represent the value of 0.0 and white cells represent the value of 1.0, so the grayscale cells are values within that range. The top sub-panel illustrates the beliefs the agent has regarding the context states, the middle sub-panel illustrates the actions the agent selected over time, and the bottom sub-panel illustrates the outcomes the agent observed over time.

explains individual differences in agential learning across the depression spectrum, and (iii) explore more complex scenarios of agential learning such as one with multiple agents and outcomes.

References

1. Gallagher, S.: Philosophical conceptions of the self: implications for cognitive science. Trends Cogn. Sci. **4**(1), 14–21 (2000)
2. Haggard, P.: Sense of agency in the human brain. Nat. Rev. Neurosci. **18**(4), 196–207 (2017)
3. Albarracin, M., Pitliya, R.J.: The nature of beliefs and believing. Front. Psychol. **3** (2022)
4. Verschure, P.F., Pennartz, C.M., Pezzulo, G.: The why, what, where, when and how of goal-directed choice: neuronal and computational principles. Philos. Trans. R. Soc. B Biol. Sci. **369**(1655), 20130483 (2014)
5. Ferster, C.B.: The use of the free operant in the analysis of behavior. Psychol. Bull. **50**(4), 263 (1953)
6. Allan, L.G., Jenkins, H.M.: The judgment of contingency and the nature of the response alternatives. Can. J. Psychol./Revue canadienne de psychologie **34**(1), 1 (1980)
7. Shanks, D.R., Dickinson, A.: Instrumental judgment and performance under variations in action-outcome contingency and contiguity. Memory Cogn. **19**, 353–360 (1991)
8. Wasserman, E.A., Chatlosh, D., Neunaber, D.: Perception of causal relations in humans: factors affecting judgments of response-outcome contingencies under free-operant procedures. Learn. Motiv. **14**(4), 406–432 (1983)
9. Wasserman, E.A., Elek, S.M., Chatlosh, D.L., Baker, A.G.: Rating causal relations: role of probability in judgments of response-outcome contingency. J. Exp. Psychol. Learn. Mem. Cogn. **19**(1), 174 (1993)
10. Vallée-Tourangeau, F., Murphy, R.A., Baker, A.: Contiguity and the outcome density bias in action-outcome contingency judgements. Q. J. Exp. Psychol. Sect. B **58**(2b), 177–192 (2005)
11. Vallee-Tourangeau, F., Murphy, R.: Action-effect contingency judgment tasks foster normative causal reasoning. In: Proceedings of the Twenty First Annual Conference of the Cognitive Science Society, pp. 820–820 (1999)
12. Msetfi, R.M., Murphy, R.A., Simpson, J., Kornbrot, D.E.: Depressive realism and outcome density bias in contingency judgments: the effect of the context and intertrial interval. J. Exp. Psychol. Gen. **134**(1), 10 (2005)
13. Cheng, P.W.: From covariation to causation: a causal power theory. Psychol. Rev. **104**(2), 367 (1997)
14. Hume, D.: A treatise of human nature: Volume 1: Texts (1739)
15. Kant, I.: Critique of pure reason. 1781, Modern Classical Philosophers, Cambridge, MA, Houghton Mifflin, pp. 370–456 (1908)
16. Michotte, A.: The perception of causality. Routledge, vol. 21 (2017)
17. Shanks, D.R., Lopez, F.J., Darby, R.J., Dickinson, A.: Distinguishing associative and probabilistic contrast theories of human contingency judgment. In: Psychology of learning and motivation, vol. 34, pp. 265–311. Elsevier (1996)
18. De Houwer, J., Beckers, T.: A review of recent developments in research and theories on human contingency learning. Q. J. Exp. Psychol. Sect. B **55**(4), 289–310 (2002)

19. Pineño, O., Miller, R.R.: Comparing associative, statistical, and inferential reasoning accounts of human contingency learning. Q. J. Exp. Psychol. **60**(3), 310–329 (2007)
20. Shanks, D.R.: Associationism and cognition: human contingency learning at 25. Q. J. Exp. Psychol. **60**(3), 291–309 (2007)
21. Mackintosh, N.J.: A theory of attention: variations in the associability of stimuli with reinforcement. Psychol. Rev. **82**(4), 276 (1975)
22. Miller, R.R., Matzel, L.D.: The comparator hypothesis: a response rule for the expression of associations. In: Psychology of learning and motivation, vol. 22, pp. 51–92. Elsevier (1988)
23. Pearce, J.M., Hall, G.: A model for pavlovian learning: Variations in the effectiveness of conditioned but not of unconditioned stimuli. Psychol. Rev. **87**(6), 532 (1980)
24. Rescorla, R.A.: A theory of pavlovian conditioning: variations in the effectiveness of reinforcement and non-reinforcement. Class. Conditioning Curr. Res. Theor. **2**, 64–69 (1972)
25. Wagner, A.R., Rescorla, R.A.: Inhibition in Pavlovian conditioning: application of a theory. In: Boakes, R.A., Halliday, M.S. (eds.) Inhibition and Learning. Academic Press, New York (1972)
26. Chapman, G.B.: Trial order affects cue interaction in contingency judgment. J. Exp. Psychol. Learn. Mem. Cogn. **17**(5), 837 (1991)
27. De Houwer, J., Beckers, T.: Higher-order retrospective revaluation in human causal learning. Q. J. Exp. Psychol. Sect. B **55**(2b), 137–151 (2002)
28. Dickinson, A.: Within compound associations mediate the retrospective revaluation of causality judgements. Q. J. Exp. Psychol. Sect. B **49**(1), 60–80 (1996)
29. Cheng, P.W., Novick, L.R.: Covariation in natural causal induction. Psychol. Rev. **99**(2), 365 (1992)
30. López, F.J., Almaraz, J., Fernández, P., Shanks, D.: Adquisición progresiva del conocimiento sobre relaciones predictivas: Curvas de aprendizaje en juicios de contingencia. Psicothema, pp. 337–349 (1999)
31. Friston, K., FitzGerald, T., Rigoli, F., Schwartenbeck, P., Pezzulo, G.: Active inference: a process theory. Neural Comput. **29**(1), 1–49 (2017)
32. Blanco, F., Matute, H., Vadillo, M.A.: Mediating role of activity level in the depressive realism effect (2012)
33. Blanco, F., Matute, H., Vadillo, M.A.: Interactive effects of the probability of the cue and the probability of the outcome on the overestimation of null contingency. Learn. Behav. **41**(4), 333–340 (2013). https://doi.org/10.3758/s13420-013-0108-8
34. Byrom, N., Msetfi, R., Murphy, R.: Two pathways to causal control: use and availability of information in the environment in people with and without signs of depression. Acta Physiol. (Oxf) **157**, 1–12 (2015)
35. Griffiths, T.L., Tenenbaum, J.B.: Structure and strength in causal induction. Cogn. Psychol. **51**(4), 334–384 (2005)
36. Waldmann, M.R.: Competition among causes but not effects in predictive and diagnostic learning. J. Exp. Psychol. Learn. Mem. Cogn. **26**(1), 53 (2000)
37. Kruschke, J.K.: Bayesian approaches to associative learning: from passive to active learning. Learn. Behav. **36**(3), 210–226 (2008). https://doi.org/10.3758/LB.36.3.210
38. Tenenbaum, J.B., Griffiths, T.L., Kemp, C.: Theory-based bayesian models of inductive learning and reasoning. Trends Cogn. Sci. **10**(7), 309–318 (2006)
39. Chater, N., Oaksford, M., Hahn, U., Heit, E.: Bayesian models of cognition. Wiley Interdisc. Rev. Cogn. Sci. **1**(6), 811–823 (2010)

40. Doya, K., Ishii, S., Pouget, A., Rao, R.P.: Bayesian brain: Probabilistic Approaches to Neural Coding. MIT Press, Cambridge (2007)
41. Von Helmholtz, H.: Handbuch der physiologischen Optik. Voss, vol. 9 (1867)
42. Nelson, J.D.: Finding useful questions: on bayesian diagnosticity, probability, impact, and information gain. Psychol. Rev. **112**(4), 979 (2005)
43. De Ath, G., Everson, R.M., Rahat, A.A., Fieldsend, J.E.: Greed is good: exploration and exploitation trade-offs in bayesian optimisation. ACM Trans. Evol. Learn. Optim. **1**(1), 1–22 (2021)
44. Friston, K.: The free-energy principle: a unified brain theory? Nat. Rev. Neurosci. **11**(2), 127–138 (2010)
45. Parr, T., Pezzulo, G., Friston, K.J.: Active Inference: the Free Energy Principle in Mind, Brain, and Behavior. MIT Press, Cambridge (2022)
46. Friston, K.J., Daunizeau, J., Kiebel, S.J.: Reinforcement learning or active inference? PLoS ONE **4**(7), e6421 (2009)
47. Friston, K., Rigoli, F., Ognibene, D., Mathys, C., Fitzgerald, T., Pezzulo, G.: Active inference and epistemic value. Cogn. Neurosci. **6**(4), 187–214 (2015)
48. Smith, R., Friston, K.J., Whyte, C.J.: A step-by-step tutorial on active inference and its application to empirical data. J. Math. Psychol. **107**, 102–632 (2022)
49. Heins, C., Millidge, B., Demekas, D., et al.: Pymdp: a python library for active inference in discrete state spaces, arXiv preprint arXiv:2201.03904 (2022)
50. Friston, K.J., Rosch, R., Parr, T., Price, C., Bowman, H.: Deep temporal models and active inference. Neurosci. Biobehav. Rev. **90**, 486–501 (2018)
51. Blanco, F., Matute, H., Vadillo, M.A.: Depressive realism: wiser or quieter? Psychol. Record **59**(4), 551–562 (2009)
52. Friston, K.J., Parr, T., de Vries, B.: The graphical brain: belief propagation and active inference. Netw. Neurosci. **1**(4), 381–414 (2017)

From Theory to Implementation

Designing Explainable Artificial Intelligence with Active Inference: A Framework for Transparent Introspection and Decision-Making

Mahault Albarracin[1,2,3(✉)], Inês Hipólito[1,4,5], Safae Essafi Tremblay[1,6], Jason G. Fox[1], Gabriel René[1], Karl Friston[1,7], and Maxwell J. D. Ramstead[1,7]

[1] VERSES AI Research Lab, Los Angeles, CA 90016, USA
mahault.albarracin@verses.ai
[2] Département d'informatique, Université du Québec à Montréal, 201, Avenue du Président-Kennedy, Montréal H2X 3Y7, Canada
[3] Institut Santé et société, UQAM, 400, rue Sainte-Catherine Est, Montréal H2L 2C5, Canada
[4] ABC Institute, Rupert-Karls-University Heidelberg, Heidelberg, Germany
[5] Department of Philosophy, Macquarie University, Sydney, NSW, Australia
[6] Département de philosophie, Université du Québec à Montréal, 455, Boulevard René-Lévesque Est, Montréal H2L 4Y2, Canada
[7] Wellcome Centre for Human Neuroimaging, University College London, London WC1N 3AR, UK

Abstract. This paper investigates the prospect of developing human-interpretable, explainable artificial intelligence (AI) systems based on active inference and the free energy principle. We first provide a brief overview of active inference, and in particular, of how it applies to the modeling of decision-making, introspection, as well as the generation of overt and covert actions. We then discuss how active inference can be leveraged to design explainable AI systems, namely, by allowing us to model core features of "introspective" processes and by generating useful, human-interpretable models of the processes involved in decision-making. We propose an architecture for explainable AI systems using active inference. This architecture foregrounds the role of an explicit hierarchical generative model, the operation of which enables the AI system to track and explain the factors that contribute to its own decisions, and whose structure is designed to be interpretable and auditable by human users. We outline how this architecture can integrate diverse sources of information to make informed decisions in an auditable manner, mimicking or reproducing aspects of human-like consciousness and introspection. Finally, we discuss the implications of our findings for future research in AI, and the potential ethical considerations of developing AI systems with (the appearance of) introspective capabilities.

Keywords: Active Inference · Explainability · Artificial intelligence

C. L. Buckley et al. (Eds.): IWAI 2023, CCIS 1915, pp. 123–144, 2024.
https://doi.org/10.1007/978-3-031-47958-8_9

1 Introduction: Explainable AI and Active Inference

Artificial intelligence (AI) systems continue to proliferate and, at the time of writing, have become an integral part of various intellectual and industrial domains, including healthcare, finance, and transportation [71,86]. Traditional AI models, such as deep learning neural networks, have been widely recognized for their ability to achieve high performance and accuracy across various tasks [46,66]. However, it is well known that these models almost invariably function as "black boxes," with limited transparency and interpretability of their decision-making processes [18,49]. This lack of explainability can lead to skepticism and reluctance to adopt AI systems—and indeed, to harm, particularly in high-stakes situations, where the consequences of a wrong decision can be severe and harmful [11,12,26,94].

The problem of explainable AI (sometimes referred to as the "black box" problem) is the problem of understanding and interpreting how these models arrive at their decisions or predictions [9,10]. While researchers and users may have knowledge of the inputs provided to the model and the corresponding outputs that it produces, comprehending the internal workings and decision-making processes of AI systems can be complex and challenging. This is in no small part because their intricate architectures and numerous interconnected layers learn to make predictions by analyzing vast amounts of training data and adjusting their internal parameters, without explicit instruction from a programmer [5]. The method by which these systems are trained thus, by design, limits their explainability. Moreover, the internal computations that are performed by these models—when they engage in decision-making—can be highly complex and nonlinear, making it difficult to extract meaningful explanations of their behavior, or insights into their decision-making process [31]. This problem is compounded by the fact that most machine learning implementations of AI fail to represent or quantify their uncertainty; especially, uncertainty about the parameters and weights that underwrite their accurate performance. This means that AI, in general, cannot evaluate (or report) the confidence in its decisions, choices or recommendations.

The lack of interpretability poses several challenges. Firstly, it hampers transparency and makes audits by third parties next to impossible, as the designers, users, and stakeholders of these systems may struggle to understand why a particular decision or prediction was made. This becomes problematic in critical domains such as healthcare or finance, where the ability to explain the reasoning behind a decision is essential for trust, accountability, and compliance with regulations [29,76]. Secondly, the black box nature of machine learning models can hinder the identification and mitigation of biases or discriminatory patterns. Without visibility into the underlying decision-making process, it becomes challenging to detect and address biases that may exist within the model's training data or architecture.

This opacity can lead to unfair or biased outcomes, perpetuating social inequalities or discriminatory practices [44,79,110]. Additionally, the lack of interpretability of the model limits its ability to provide meaningful explanations to end-users. Individuals interacting with machine learning systems often seek explanations for the decisions made by these systems [61,109]. For instance, in medical diagnosis, patients and healthcare professionals may want to understand why a particular diagnosis or treatment recommendation was given [80,81]; or consider automated suggestions in practical industrial settings [65]. Without explainability, users may be hesitant to trust the system's recommendations or may feel apprehensive (not without good reason) about relying on the outputs of such models.

Accordingly, the need for explainable AI has become increasingly important [1]. "Explainable AI" refers to the development of AI systems that can provide human-understandable explanations for their decisions and actions [48]. This level of transparency is crucial for fostering trust [17], ensuring accountability [97], and facilitating inclusive collaboration between humans and AI systems [13,52,58]. Recent efforts to regulate AI may turn explainability into a requirement for the deployment of any AI system at scale. For instance, in the United States, the National Institute of Standards and Technology (NIST) released its Artificial Intelligence Risk Management Framework (RMF) in 2023 [107], which includes explainability and interpretability as crucial characteristics of a trustworthy AI system. The RMF is envisioned as a guide for tech companies to manage the risks of AI and could eventually be adopted as an industry standard. In a similar vein, US Senator Chuck Schumer has led a congressional effort to establish US regulations on AI, with one of the key aspects being the availability of explanations for how AI arrives at its responses [27].

In the European Union, a proposed Regulation Laying Down Harmonized Rules on Artificial Intelligence (better known as the "AI Act") is set to increase the transparency required for the use of so-called "high-risk" AI systems [20]. For instance, groups that deploy automated emotion recognition systems may be obligated to inform those on whom the system is being deployed that they are being exposed to such a system. The AI Act is expected to be finalized and adopted in 2023, with its obligations likely to apply within three years' time. The Council of Europe is also in the process of developing a draft convention on artificial intelligence, human rights, democracy, and the rule of law, which will be the first legally binding international instrument on AI. This convention seeks to ensure that research, development, and deployment of AI systems are consistent with the values and interests of the EU, and that they remain compatible with the AI Act and the proposed AI Liability Directive, which includes a risk-based approach to AI. In addition, the US-EU Trade and Technology Council published a joint Roadmap for Trustworthy AI and Risk Management in 2022, which aims to advance collaborative approaches in international standards bodies related to AI, among other objectives [101]. Therefore, explainability is clearly a major issue in research, development, and deployment of AI systems, and will remain so for the foreseeable future.

Explainable AI aims to bridge the gap between the complexity and lack of auditability of contemporary AI systems and the need for human interpretability and auditability [1,14,48]. It seeks to provide insights into the factors that influence AI decision-making, enabling users to understand the explicit reasoning and other factors driving the output of AI systems. Understanding the performance and potential biases of AI systems is crucial for their ethical and responsible deployment [93,95]. This understanding, however, must extend beyond the performance of AI systems on academic benchmarks and tasks to include a deep understanding of what the models represent or learn, as well as the algorithms that they instantiate [47].

Transparency considerations are embedded in the design, development, and deployment of AI systems, from the societal problems that arise worth developing a solution for, to the data collection stage, and still at the point where the AI system is deployed in the real world and iteratively improved [51,52]. This transparency may enable the implementation of other ethical AI dimensions like interpretability, accountability, and safety [19].

Researchers have been exploring various approaches to develop more explainable AI systems [6,26]. However, these efforts have yet to yield a principled and widely accepted path method for, or path to, explainability. One promising direction is to draw inspiration from research into human introspection and decision-making processes [24]. Furthermore, a two-stage decision-making process, which includes a reflection stage where the network reflects on its feed-forward decision, can enhance the robustness and calibration of AI systems [85]. It has been suggested that explainability in AI systems can be further enhanced through techniques such as layer-wise relevance propagation [7] and saliency maps [116], which aid in visualizing the model's reasoning process. By translating the internal models of AI systems into human-understandable explanations, we can foster trust and collaboration between AI systems and their human users [63]. However, as [47] argue, we must also consider the metatheoretical calculus that underpins our understanding and use of these models. This involves not only considering the performance of the model on a task, but also the implications of the performance of the model for our understanding of the mind and brain.

In this paper, we investigate the potential of active inference, and the free energy principle (FEP) upon which is based [42,89], to enhance explainability in AI systems, notably by capturing core aspects of introspective processes, hierarchical decision-making processes, and (cover and overt) forms of action in human beings [54,90,91]. The FEP is a variational principle of information physics that can be used to model the dynamics of self-organizing systems like the brain. Active inference is an application of the FEP to model the perception-action loops of cognitive systems: it provides us with the basis of a unified theory of the structure and function of the brain (and indeed, of living and self-organizing systems more generally; [87,92]. Active inference allows us to model self-organizing systems like brains as being driven by the imperative to minimize surprising encounters with the environment; where this surprise scores how far a thing or system deviates from its characteristic states (e.g., a fish out of water). By doing so, the brain continually updates and refines its world model, allowing the agent to act adaptively and in situationally appropriate ways.

The relevance of using active inference is that the models of cognitive dynamics—and in particular, introspection—that have been developed using its tools can be adapted to enable the design of human interpretable and auditable (and indeed, self-auditable) AI systems. The ethical and epistemological or epistemic gains that this enables are notable. The proposed active inference based AI system architecture would enable artificial agents to access and analyze their own internal states and decision-making processes, leading to a better understanding of their decision-making processes, and the ability to report on themselves. Proof of concept for this kind of "self report" is already at hand [83] and, in principle, is supported in any application of active inference. At one level, committing to a generative model—implicit in any active inference scheme—dissolves the explainability problem. This is because one has direct access to the beliefs and belief-updating of the agent in question.

Indeed, this is why active inference has been so useful in neuroscience to model and explain behavioral and neuronal responses in terms of underlying belief states: e.g., [2,3,102,104,108]. As demonstrated in [83] it is a relatively straightforward matter to augment generative models to self-report their belief states. In this paper, we address a slightly more subtle aspect of explainability that rests upon "self-access"; namely, when an agent infers its own "states of mind"—states of mind that underwrite its sense-making and choices. Crucially, this kind of meta-inference [34,43,98,115] may rest on exactly the representations of uncertainty (a.k.a., precision) that are absent in conventional AI.

This paper is organized as follows. We first introduce essential aspects of active inference. We then discuss how active inference can be used to design explainable AI systems. In particular, we propose that active inference can be used as the basis for a novel AI architecture—based on explicit generative models—that both endows AI systems with a greater degree of explainability and audibility from the perspective of users and stakeholders, and allows AI systems to track and explain their own decision-making processes in a manner understandable to users and stakeholders. Finally, we discuss the implications of our findings for future research in auditable, human-interpretable AI, as well as the potential ethical considerations of developing AI systems with the appearance of introspective capabilities.

2 Active Inference and Introspection

2.1 A Brief Introduction to Active Inference

Active inference offers a comprehensive framework for naturalizing, explaining, simulating, and understanding the mechanisms that underwrite decision-making, perception, and action [21,23]. The free energy principle (FEP) is a variational principle of information physics [89]. It has gained considerable attention and traction since it was first introduced in the context of computational neuroscience and biology [36,37]. Active inference denotes a family of models premised on the FEP, which are used to understand and predict the behavior of self-organizing systems. The tools of active inference allow us to model self-organizing systems as driven by the imperative to minimize surprise, which quantifies the degree to which a given path or trajectory deviates from its inertial or characteristic path—or its upper bound, variational free energy, which scores the difference between its predictions and the actual sensory inputs it receives [87].

Active inference modeling work suggests that decision-making, perception, and action involve the optimization of a world model that represents the causal structure of the system generating outcomes of observations [89]. In particular, active inference models the way that latent states or factors in the world cause sensory inputs, and how those factors cause each other, thereby capturing the essential causal structure of the measured or sensed world [59]. Minimizing surprise or free energy on average and over time allows the brain to maintain a consistent and coherent internal model of the world—one that maximizes predictive accuracy while minimizing model complexity—which, in turn, enables agents to adapt and survive in their environments [37,38]. (Strictly speaking, this is the other way around. In other words, agents who "survive" can always be read as minimizing variational free energy or maximizing their marginal likelihood (a.k.a., model evidence). This is often called self-evidencing [55].)

Active inference has instrumental value in allowing us to model, and thereby hopefully help to understand, core aspects of human consciousness (for a review, see [37]). Of particular interest to us here, it enables us to model the processes involved in introspective self-access (see [90,91]. Active inference modeling deploys the construct of generative models to make sense of the dynamics of self-organizing systems. In this context, a generative model is a joint probability density over the hidden or latent causes of observable outcomes; see [89] for a discussion of how to interpret these models philosophically and [98] for a gentle introduction to the technical implementation of these models.

We depict a simple generative model, apt for perceptual inference, in Fig. 1, and a more complex generative model, apt for the selection of actions (a.k.a. policy selection) in Fig. 2. These models specify the way in which observable outcomes are generated by (typically non-observable) states or factors in the world.

The main advantage of using generative models over current state of the art black box approaches is interpretability and auditability. Indeed, the factors that figure in the generative model are explicitly labeled, such that their contributions to the operations of the model can be read directly off its structure. This lends the generative model a degree of auditability that other approaches do not have.

2.2 Active Inference, Introspection, and Self-modeling

Active inference modeling has been deployed in the context of the scientific study of introspection, self-modeling, and self-access, which has led to the development of several leading theories of consciousness (for a review, see [90,100]). Introspection, which is defined as the ability to access and evaluate one's own mental states, thoughts, and experiences, plays a pivotal role in self-awareness, learning, and decision-making and is a pillar of human consciousness [69]. Self-modeling and self-access can be defined as interconnected processes that contribute to the development of self-awareness and to the capacity for introspection.

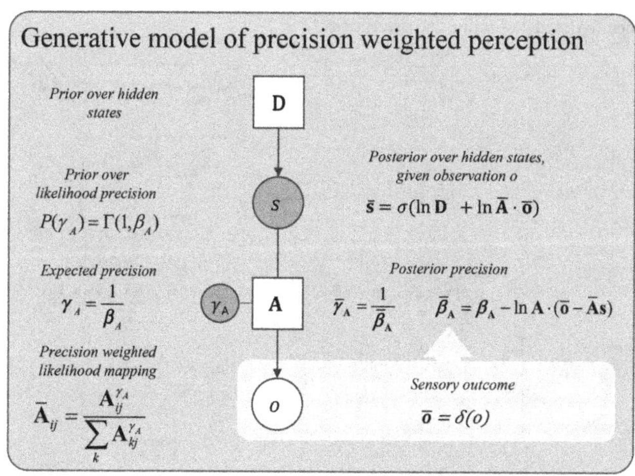

Fig. 1. A basic generative model for precision-weighted perceptual inference. This figure depicts an elementary generative model that is capable of performing precision-weighted perceptual inference. States are depicted as circles and denoted in lowercase: observable states or outcomes are denoted o and latent states (which need to be inferred) are denoted s. Parameters are depicted as squares and denoted as uppercase. The likelihood mapping **A** relates outcomes to the states that cause them, whereas **D** harnesses our prior beliefs about states, independent of how they are sampled. The precision term γ controls the precision or weighting assigned to elements of the likelihood, and implements attention as precision-weighting. Figure from [98].

Self-modeling involves the creation of internal representations of oneself, while self-access refers to the ability to access and engage with these representations for self-improvement and learning [8,78]. These processes, in conjunction with introspection, form a complex dynamic system that enriches our understanding of consciousness and the self-and indeed, may arguably form the causal basis of our capacity to understand ourselves and others.

Introspective self-access has been modeled using active inference by deploying a hierarchically structured generative model [70]. The basic idea is that for a system to report or evaluate its own inferences, it must be able to enact some form of self-access, where some parts of the system can take the output of other parts as their own input, for further processing. This has been discussed in computational neuroscience under the rubric of "opacity" and "transparency" [72–74,98]. The idea is that some cognitive processes are "transparent": like a (clean, transparent) window, they enable us to access some other thing (say, a tree outside) while not themselves being perceivable. Other cognitive processes are "opaque": they can be assessed per se, as in introspective self-awareness (i.e., aware that you are looking at a tree as opposed to seeing a tree). The idea, then, is that introspective processes make other cognitive processes accessible to the system as such, rendering them opaque.

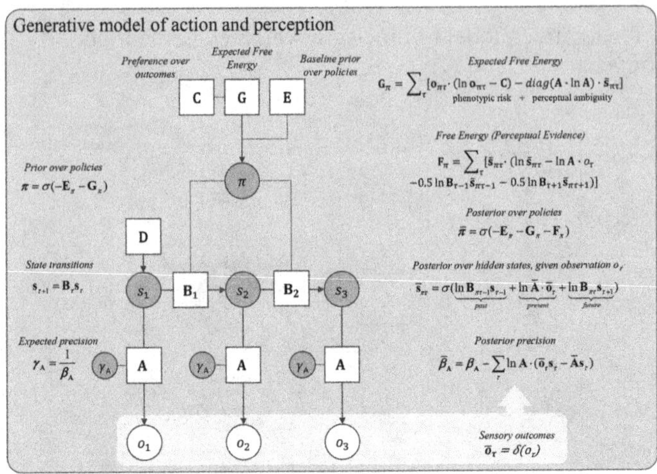

Fig. 2. A generative model for policy selection. This figure depicts a more sophisticated generative model that is apt for planning and the selection of actions in the future. The basic model depicted in Fig. 1 has now been expanded to include beliefs about the current course of action or policy (denoted $\bar{\pi}$), as well as **B**, **C**, **E**, **F** and **G** parameters. This kind of model generates a time series of states (s_1, s_2, etc.) and outcomes (o_1, o_2, etc.). The state transition (**B**) parameter encodes the transition probabilities between states over time, independently of the way they are sampled. **B**, **C**, **E**, **F** and **G** enter into the selection of beliefs about courses of action, a.k.a. policies. The **C** vector specifies preferred or expected outcomes and enters into the calculation of variational (**F**) and expected (**G**) free energies. The **E** vector specifies a prior preference for specific courses of action. Figure from [98].

In the context of self-access, the transparency and opacity of introspective processes has been modeled using a three-level generative model [98]. The model is depicted in Fig. 3. This model provides a framework for understanding how we access and interpret our internal states and experiences. The first level of the model (in blue), which implements the selection of overt actions, can be seen as a transparent process. The second, hierarchically superordinate level (in orange), which implements attention and covert action [74,91], represents more opaque processes, which make processes in the first layer accessible to the system. This layer models mental actions and shifts in attention that we may not be consciously aware of, or able to report. The second level takes as its input the inferences (posterior state estimations) ongoing at the first level, as data for further inference—about the system's inferences. Attentional processes are of this sort: they are about cognitive processes and action, and they modulate the activity of the first level. The third, final level (in green) implements the awareness of where one's attention is deployed. In other words, it both recognizes and instantiates a particular attentional set via bottom-up and top-down messages between levels, respectively. On the whole, this three-level architecture models our self-access and introspective abilities in terms of the processes regulating

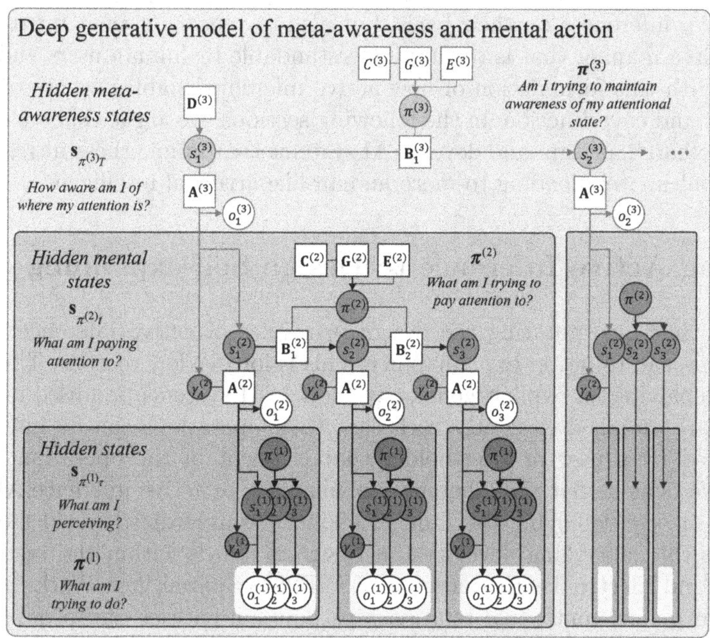

Fig. 3. A hierarchical generative model capable of self-access. Here, the generative model depicted in Fig. 2 (in blue) has been augmented with two superordinate hierarchical layers. In this architecture, posterior state estimates at one level are passed onto the next level as data for further inference. Note that this induces an architecture where the system is able to make inferences about its own inferences. Figure from [98].

transparency and opacity at a phenomenal level of description, or attentional selection at a psychological level.

Ramstead, Albarracin et al. (2023) recently discussed how active inference enables us to model both overt and covert action (also see [34,69,70,74,115]). Overt actions—observable behaviors such as physical movements or verbal responses—are directly influenced by the brain's hierarchical organization and can be modeled using active inference [39–41]. In contrast, covert actions refer to internal mental processes, such as attention and imagination, which involve the manipulation and processing of internal representations in the absence of observable behaviors [4,15,28,32,53,57,67,82,84,112]—of the sort discussed as "mental action" [68,69,74,98]. These actions are essential for higher cognitive functions, which rely on the brain's capacity to explore and manipulate abstract concepts and relationships.

In a significant new body of work in the active inference tradition [103–105,113,114], a hierarchical architecture of this type was deployed that was augmented with the capacity to report on its emotional states. Thus, it is possible to use active inference to design systems that can not only access their own states and perform inferences on their basis, but also to report on their introspective processes in a manner that is readily understandable by human users and stakeholders. With this formulation of how active inference enables agents to model their overt and covert action, in the following sections, we argue that we can and ought to research, design, and develop AI systems that mimic these introspective processes, ultimately leading to more human-like artificial intelligence.

3 Using Active Inference to Design Self-explaining AI

We argue that incorporating the design principles of active inference into AI systems can lead to better explainability. This is for two key reasons. The first is that, by deploying an explicit generative model, AI systems premised on active inference are designed explicitly such that their operations can be interpreted and audited by a user or stakeholder that is fluent in the operation of such models. We believe that the inherent explainability of active inference AI might be scaled up, by deploying the kind of explicit, standardized world modelling techniques that are being developed as open standards within the Institute of Electrical and Electronics Engineers (IEEE) P2874 Spatial Web Working Group [106], to formalize contextual relationships between entities and processes and to create digital twins of environments that are able to update in real time.

The second is that, by implementing an architecture inspired by active inference models of introspection, we can build systems that are able to access—and report on—the reasons for their decisions, and their state of mind when reaching these decisions.

AI systems designed using active inference can incorporate the kind of hierarchical self-access described by [98,103–105,113,114], to enhance their introspection during decision-making. As discussed, in the active inference tradition, introspection can be understood in the context of the (covert and overt) actions that AI systems perform. Covert actions, which are internal computations and decision-making processes that are not directly observable to users and stakeholders, can be recorded or explained to make the system more explainable. Overt actions, which are actions that an AI system takes based on its internal computations, such as making a recommendation or decision, can be explained to help users understand why the AI system acted as it did. This kind of deep inference promotes introspection, adaptability, and responses to environmental changes [25,99].

The proposed AI architecture includes components that continuously update and maintain an internal model of its own states, beliefs, and goals. This capacity for self-access (and implicitly self-report) enables the AI system to optimize (and report on) its decision-making processes, fostering introspection (and enhanced explainability). It incorporates metacognitive processing capabilities, which involve the ability to monitor, control, and evaluate its own cognitive processes. The AI system can thereby better explain the factors that contribute to its decisions, as well as identify potential biases or errors, ultimately leading to improved decision-making and explainability.

The proposed AI architecture would include introspection and a self-report interface, which translates the AI system's internal models and decision-making processes into human-understandable (natural) language (using, e.g., large language models). In effect, the agent would be talking to itself, describing its current state of mind and beliefs. This interface bridges the gap between the AI system's internal workings and human users, promoting epistemic trust and collaboration. In this way, the system can effectively mimic human-like consciousness and transparent introspection, leading to a deeper understanding of its decision-making processes and explainability. This advancement may be essential in fostering trust and collaboration between AI systems and their human users, paving the way for more effective and responsible AI applications.

Augmenting a generative model with black box systems—like large language models—may be a useful strategy to help AI systems articulate their "understanding" of the world. Using large language models to furnish an introspective interface may be relatively straightforward, leveraging their powerful natural language processing capabilities to create explanations of belief updating. This architecture—with a hierarchical generative model at its core—may contribute to the overall performance and explainability of hybrid AI systems. Attention mechanisms also achieve this purpose by enhancing the explainability of the AI system's decision making, emphasizing important factors in the hierarchical generative model that contribute to its decisions and actions.

These ideas are not new. Attentional mechanisms, particularly those at the word-level, have been identified as crucial components in AI architecture, specifically in the context of hierarchical generative models—and in generative AI, in the form of transformers. They function by focusing on relevant aspects during decision-making processes, thereby allowing the system to effectively process and prioritize information [64]. In fact, the performance of hierarchical models, which are a type of AI architecture, can be significantly improved by integrating word-level attention mechanisms. These mechanisms are powerful because they can leverage context information more effectively, especially fine-grained information.

The AI architecture that we propose employs a soft attention mechanism, which uses a weighted combination of hierarchical generative model components to focus on relevant information. The attention weights are dynamically computed based on the input data and the AI system's internal state, allowing the system to adaptively focus on different aspects of the hierarchical generative model [60]. This approach is similar to the use of deep learning models for global coordinate transformations that linearize partial differential equations, where the model is trained to learn a transformation from the physical domain to a computational domain where the governing partial differential equations are simpler, or even linear [45].

The AI architecture that we describe here effectively integrates diverse information sources for decision-making, mirroring the complex information processing capabilities observed in the human brain. The hierarchical structure of the generative model facilitates the exchange of information between different levels of abstraction. This exchange allows the AI system to refine and update its internal models based on both high-level abstract knowledge and low-level detailed information.

In conclusion, the integration of introspective processes in AI systems may represent a significant step towards achieving more explainable AI. By leveraging explicit generative models, as well as attention and introspection mechanisms, we can design AI systems that are not only more efficient and robust, but also more understandable and trustworthy. This approach allows us to bridge the gap between the complex internal computations of AI systems and the human users who interact with them. Ultimately, the goal is to create AI systems that can effectively communicate the reasons that drive their decision-making processes, adapt to environmental changes, and collaborate seamlessly with human users. As we continue to advance in this field, the importance of introspection in AI will only become more apparent, paving the way for more sophisticated and ethically sound AI systems.

4 Discussion

4.1 Directions for Future Research

The problem of explainable AI is the problem of understanding how AI models arrive at their decisions or predictions. This problem is especially relevant to avoid biases and harm in the design, implementation, and use of AI systems. By incorporating explicit generative models and introspective processing into the proposed AI architecture, we can create a system that is or seems capable of introspection and, thereby, that displays greatly enhanced explainability and auditability. This approach to AI design paves the way for more effective AI deployment across various real-world applications, by shedding light upon the problem of explainability, thereby offering opportunities for fostering trust, fairness, and inclusivity.

The development of the AI architecture based on active inference opens several potential avenues for future research. One possible direction is to further investigate the role of attention and introspection mechanisms in both AI systems and human cognition, as well as the development of more efficient attentional models to improve the AI system's ability to focus on salient information during decision-making. The approach that we propose bridges the gap between AI and cognitive neuroscience by incorporating biologically-inspired mechanisms into the design of AI systems. As a result, the proposed architecture promotes a deeper understanding of the nature of cognition and its potential applications in artificial intelligence, thus paving the way for more human-like AI systems capable of introspection and enhanced collaboration with human users.

Future work could explore more advanced data fusion techniques, such as deep learning-based fusion or probabilistic fusion, to improve the AI system's ability to combine and process multimodal data effectively. Evaluating the effectiveness of these techniques in diverse application domains will also be a valuable avenue for research [62,75]. Furthermore, the explanation dimension of these AI systems has been a significant topic in recent years, particularly in decision-making scenarios. These systems provide more awareness of how AI works and its outcomes, building a relationship with the system and fostering trust between AI and humans [33].

In addition to the aforementioned avenues for future research, another promising direction lies in the realm of computational phenomenology (for a review and discussion, see [88]. Beckmann, Köstner, & Hipólito (2023) have proposed a framework that deploys phenomenology—the rigorous descriptive study of first-person experience—for the purposes of machine learning training. This approach conceptualizes the mechanisms of artificial neural networks in terms of their capacity to capture the statistical structure of some kinds of lived experience, offering a unique perspective on deep learning, consciousness, and their relation. By grounding AI training in socioculturally situated experience, we can create systems that are more aware of sociocultural biases and capable of mitigating their impact. Ramstead et al. (2022) propose a similar methodology based on explicit generative models as they figure in the active inference tradition. This connection to first-person experience, of course, does not guarantee unbiased AI. But by moving away from traditional black box AI systems, we shift towards human-interpretable models that enable the identification and correction of biases in the AI system. This approach aligns with our goal of creating AI systems that are not only efficient and effective, but also ethically sound and socially responsible.

The incorporation of computational phenomenology into our proposed AI architecture could further enhance its introspective capabilities and its ability to understand and navigate the complexities of human sociocultural contexts. This could lead to AI systems that are more adaptable, more trustworthy, and more capable of meaningful collaboration with human users. As we continue to explore and integrate such innovative approaches, we move closer to our goal of creating AI systems that truly mirror the richness and complexity of human cognition and consciousness.

4.2 Ethical Considerations of Introspective AI Systems

Ethical AI starts with the development of AI systems that are ethically designed; AI systems must be designed in such a way as to be transparent, auditable, explainable, and to minimize harm. The experience of AI may improve our ethical intuitions and self-understanding, potentially helping our societies make better-informed decisions on serious ethical dilemmas [16]. But as these systems become increasingly integrated into our daily lives, research on the ethical implications of introspective AI systems, as well as the development of regulatory frameworks and guidelines for responsible AI use, become crucial.

The development of introspective AI systems raises several ethical considerations. Even if these systems provide more human-like decision-making capabilities and enhanced explainability, it is and will remain crucial to ensure that their decisions are transparent, fair, and unbiased, and that their designers and users can be held accountable for harm that their use may cause. The lack of ethics and interpretability of AI decisions are critical issues, leading to the proposal of two scenarios for the future development of ethical AI: more external regulation or more liberalization of AI explanations [56].

To address these concerns, future research should focus on developing methods to audit and evaluate the AI system's decision-making processes, as well as identify and mitigate potential biases within the system. The development of laws, policies, and best practices for seizing the opportunities and minimizing the risks posed by AI technologies would benefit from building on ethical frameworks such as the one offered by Cowls & Floridi [22]. This framework emphasizes the importance of transparency, accountability, and the alignment of AI with human values.

Additionally, the AI4People initiative presents five ethical principles and 20 recommendations to establish a Good AI Society. These principles and recommendations, if adopted, would provide a strong foundation for achieving this goal [35]. The recommendations are structured around five key principles: Beneficence (promoting good), Non-Maleficence (preventing harm), Autonomy (protecting human intervention), Justice (ensuring fairness), and Explicability (ensuring transparency). These principles guide the development of AI in a way that aligns with societal values and ethical considerations, fostering responsible innovation and deployment.

Moreover, as introspective AI systems become more prevalent, issues related to agency, privacy, and data security may arise. Ensuring that these systems protect sensitive information by abiding by data protection regulations, thereby safeguarding agency, will be of paramount importance. The results from a survey study by Esmaeilzadeh [30] show that technological, ethical, and regulatory concerns significantly contribute to the perceived risks of using AI applications in healthcare, highlighting the need for robust data protection measures.

In terms of the implications of developing sentient/introspective AI, beyond the human-centric ethics, ethical frameworks such as the one offered by Cowls & Floridi [22] are meaningful and useful when non-human agents (animals as well as AI agents) may also be deserving of ethical consideration and care. The conceptual analysis reveals interdependencies and tensions between ethical principles, advocating the need for a basic understanding of AI inputs, functioning, agency, and outcomes [50]. The AI4People initiative also presents five ethical principles and 20 recommendations to establish a Good AI Society, providing a strong foundation for achieving this goal [35].

The ethics of doing cognitive modeling on/about humans require a wider range of driving ethical principles for designing more socially responsible AI agents [111]. An embedded ethics approach, such as embedding ethicists into the development team, can improve the consideration of ethical issues during AI development [96].

The development of AI systems based on active inference has broad implications for both the fields of AI and consciousness studies. As future research explores the potential of this novel approach, ethical considerations and responsible use of introspective AI systems must remain at the forefront of these advancements, ultimately leading to more transparent, effective, and user-friendly AI applications. The dearth of literature on the ethics of AI within LMICs, as well as in public health, also points to a critical need for further research into the ethical implications of AI within both global and public health, to ensure that its development and implementation is ethical for everyone, everywhere [77].

5 Conclusion

We have argued that active inference has demonstrated significant potential in advancing the field of explainable AI. By incorporating design principles from active inference, the AI system can better tackle complex real-world problems with improved auditability of decision-making, thereby increasing safety and user trust.

Throughout our discussions and analysis, we have highlighted the importance of active inference models as a foundation for designing more human-like AI systems, seemingly capable of introspection and finessed (epistemic) collaboration with human users. This novel approach bridges the gap between AI and cognitive neuroscience by incorporating biologically-inspired mechanisms into the design of AI systems, thus promoting a deeper understanding of the nature of consciousness and its potential applications in artificial intelligence.

As we move forward in the development of AI systems, the importance of advancing explainable AI becomes increasingly apparent. By designing AI systems that can not only make accurate and efficient decisions, but also provide understandable explanations for their decisions, we foster (epistemic) trust and collaboration between AI systems and human users. This advancement ultimately leads to more transparent, effective, and user-friendly AI applications that can be tailored to a wide range of real-world scenarios.

Acknowledgements. The authors are grateful to VERSES for supporting the open access publication of this paper. SET is supported in part by funding from the Social Sciences and Humanities Research Council of Canada (Ref: 767-2020-2276). KF is supported by funding for the Wellcome Centre for Human Neuroimaging (Ref: 205103/Z/16/Z) and a Canada-UK Artificial Intelligence Initiative (Ref: ES/T01279X/1). The authors are grateful to Brennan Klein for assistance with typesetting.

Conflict of interest statement. The authors disclose that they are contributors to the Institute of Electrical and Electronics Engineers (IEEE) P2874 Spatial Web Working Group.

References

1. Adadi, A., Berrada, M.: Peeking inside the black-box: a survey on explainable artificial intelligence (XAI). IEEE Access **6**, 52138–52160 (2018). https://doi.org/10.1109/ACCESS.2018.2870052
2. Adams, R.A., Shipp, S., Friston, K.J.: Predictions not commands: active inference in the motor system. Brain Struct. Funct. **218**(3), 611–643 (2013). https://doi.org/10.1007/s00429-012-0475-5

3. Adams, R.A., et al.: Everything is connected: inference and attractors in delusions. Schizophrenia Res. **245**, 5–22 (2022). https://doi.org/10.1016/j.schres.2021.07.032
4. Ainley, V., et al.: Bodily precision: a predictive coding account of individual differences in interoceptive accuracy. Philos. Trans. R. Soc. B Biol. Sci. **371**(1708), 20160003 (2016). https://doi.org/10.1098/rstb.2016.0003
5. Ali, S., et al.: Explainable artificial intelligence (XAI): what we know and what is left to attain trustworthy artificial intelligence. Inf. Fusion **99**, 101805 (2023). https://doi.org/10.1016/j.inffus.2023.101805
6. Arrieta, A.B., et al.: Explainable artificial intelligence (XAI): concepts, taxonomies, opportunities and challenges toward responsible AI. Inf. Fusion **58**, 82–115 (2020). https://doi.org/10.1016/j.inffus.2019.12.012
7. Bach, S., et al.: On pixel-wise explanations for non-linear classifier decisions by layer-wise relevance propagation. PLoS ONE **10**(7), e0130140 (2015). https://doi.org/10.1371/journal.pone.0130140
8. Baker, J.R.: Going beyond brick and mortar self-access centers: Establishing a satellite activity self-access program. Stud. Self-Access Learn. J. **13**(1), 129–141 (2022). https://doi.org/10.37237/130107
9. Bauer, K., von Zahn, M., Hinz, O.: Expl(AI)ned: the impact of explainable artificial intelligence on cognitive processes. Inf. Syst. Res. (2021). https://doi.org/10.1287/isre.2023.1199
10. Bélisle-Pipon, J.-C., Monteferrante, E., Roy, M.-C., Couture, V.: Artificial intelligence ethics has a black box problem. AI Soc. 1–16 (2022). https://doi.org/10.1007/s00146-021-01380-0
11. Birhane, A.: The impossibility of automating ambiguity. Artif. Life **27**(1), 44–61 (2021). https://doi.org/10.1162/artl_a_00336
12. Birhane, A., et al.: Frameworks and challenges to participatory AI. In: Proceeding of the Second Conference on Equity and Access in Algorithms, Mechanisms, and Optimization (EAAMO 2022) (2022). https://doi.org/10.48550/arXiv.2209.07572
13. Birhane, A., et al.: The forgotten margins of AI ethics. In: 2022 ACM Conference on Fairness, Accountability, and Transparency, pp. 948–958 (2022). https://doi.org/10.1145/3531146.3533157
14. Brennen, A.: What do people really want when they say they want "Explainable AI?" we asked 60 stakeholders. In: Extended Abstracts of the 2020 CHI Conference on Human Factors in Computing Systems, pp. 1–7 (2020). https://doi.org/10.1145/3334480.3383047
15. Brown, H., Adams, R.A., Parees, I., Edwards, M., Friston, K.: Active inference, sensory attenuation and illusions. Cogn. Process. **14**(4), 411–427 (2013). https://doi.org/10.1007/s10339-013-0571-3
16. Bryson, J., Kime, P.P.: Just an artifact: why machines are perceived as moral agents (2011)
17. Burrell, J.: How the machine 'thinks': understanding opacity in machine learning algorithms. Big Data Soc. **3**(1), 2053951715622512 (2016). https://doi.org/10.1177/2053951715622512
18. Castelvecchi, D.: Can we open the black box of AI?" Nat. News **538**(7623) 20 (2016). https://doi.org/10.1038/538020a
19. Chaudhry, M.A., Cukurova, M., Luckin, R.: A transparency index framework for AI in education. In: Artificial Intelligence in Education. Posters and Late Breaking Results, Workshops and Tutorials, Industry and Innovation Tracks, Practitioners' and Doctoral Consortium: 23rd International Conference, AIED 2022, Durham,

UK, Proceedings, Part II. 2022, pp. 195–198, 27–31 July 2022. https://doi.org/10.1007/978-3-031-11647-6_33
20. European Commission. Proposal for a Regulation laying down harmonised rules on artificial intelligence. In: Shaping Europe's digital future. The Commission has proposed the first ever legal framework on AI, which addresses the risks of AI and positions Europe to play a leading role globally, April 2021. https://digital-strategy.ec.europa.eu/en/library/proposal-regulation-laying-down-harmonised-rulesartificial-intelligence
21. Constant, A., et al.: Regimes of expectations: an active inference model of social conformity and human decision making. Front. Psychol. **10**, 679 (2019). https://doi.org/10.3389/fpsyg.2019.00679
22. Cowls, J., Floridi, L.: Prolegomena to a white paper on an ethical framework for a good AI society (2018)
23. Da Costa, L., et al.: Bayesian mechanics for stationary processes. In: Proceedings of the Royal Society A, vol. 477, p. 2256 (2021). https://doi.org/10.1098/rspa.2021.0518
24. Da Costa, L., et al.: How active inference could help revolutionise robotics. Entropy **24**(3), 361 (2022)
25. Dhulipala, S.L.N., Hruska, R.C.: Efficient interdependent systems recovery modeling with DeepONets. In: arXiv, pp. 1–6 (2022). https://doi.org/10.48550/arXiv.2206.10829
26. Doshi-Velez, F., Kim, B.: Towards a rigorous science of interpretable machine learning. In: arXiv (2017). https://doi.org/10.48550/arXiv.1702.08608
27. Drake, M., et al.: EU AI policy and regulation: what to look out for in 2023. In: Inside Privacy (2023). https://www.insideprivacy.com/artificial-intelligence/eu-ai-policy-and-regulationwhat-to-look-out-for-in-2023/
28. Edwards, M.J., et al.: A Bayesian account of 'hysteria. Brain **135**(11), 3495–3512 (2012). https://doi.org/10.1093/brain/aws129
29. Eschenbach, W.J.: Transparency and the black box problem: why we do not trust AI. Philos. Technol. **34**(4), 1607–1622 (2021). https://doi.org/10.1007/s13347-021-00477-0
30. Esmaeilzadeh, P.: Use of AI-based tools for healthcare purposes: a survey study from consumers perspectives. In: BMC Medical Informatics and Decision Making (2020)
31. Esterhuizen, J.A., Goldsmith, B.R., Linic, S.: Interpretable machine learning for knowledge generation in heterogeneous catalysis. Nat. Catal. **5**(3), 175–184 (2022). https://doi.org/10.1038/s41929-022-00744-z
32. Feldman, H., Friston, K.J.: Attention, uncertainty, and freeenergy. Front. Hum. Neurosci. **4** (2010). https://doi.org/10.3389/fnhum.2010.00215
33. Ferreira, J.J., Monteiro, M.: The human-AI relationship in decision-making: AI explanation to support people on justifying their decisions. In: arXiv (2021). https://doi.org/10.48550/arXiv.2102.05460
34. Fleming, S.M.: Awareness as inference in a higher-order state space. Neurosci. Conscious. **2020**(1) niz020 (2020). https://doi.org/10.1093/nc/niz020
35. Floridi, L., et al.: AI4People–an ethical framework for a good ai society: opportunities, risks, principles, and recommendations. Minds Mach. (2018)
36. Friston, K.J.: A theory of cortical responses. Philos. Trans. R. Soc. B: Biol. Sci. **360**(1456), 815–836 (2005). https://doi.org/10.1098/rstb.2005.1622
37. Friston, K.J.: Is the free-energy principle neurocentric?". Nat. Rev. Neurosci. **11**(8), 605–605 (2010). https://doi.org/10.1038/nrn2787-c2

38. Friston, K.J.: Life as we know it. J. R. Soc. Interface **10**(86), 20130475 (2013). https://doi.org/10.1098/rsif.2013.0475
39. Friston, K.J., Mattout, J., Kilner, J.: Action understanding and active inference. Biol. Cybern. **104**, 137–160 (2011). https://doi.org/10.1007/s00422-011-0424-z
40. Friston, K.J., Parr, T., de Vries, B.: The graphical brain: Belief propagation and active inference. Netw. Neurosci. **1**(4), 381–414 (2017). https://doi.org/10.1162/NETN_a_00018
41. Friston, K.J., et al.: Deep temporal models and active inference. Neurosci. Biobehav. Rev. **77**, 388–402 (2017). https://doi.org/10.1016/j.neubiorev.2017.04.009
42. Friston, K.J., et al.: Designing ecosystems of intelligence from first principles. In: arXiv (2022). https://doi.org/10.48550/arXiv.2212.01354
43. Frith, C.D.: Consciousness, (meta) cognition, and culture. Q. J. Exp. Psychol. 17470218231164502 (2023). https://doi.org/10.1177/17470218231164502
44. van Giffen, B., Herhausen, D., Fahse, T.: Overcoming the pitfalls and perils of algorithms: a classification of machine learning biases and mitigation methods. J. Bus. Res. **144**, 93–106 (2022). https://doi.org/10.1016/j.jbusres.2022.01.076
45. Gin, C., et al.: Deep learning models for global coordinate transformations that linearise PDEs. Eur. J. Appl. Math. **32**(3), 515–539 (2021). https://doi.org/10.1017/S0956792520000327
46. Goodfellow, I., Bengio, Y., Courville, A.: Deep Learning. MIT Press, Cambridge (2016)
47. Guest, O., Martin, A.E.: On logical inference over brains, behaviour, and artificial neural networks. Comput. Brain Behav. 1–15 (2023). https://doi.org/10.1007/s42113-022-00166-x
48. Guidotti, R., et al.: A survey of methods for explaining black box models. ACM Comput. Surv. **51**(5), 1–42 (2018). https://doi.org/10.1145/3236009
49. Gunning, D.: Explainable artificial intelligence (XAI). Def. Sci. Res. Projects Agency **2**(2), 1 (2017). https://doi.org/10.1609/aimag.v40i2.2850
50. Hermann, E.: Artificial intelligence and mass personalization of communication content-an ethical and literacy perspective. New Media Soc. **24**(5), 1258–1277 (2021)
51. Hipólito, I.: The human roots of artificial intelligence (2023). https://doi.org/10.31234/osf.io/cseqt
52. Hipólito, I., Winkle, K., Lie, M.: Enactive artificial intelligence: subverting gender norms in robot-human interaction. Front. Neurorobot. **17** 77 (2023). https://doi.org/10.48550/arXiv.2301.08741
53. Hohwy, J.: Attention and conscious perception in the hypothesis testing brain. Front. Psychol. **3**, 96 (2012). https://doi.org/10.3389/fpsyg.2012.00096
54. Hohwy, J.: The Predictive Mind. Oxford University Press, Oxford (2013). https://doi.org/10.1093/acprof:oso/9780199682737.001.0001
55. Hohwy, J.: The self-evidencing brain. Nous **50**(2), 259–285 (2016). https://doi.org/10.1111/nous.12062
56. John-Mathews, J.-M.: Some critical and ethical perspectives on the empirical turn of AI interpretability (2021)
57. Kanai, R., et al.: Cerebral hierarchies: predictive processing, precision and the pulvinar. Philos. Trans. R. Soc. B: Biol. Sci. **370**(1668), 20140169 (2015). https://doi.org/10.1098/rstb.2014.0169
58. Kokciyan, N., et al.: Sociotechnical perspectives on AI ethics and accountability. IEEE Internet Comput. **25**(6), 5–6 (2021). https://doi.org/10.1109/MIC.2021.3117611

59. Konaka, Y., Naoki, H.: Decoding reward-curiosity conflict in decision-making from irrational behaviors. Nat. Computat. Sci. **3**(5), 418–432 (2023). https://doi.org/10.1038/s43588-023-00439-w
60. Kulkarni, M., Abubakar, A.: Soft attention convolutional neural networks for rare event detection in sequences (2020). https://doi.org/10.48550/arXiv.2011.02338
61. Laato, S., et al.: How to explain AI systems to end users: a systematic literature review and research agenda. Internet Res. **32**(7), 1–31 (2022). https://doi.org/10.1108/INTR-08-2021-0600
62. Lahat, D., Adali, T., Jutten, C.: Multimodal data fusion: an overview of methods, challenges, and prospects. In: Proceedings of the IEEE, vol. 103, no. 9, pp. 1449–1477 (2015). https://doi.org/10.1109/JPROC.2015.2460697
63. Lamberti, W.F.: An overview of explainable and interpretable AI. In: AI Assurance, pp. 55–123 (2023). https://doi.org/10.1016/B978-0-32-391919-7.00015-9
64. Lan, T., et al.: Which kind is better in open-domain multi-turn dialog, hierarchical or non-hierarchical models? An empirical study. In: arXiv (2020). https://doi.org/10.48550/arXiv.2008.02964
65. Le, T.-T.-H., et al.: Exploring local explanation of practical industrial AI applications: a systematic literature review. Appl. Sci. **13**(9), 5809 (2023). https://doi.org/10.3390/app13095809
66. LeCun, Y., Bengio, Y., Hinton, G.: Deep learning. Nature **521**(7553), 436–444 (2015). https://doi.org/10.1038/nature14539
67. Limanowski, J.: (Dis-)Attending to the body – action and self- experience in the active inference framework. In: Metzinger, T.,. Wiese, W. (ed). Philosophy and Predictive Processing. Frankfurt am Main: MIND Group (2017). https://doi.org/10.15502/9783958573192
68. Limanowski, J.: Precision control for a flexible body representation. Neurosci. Biobehav. Rev. **134**, 104401 (2022). https://doi.org/10.1016/j.neubiorev.2021.10.023
69. Limanowski, J., Friston, K.J.: Seeing the dark: grounding phenomenal transparency and opacity in precision estimation for active inference. Front. Psychol. **9**, 643 (2018). https://doi.org/10.3389/fpsyg.2018.00643
70. Limanowski, J., Friston, K.J.: Attenuating oneself: an active inference perspective on "selfless" experiences. Philos. Mind Sci. **1**(I), 1–16 (2020). https://doi.org/10.33735/phimisci.2020.I.35
71. Mascarenhas, M., et al.: The promise of artificial intelligence in digestive healthcare and the bioethics challenges it presents. Medicina **59**(4), 790 (2023). https://doi.org/10.3390/medicina59040790
72. Metzinger, T.: Empirical perspectives from the self-model theory of subjectivity: a brief summary with examples. Prog. Brain Res. **168**, 215–278 (2007). https://doi.org/10.1016/S0079-6123(07)68018-2
73. Metzinger, T.: Phenomenal transparency and cognitive self-reference. Phenomenol. Cogn. Sci. **2**, 353–393 (2003). https://doi.org/10.1023/b:phen.0000007366.42918.eb
74. Metzinger, T.: The problem of mental action. In: Metzinger, T., Wiese, W. (ed.). Philosophy and Predictive Processing. Frankfurt am Main: MIND Group (2017). https://doi.org/10.15502/9783958573208
75. Microsoft Defender Security Research Team. Seeing the big picture: Deep learning-based fusion of behavior signals for threat detection (2020). https://tinyurl.com/3kpzvk9d
76. Mishra, A.: Transparent AI: reliabilist and proud. J. Med. Ethics **47**(5), 341–342 (2021). https://doi.org/10.1136/medethics-2021-107352

77. Murphy, A. et al.: Ethics of AI in Low- and Middle-Income Countries and Public Health. Glob. Public Health (2021)
78. Murray, G.: Self-access environments as self-enriching complex dynamic ecosocial systems. Stud. Self-Access Learn. J. **9**(2) (2018). https://doi.org/10.37237/090204
79. Nascimento, N., Alencar, P., Cowan, D.: Comparing software developers with ChatGPT: an empirical investigation. In: arXiv (2023). https://doi.org/10.48550/arXiv.2305.11837
80. Neri, E., et al.: Explainable AI in radiology: a white paper of the Italian society of medical and interventional radiology. In: La Radiologia Medica, pp. 1–10 (2023). https://doi.org/10.1007/s11547-023-01634-5
81. Oberste, L., et al.: Designing user-centric explanations for medical imaging with informed machine learning. In: Design Science Research for a New Society: Society 5.0: 18th International Conference on Design Science Research in Information Systems and Technology, DESRIST 2023, Pretoria, South Africa, May 31-June 2, 2023, Proceedings, pp. 470–484 (2023). https://doi.org/10.1007/978-3-031-32808-4_29
82. Parr, T., Friston, K.J.: Attention or salience? Curr. Opin. Psychol. **29**, 1–5 (2019). https://doi.org/10.1016/j.copsyc.2018.10.006
83. Parr, T., Pezzulo, G.: Understanding, explanation, and active inference. Front. Syst. Neurosci. **15**, 772641 (2021). https://doi.org/10.3389/fnsys.2021.772641
84. Pezzulo, G.: An active inference view of cognitive control. Front. Psychol. **3**, 478 (2012). https://doi.org/10.3389/fpsyg.2012.00478
85. Prabhushankar, M., AlRegib, G.: Introspective learning: a two-stage approach for inference in neural networks. In: arXiv (2022). https://openreview.net/forum?id=in1ynkrXyMH
86. Raghupathi, W., Raghupathi, V.: Big data analytics in healthcare: promise and potential. Health Inf. Sci. Syst. **2**, 1–10 (2014). https://doi.org/10.1186/2047-2501-2-3
87. Ramstead, M.J.D., Badcock, P.B., Friston, K.J.: Answering Schrödinger's question: a free-energy formulation. Phys. Life Rev. **24**, 1–16 (2018). https://doi.org/10.1016/j.plrev.2017.09.001
88. Ramstead, M.J.D., et al.: From generative models to generative passages: a computational approach to (Neuro) Phenomenology. Rev. Philos. Psychol. **13**(4) (2022). https://doi.org/10.1007/s13164-021-00604-y
89. Ramstead, M.J.D., et al.: On Bayesian mechanics: a physics of and by beliefs. Interface Focus **13**, 20220029 (2023). https://doi.org/10.1098/rsfs.2022.0029
90. Ramstead, M.J.D., et al.: Steps towards a minimal unifying model of consciousness: an integration of models of consciousness based on the free energy principle (2023)
91. Ramstead, M.J.D., et al.: The inner screen model of consciousness: applying the free energy principle directly to the study of conscious experience. In: PsyArXiv (2023). https://doi.org/10.31234/osf.io/6afs3
92. Ramstead, M.J.D., et al.: Variational ecology and the physics of sentient systems. Phys. Life Rev. **31**, 188–205 (2019). https://doi.org/10.1016/j.plrev.2018.12.002
93. Ratti, E., Graves, M.: Explainable machine learning practices: opening another black box for reliable medical AI. AI Ethics **2**(4), 801–814 (2022). https://doi.org/10.1007/s43681-022-00141-z
94. Ribeiro, M.T., Singh, S., Guestrin, C.: Why should I trust you? Explaining the predictions of any classifier. In: Proceedings of the 22nd ACM SIGKDD International Conference on Knowledge Discovery and Data Mining, pp. 1135–1144 (2016). https://doi.org/10.1145/2939672.2939778

95. Ridley, M.: Explainable artificial intelligence (XAI). Inf. Tech. Libr. **41**(2), 14683 (2022). https://doi.org/10.6017/ital.v41i2
96. McLennan, S., et al.: An embedded ethics approach for AI development. Nat. Mach. Intell. **2**(9), 488–490 (2020)
97. Miguel, B.S., Naseer, A., Inakoshi, H.: Putting accountability of AI systems into practice. In: Proceedings of the Twenty- Ninth International Conference on International Joint Conferences on Artificial Intelligence, pp. 5276–5278 (2021). https://doi.org/10.24963/ijcai.2020/768
98. Sandved-Smith, L., et al.: Towards a computational phenomenology of mental action: modelling meta-awareness and attentional control with deep parametric active inference. Neurosci. Conscious. **2021**, 1 (2021). https://doi.org/10.1093/nc/niab018
99. Schoeffer, J., et al.: On the interdependence of reliance behavior and accuracy in AI-assisted decision-making. In: arXiv (2023). https://doi.org/10.48550/arXiv.2304.08804
100. Seth, A.K., Bayne, T.: Theories of consciousness. Nat. Rev. Neurosci. **23**(7), 439–452 (2022). https://doi.org/10.1038/s41583-022-00587-4
101. Skeath, C., Tonsager, L., Zhang, J.: FTC Announces COPPA Settlement against Ed tech provider including strict data minimization and data retention requirements. Inside Priv. (2023). https://www.insideprivacy.com/childrens-privacy/ftcannounces-coppa-settlement-against-ed-tech-provider-includingstrict-data-minimization-and-data-retention-requirements
102. Smith, R., Khalsa, S.S., Paulus, M.P.: An active inference approach to dissecting reasons for nonadherence to antidepressants. Biol. Psychiatry Cogn. Neurosci. Neuroimaging **6**(9), 919–934 (2021). https://doi.org/10.1016/j.bpsc.2019.11.012
103. Smith, R., Parr, T., Friston, K.J.: Simulating emotions: an active inference model of emotional state inference and emotion concept learning. Front. Psychol. **10**, 2844 (2019). https://doi.org/10.3389/fpsyg.2019.02844
104. Smith, R., Taylor, S., Bilek, E.: Computational mechanisms of addiction: recent evidence and its relevance to addiction medicine. Curr. Addict. Rep. **8**(4), 509–519 (2021). https://doi.org/10.1007/s40429-021-00399-z
105. Smith, R., et al.: Neurocomputational mechanisms underlying emotional awareness: insights afforded by deep active inference and their potential clinical relevance. Neurosci. Biobehav. Rev. **107**, 473–491 (2019). https://doi.org/10.1016/j.neubiorev.2019.09.002
106. Standard for Spatial Web Protocol, Architecture and Governance (2020). https://standards.ieee.org/ieee/2874/10375/
107. National Institute of Standards and Technology (NIST). AI Risk Management Framework. In: On January 26, 2023, NIST released the AI Risk Management Framework (AI RMF 1.0) along with various resources. In collaboration with the private and public sectors, NIST has developed a framework to better manage risks associated with artificial intelligence (AI). The NIST AI Risk Management Framework is intended for voluntary use and aims to improve trustworthiness considerations in the design, development, use, and evaluation of AI products, services, and systems, January 2023. https://www.nist.gov/itl/ai-riskmanagement-framework
108. Sterzer, P., et al.: The predictive coding account of psychosis. Biol. Psychiatry **84**(9), 634–643 (2018). https://doi.org/10.1016/j.biopsych.2018.05.015
109. Stiglic, G., et al.: Interpretability of machine learning-based prediction models in healthcare. Wiley Interdisciplinary Rev. Data Min. Knowl. Disc. **10**(5), e1379 (2020). https://doi.org/10.1002/widm.1379

110. Veale, M., Binns, R.: Fairer machine learning in the real world: mitigating discrimination without collecting sensitive data. Big Data Soc. **4**(2) (2017). https://doi.org/10.1177/2053951717743530
111. Vetrò, A., et al.: AI: from rational agents to socially responsible agents. In: Digital Policy, Regulation and Governance (2019)
112. Vossel, S., et al.: Cortical coupling reflects Bayesian belief updating in the deployment of spatial attention. J. Neurosci. **35**(33), 11532–11542 (2015). https://doi.org/10.1523/JNEUROSCI.1382-15.2015
113. Whyte, C.J., Hohwy, J., Smith, R.: An active inference model of conscious access: how cognitive action selection reconciles the results of report and no-report paradigms. Curr. Res. Neurobiol. **3**, 100036 (2022). https://doi.org/10.1016/j.crneur.2022.100036
114. Whyte, C.J., Smith, R.: The predictive global neuronal workspace: a formal active inference model of visual consciousness. Prog. Neurobiol. **199**, 101918 (2021). https://doi.org/10.1016/j.pneurobio.2020.101918
115. Yon, D., Frith, C.D.: Precision and the Bayesian brain. Curr. Biol. **31**(17), R1026–R1032 (2021). https://doi.org/10.1016/j.cub.2021.07.044
116. Zhang, Q., Wu, Y.N., Zhu, S.-C.: Interpretable convolutional neural networks. In: Proceedings of the IEEE Conference on Computer Vision and Pattern Recognition, pp. 8827–8836 (2018). https://doi.org/10.1109/CVPR.2018.00920

An Analytical Model of Active Inference in the Iterated Prisoner's Dilemma

Daphne Demekas[1,2](✉), Conor Heins[1,3,4,5], and Brennan Klein[1,5,6]

[1] Network Science Institute, Northeastern University, Boston, MA, USA
daphnedemekas@gmail.com, b.klein@northeastern.edu
[2] Wheeler Lab, University of Arizona, Tucson, AZ, USA
[3] Department of Collective Behaviour, Max Planck Institute of Animal Behavior, 78464 Konstanz, Germany
[4] Department of Biology and the Centre for the Advanced Study of Collective Behaviour, University of Konstanz, 78464 Konstanz, Germany
[5] VERSES AI Research Lab, Los Angeles, CA 90016, USA
[6] Institute for Experiential AI, Northeastern University, Boston, MA, USA

Abstract. This paper addresses a mathematically tractable model of the Prisoner's Dilemma using the framework of active inference. In this work, we design pairs of Bayesian agents that are tracking the joint game state of their and their opponent's choices in an Iterated Prisoner's Dilemma game. The specification of the agents' belief architecture in the form of a partially-observed Markov decision process allows careful and rigorous investigation into the dynamics of two-player gameplay, including the derivation of optimal conditions for phase transitions that are required to achieve certain game-theoretic steady states. We show that the critical time points governing the phase transition are linearly related to each other as a function of learning rate and the reward function. We then investigate the patterns that emerge when varying the agents' learning rates, as well as the relationship between the stochastic and deterministic solutions to the two-agent system.

Keywords: Game Theory · Bounded Rationality · Multi-Agent Systems · Prisoner's Dilemma

1 Introduction

Studies of behavioural science, be it in biology, psychology, or machine learning, often rely on the concept of rational thinking and decision making [3,23,24,30]. Game theory has had wide success in precisely formulating contexts in which players or agents are challenged to converge to an optimal yet counter-intuitive strategy that maximises reward. In particular, game theory models communication among agents that can result in bounded-complex emergent behaviour [6,31]. The Iterated Prisoner's Dilemma (IPD) is a quintessential game, in which the 'dilemma' is that the highest reward is attributed to the action of defection, but the optimal behaviour in the long run is to cooperate, because of the

C. L. Buckley et al. (Eds.): IWAI 2023, CCIS 1915, pp. 145–172, 2024.
https://doi.org/10.1007/978-3-031-47958-8_10

'Shadow of the Future' phenomenon [11][1]. When played iteratively, agents learn each other's predictable behaviour and can form an optimal strategy, away from the Nash equilibrium of the one-shot game. To do so, agents need to be aware of what their opponent is likely to do, which is why the IPD is widely used to study the evolution of cooperation for selfish agents [20].

This work addresses a computational model of the (memory-one) Iterated Prisoner's Dilemma under the framework of active inference (AIF) [12,25,28]. AIF is an agent-based modelling framework derived from theoretical neuroscience, where cognitive processes like action, perception, and learning are seen as solutions to an inference problem. As an explicitly model-based, Bayesian framework for simulating behaviour, AIF provides cognitively 'transparent' agents, whose posterior beliefs about the world and associated uncertainties are accessible and interpretable. This enables careful investigation into the Bayesian basis of behaviour in these simple models, in turn allowing us to identify the conditions under which optimal behaviour is possible.

When two identical and deterministic AIF agents play against one another, we show that the equation governing across-trial learning dynamics is mathematically tractable given one approximation. This enables us to derive functions that model the specific conditions under which convergence to an optimal strategy—namely the Pavlov Strategy [20]—for the IPD can occur, given a multi-agent AIF model. The Pavlov strategy is win-stay-lose-change, where agents will cooperate if the agent's and opponent's moves are the same in the previous round and defect otherwise. We explore how these dynamics vary across different configurations of the agents' learning rates, as well as how stochasticity in the agent network determines the probabilities of agents reaching the optimal outcome.

1.1 Iterated Prisoner's Dilemma

In the Prisoner's Dilemma, at each round, both players can either defect or cooperate, leading to 4 possible outcomes [16] (see Table 1 with different reward levels). The outcome with the highest reward is if the player defects and its opponent cooperates (DC), which is also the outcome with the lowest reward for the opponent (CD). The second-best outcome is if both cooperate (CC), and the third-best outcome for both players is if they defect (DD). In this model, the four reward levels are respectively [3, 1, 4, 2]. This work specifically models the memory-one IPD, where each player only considers the previous move of their opponent when making their decision for the current round.

There are several notable strategies in the IPD, which have been categorised in different ways [17]. First, a dominant strategy produces the best possible payoff for an agent, regardless of the strategies used by opponents. The most commonly cited dominant outcome is when both players defect (choose to betray) in every round. From an individual player's perspective, defecting in every round provides

[1] This is when agents in repeated play—without awareness of when the play will end—will be more cooperative because they are made to learn about the possibility of being punished and plan accordingly [16].

Table 1. Example payout matrix in a Prisoner's Dilemma game.

		Player 2	
		Cooperate (**C**)	Defect (**D**)
Player 1	**C**	(3, 3)	(1, 4)
	D	(4, 1)	(2, 2)

a higher immediate payoff compared to cooperation, especially when the other player cooperates. However, defecting in every round is not socially optimal as it leads to a lower overall payoff compared to mutual cooperation. The challenge is to find strategies that can foster cooperation and lead to better outcomes for both players in the long run, rather than succumbing to the dominant outcome of mutual defection [14].

In order to reach the social optimum of cooperation, new heuristics or bounds on the agents need to emerge in order for them to look beyond the reward function when deciding their actions. This makes the IPD a good arena to study *bounded* rationality, in which agents do not have access to the full generative process (encompassing both themselves and their opponent), and therefore must make decisions given a bound on their awareness or knowledge, of, for instance, the other player, or any external environmental factors [30]. Agents playing the IPD has been studied in the context of reinforcement learning already [18,29], and the idea of bounded rationality serves as a motivation for using active inference agents to model the IPD, as the AIF is a transparent and interpretable framework in which agents infer actions and quantify uncertainty under the constraints of their generative model.

There are several ways to train the agents to converge to the social optimum, which we will refer to as the cooperative steady state. When agents sample their actions deterministically, our model shows that active inference agents parameterised with a constrained set of learning rates can converge to the cooperative steady state by learning the Pavlov Strategy [20], and it also demonstrates learning rate configurations that get trapped in the Nash equilibrium, in which agents converge to Unconditional Defection [26,29].

1.2 Active Inference

Active inference (AIF) agents are able to plan and learn about their state space and transition probabilities through observed experience. They infer which actions to take by minimising the expected free energy anticipated to accrue from their actions [25]. This often allows these agents to solve complex tasks often seen in reinforcement learning or neuroscience, such as the Multi-Armed Bandit [19] and other Monte-Carlo based tasks [7].

Advances in the ability to quickly build and scale models of AIF agents, particularly in Python using the `pymdp` library [13], have allowed for a much more scalable and accessible means to model these agents in different and flexible

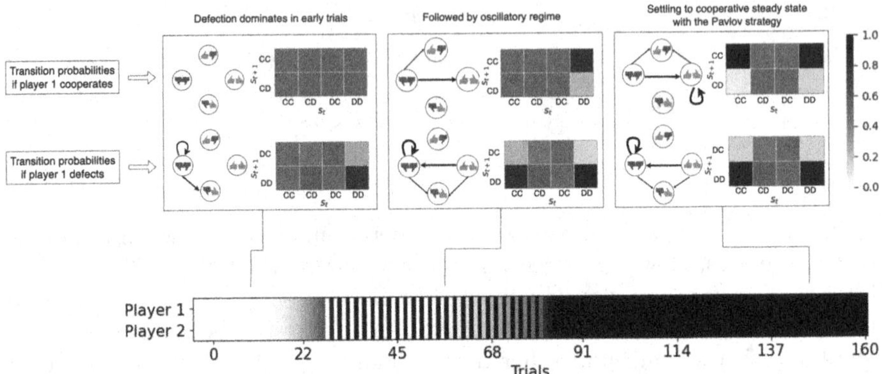

Fig. 1. Beliefs about transition probabilities over trials. Top: A representation of Player 1's beliefs at three phases of the simulation ($t = 10, 20, 150$). Each box contains a graph representation of the transition probabilities, and histograms of the cooperate-conditioned (top row) or defection-conditioned (bottom row) transition distributions at the displayed trial indices. Darker values represent a higher probability. **Bottom:** The inferred probabilities of cooperation in each trial. Agents select the action with the highest posterior probability. The agents begin by continuously defecting, then undergo an oscillatory period of defection and cooperation, and eventually reach a cooperative steady state. After this period of training, they will have learned the Pavlov strategy, i.e. they will cooperate if the agent's and opponent's moves are the same in the previous round and defect otherwise [20].

environments, as well as to connect them in networks and allow them to observe each other's actions. This has allowed researchers to ask more interesting questions about how relevant AIF is in terms of modelling rational decision making, such as those observed in game theory. In this paper, we show that not only can AIF agents effectively learn optimal strategies to the IPD, but the framework of active inference enables us to derive the exact conditions for when this will occur and have a layered understanding of the agents' 'mental process' throughout the game.

The agents in this model actively entertain beliefs about the dynamics of the game and iteratively update their beliefs about the game dynamics (i.e., a 'transition model') as they play multiple rounds against their opponent. In the context of the discrete-time and -space models used in the present work, this amounts to updating the elements of transition probability matrices that represent each agent's beliefs about game states from one trial to the next. After every trial of iterated play, the agents update these state transition probability distributions based on their actions and the outcomes that they observed. In doing so, the agents have the capacity to learn strategies, manifested as patterns of learned probabilities of transition from each state to each other state.

Table 2. Generative model variables and notation.

Variable Name	Notation
Hidden States	$\mathbf{s} \in \{\mathrm{CC}, \mathrm{CD}, \mathrm{DC}, \mathrm{DD}\}$
Observations	$\mathbf{o} \in \{\mathrm{CC}, \mathrm{CD}, \mathrm{DC}, \mathrm{DD}\}$
Actions	$\mathbf{u} \in \{u^C, u^D\}$
Observation Model	$P(\mathbf{o}_t \vert \mathbf{s}_t; A) = Cat(\mathbf{A})$
Transition Model	$P(\mathbf{s}_{t+1} \vert \mathbf{s}_{t-1}, \mathbf{u}_{t-1}; B) = Cat(\mathbf{B})$
Transition Model Parameter	$P(B) = \prod_{ju} P(B_{\bullet ju}), \quad P(B_{\bullet ju}) = Dir(\mathbf{b}_{\bullet ju})$
Initial State Prior	$P(\mathbf{s}_1; D) = Cat(\mathbf{D})$
'Biased' State Prior (Reward)	$\tilde{P}(\mathbf{s}; C) = Cat(\mathbf{C})$, s.t. $\ln \mathbf{C} = [3, 1, 4, 2]$

Our hypothesis is that throughout iterative play, the bounded-rational agents will learn to infer actions based on learned patterns of their opponent's behaviour (i.e., the ability to predict revenge from defection), and this will result in a strategy leading to the social optimum steady state in which both agents cooperate. Further, given the interpretability of the AIF, we will be able to analytically derive the process that the agents undergo during this learning process and thus predict how it might change with different parameters.

2 Simulation Dynamics

Here, we explore the long-term dynamics of the IPD. Agents play in turns for a finite set of trials, updating their transition model beliefs $Q(B; \phi_{\mathrm{b}})$ at each trial. Unless otherwise specified, agents are configured exactly the same (same priors, same learning rate) and sample their actions deterministically as described in Eq. (A.21). In this model, agents always converge to the cooperative steady state and remain there indefinitely. The magnitude of the learning rate η affects the rate of convergence by scaling the update to the transition matrix at each timestep, as shown in Eq. (A.25). In Fig. 1 we show the simulation dynamics for agents configured with learning rate $\eta = 0.3$, but it's important to note that at different learning rates, the nature of these dynamics would not change - rather the critical time points would only occur either sooner (for larger η) or later (for smaller η). Therefore, the amount of time taken in order to converge is not representative of the performance of this model, but rather a parameter that can be tweaked. Given the transparency of this deterministic system, it is possible to explain exactly how these agents are 'thinking', given their posteriors over time.

Agents are initialised with uniform transition matrices as in Eq. (A.7). Upon the first observation, they infer the game state and calculate the expected free energies (EFEs, or $\mathbf{G}$) of cooperating and defecting. They take the action that has smaller EFE, i.e., $\arg\min_u \mathbf{G}_0(u)$. At first, because of the reward parameterization and the uniformity in the transition prior $P(B; \mathbf{b})$, defection will minimise the EFE (i.e., predicts the highest reward), according to:

$$\mathbf{G}_0(u = \mathrm{C}, \phi^{\mathrm{C}}) = -(\mathbf{B}_0^{\mathrm{C}} \cdot \phi_0^{\mathrm{C}}) \cdot (\ln \mathbf{B}_0^{\mathrm{C}} \cdot \phi_0^{\mathrm{C}} - \ln \mathbf{C}) = \frac{1}{2} \ln(\mathbf{C}_1 \mathbf{C}_2) - \ln \frac{1}{2} \quad (1)$$

$$\mathbf{G}_0(u = \mathrm{D}, \phi^{\mathrm{D}}) = -(\mathbf{B}_0^{\mathrm{D}} \cdot \phi_0^{\mathrm{D}}) \cdot (\ln \mathbf{B}_0^{\mathrm{D}} \cdot \phi_0^{\mathrm{D}} - \ln \mathbf{C}) = \frac{1}{2} \ln(\mathbf{C}_3 \mathbf{C}_4) - \ln \frac{1}{2} \quad (2)$$

Therefore, as long as $\ln(\mathbf{C}_3\mathbf{C}_4) < \ln(\mathbf{C}_1\mathbf{C}_2)$, the agent always defects on the first timestep. Agents will then continue to defect, because the expected reward from realising the state DC still outweighs that of any other predicted state. As the agents continue to defect, their beliefs about $P(\mathbf{s}_t = \mathrm{DC}|s_{t-1} = \mathrm{DD}, u = \mathrm{D})$ will be decreasing with a proportional increase in $P(\mathbf{s}_t = \mathrm{DD}|\mathbf{s}_{t-1} = \mathrm{DD}, u = \mathrm{D})$, meaning $\mathbf{G}(u = D)$ will increase as the probability of getting their desired reward decreases.

At a critical time, which we denote τ_1[2], the agents will begin assigning more probability to cooperation than defection $\phi^C > \phi^D$, because the transition probabilities have decreased sufficiently for the EFE of cooperation to outweigh that of defection. Once the agents begin cooperating, they undergo an oscillatory period during which their actions fluctuate from cooperation to defection. This is because at τ_1, the transition probabilities $P(\mathbf{s}_t|s_{t-1} = \mathrm{CC})$ are fixed at their initial value, since the agents have yet observed the previous state being CC. Thus the agents will still be optimistic about realising the highest reward state DC via the transition probability $P(\mathbf{s}_{t+1} = \mathrm{DC}|\mathbf{s}_t = \mathrm{CC}, u = \mathrm{D})$.

The agents will eventually learn that inferring to defect will inevitably lead to observing DD, and inferring to cooperate will inevitably lead to CC. The oscillatory period is crucial to this because it teaches the agent that defecting in response to cooperation will only ever lead to DD. The oscillation continues until the critical time point τ_2, in which the probability $p(\mathbf{s}_{t+1} = \mathrm{DC}|\mathbf{s}_t = \mathrm{CC}, \mathbf{u}_t = \mathrm{D})$ becomes smaller than $p(\mathbf{s}_{t+1} = \mathrm{DD}|\mathbf{s}_t = \mathrm{CC}, \mathbf{u}_t = \mathrm{D})$, at which point the agents will cooperate for all remaining rounds.

2.1 The Analytic Transition Function

In the above model of AIF agents, an analytic solution for the evolution of each agent's beliefs about the transition likelihood $Q(B; \phi_{\mathrm{b}}^*)$ is available. This is formulated by deriving approximations to τ_1—the critical trial in which the agents transition to an oscillatory period between defection and cooperation—and τ_2, the second phase transition in which the agents converge to the cooperative steady state. Given the expressions for τ_1 and τ_2 in Eq. (3), we can write down the evolution of the Dirichlet parameters of the transition probability matrix. The derivations for the following expressions are in Appendix A.4 and A.5. Here, $\mathbf{C}$ corresponds to the 'biased' state reward prior, and each entry of $\mathbf{C}$ corresponds to the reward value of that observation $(r_{\mathrm{CC}}, r_{\mathrm{CD}}, r_{\mathrm{DC}}, r_{\mathrm{DD}})$. For the full definition see Eq. (A.5).

[2] Whose solution in terms of generative model parameters we derive in the next section.

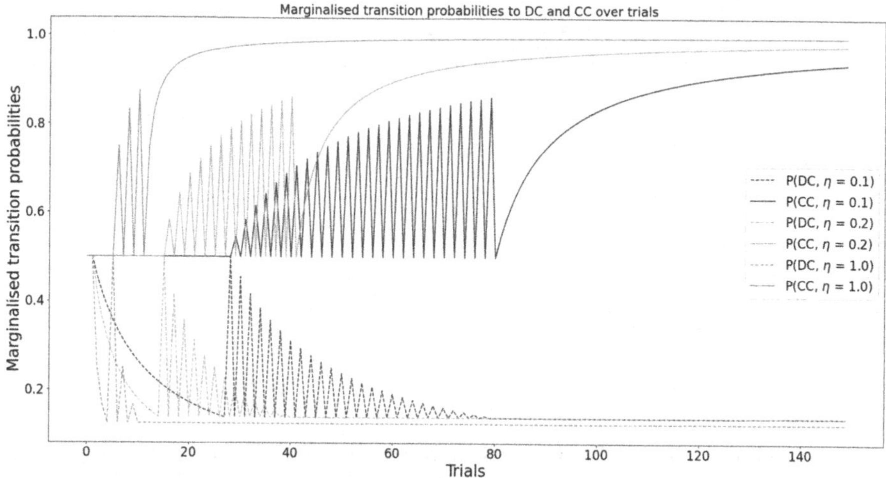

Fig. 2. Marginalised transition probabilities under different η. The dotted lines represent the marginalised probabilities from all states to the highest reward state DC, and the solid lines represent the marginalised probabilities from all states to the socially optimal state CC. The transition probabilities to DC decrease initially during the period of defection, then fluctuate during the period of oscillation and steady out close to 0 once the agents reach the cooperate steady state, and the probabilities to state CC take the same pattern in the opposite direction. This happens more rapidly for larger η, because the updates to the parameters of the transition likelihood distribution are larger at every trial.

$$\tau_1 \approx \frac{R_1(\beta)}{\eta} \qquad \tau_2 \approx \frac{R_2(\beta)}{\eta} \tag{3}$$

where

$$R_1 = \frac{2}{\ln \frac{\mathbf{C}_3}{\mathbf{C}_4} + 2 - \sqrt{(\ln \frac{\mathbf{C}_4}{\mathbf{C}_3} - 2)^2 - 8(-\ln \frac{\mathbf{C}_4}{2\sqrt{\mathbf{C}_1\mathbf{C}_2}} - \frac{1}{5})}} - 1 \qquad R_2 = \frac{3}{2}R_1 \tag{4}$$

This means that τ_2, the number of trials it takes the system to reach the steady state, can be precisely approximated as a linear function of τ_1, the number of trials it takes to start the oscillatory period (see Fig. 3)—i.e. the critical time points governing the phase transition are linearly related to each other as a function of learning rate and the reward function. Given these expressions, our analytic form of the transition rule for the posterior Dirichlet parameters over the transition model is:

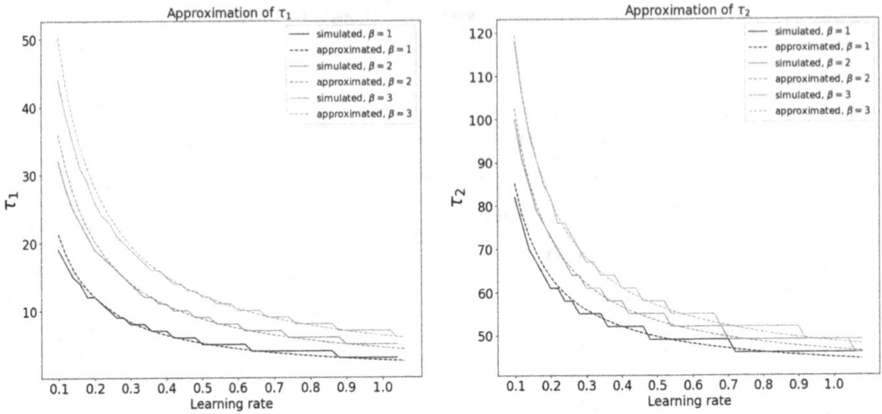

Fig. 3. Simulated vs. derived relation between reward and learning rate. Simulated and approximated τs for three values of β parameterizing the reward function. On the left, we approximate τ_1 with the equation $\tau_1 = \frac{R_1}{\eta}$ where R_1 depends on the reward parameter of β. On the right, we approximate τ_2 with $\tau_2 = \tau_1 + \frac{1}{\eta}R_2$ where again, R_2 depends on β. With a larger β, meaning a higher predicted reward for the state DC, the values of τ increase as it will take more trials for the players to update their transition probabilities away from having a preference to defect.

$$\phi^{\mathbf{C}}_{\mathbf{b}_{t+1}} = \begin{cases} \mathbf{b}^{\mathbf{C}}_0 \\ \mathbf{b}^{\mathbf{C}}_0 + \frac{\eta}{2} s^{\mathbf{CC}} \otimes s^{\mathbf{DD}}(t - \frac{R_1}{\eta}) \\ \mathbf{b}^{\mathbf{C}}_{\tau_2} + \eta s^{\mathbf{CC}} \otimes s^{\mathbf{CC}}(t - \frac{R_2}{\eta}) \end{cases} \quad \phi^{\mathbf{D}}_{\mathbf{b}_{t+1}} = \begin{cases} \mathbf{b}^{\mathbf{D}}_0 + \eta s^{\mathbf{DD}} \otimes s^{\mathbf{DD}} t \\ \mathbf{b}^{\mathbf{D}}_{\tau_1} + \frac{\eta}{2} s^{\mathbf{DD}} \otimes s^{\mathbf{CC}}(t - \frac{R_1}{\eta}) \\ \mathbf{b}^{\mathbf{D}}_{\tau_2} \end{cases} \left| \begin{array}{c} t < \frac{R_1}{\eta} \\ \frac{R_1}{\eta} < t < \frac{R_2}{\eta} \\ t > \frac{R_2}{\eta} \end{array} \right. \tag{5}$$

which can be used to exactly replicate the trajectory of $Q(s_{t+1}|s_t, u_t)$ over time (Fig. 3).

We conclude by noting that the agents in this model, after undergoing these two phase transitions and converging to CC, have learned the well-known Pavlov (also known as the "Win-Stay Lose-Shift") strategy from IPD literature [20]. Agents learned during $0 < t < \tau_1$ that given the observation DD, the best strategy is to cooperate, and during $\tau_1 < t < \tau_2$ they learned that cooperating is the best outcome given the observation CC—therefore, having reached τ_2, they continue cooperating. To show that the agents learned the Pavlov strategy, we performed an experiment where once an agent converged to the steady state, we disabled additional learning and had this agent play against an agent that behaves completely randomly. When playing against this random agent, they observe the new asymmetric states DC or CD. The desire to maximise expected utility (via the drive to minimise KL risk, a.k.a., the expected free energy) will lead them to perform the 'greedy' strategy of defection, which is how their behaviour is consistent with the Pavlov strategy[3]. Future work will further characterise the space of learnable strategies under this framework.

3 Generalizing the Model

In the previous section, we found an approximate solution for the belief-, action-, and learning-dynamics, which completely describes the case of two symmetrically-parameterised agents playing IPD. For any given parameterisation of the prior preferences **C**, we derived the trials at which the critical transitions take place in the two-agent system, steering it away from the Nash equilibrium and towards the cooperative steady state.

The simplicity of this model is that these agents are configured exactly alike, and therefore there is complete symmetry in the state space. This means that the agents will only ever observe two out of four possible states in the space. However, this case no longer holds when either the agents are parameterised with different learning rates, or when they sample their actions stochastically, according to Eq. (A.22). These cases open the space of possible strategies that the agents can learn, some of which will lead the agents to fall into the Nash equilibrium, and others which will allow them to reach the optimal outcome.

3.1 Different Learning Rates

We now assume agents parameterised with different η and the same β, performing actions deterministically. We denote the agent with larger η_1 as a_1, and the agent with smaller η_2 as a_2. According to Eq. (A.38), the critical value τ_1 depends on η, and since $\eta_1 > \eta_2$, this means $\tau_1^{a_1} < \tau_2^{a_2}$. Thus, a_1 will cooperate at $\tau_1^{a_1} = \frac{R_1}{\eta_1}$, but a_2 will not yet deem cooperation a better policy than defection (namely, the EFE of defection will remain below that of cooperation). Therefore, at $\tau_1^{a_1}$, the game state will be CD from a_1's perspective and DC from a_2's perspective. This symmetry-breaking means that the system will not enter into the typical oscillation phase triggered by mutual cooperation (as is guaranteed when $\eta_1 = \eta_2$ and thus $\tau_1^{a_1} = \tau_1^{a_2}$).

The nonidentical observations imply that after $\tau_1^{a_1}$, a_1 believes $P(\mathbf{s}_{t+1} = \text{CD}|\text{DD})$ is more probable, thereby being disincentivised to continue cooperating, and a_2 believes $P(\mathbf{s}_{t+1} = \text{DC}|\text{DD})$ is more probable, being incentivised to continue defecting. The degree of disincentivisation (or incentivisation) will increase in proportion to η_1 or η_2, respectively, due to a corresponding η_1-scaled increase in $\mathbf{G}^{a_1}(\mathbf{u} = C)$ and an η_2-scaled decrease in $\mathbf{G}^{a_2}(\mathbf{u} = D)$. This growing

[3] An agent exhibiting the Pavlov strategy will only cooperate if in the previous trial, both agents performed the same action (i.e., the state was either CC or DD, otherwise they will defect).

asymmetry in the agents' beliefs means that Eq. (5) no longer holds. At this point, the agents will return to continuous defection until another instance of $\mathbf{G}(\mathbf{u} = D) = \mathbf{G}(\mathbf{u} = C)$ occurs; the duration of this depends on η.

In sum, the conditions under which the joint-agent system converges to the optimal steady state is determined by whether or not the agents' learning rates are configured such that there will be some time point t less than some threshold T_{max} in which both agents cooperate simultaneously. If this is not the case, then as defection continues, the rate of increase of $\mathbf{G}(\mathbf{u} = D)$ slows, and after a certain amount of time (governed by η) it will become too slow and never catch up to $\mathbf{G}(\mathbf{u} = C)$ (see Fig. 4). In other words, if at any point, for either agent, $\mathbf{G}_t(\mathbf{u} = D) < \mathbf{G}_t(\mathbf{u} = C)$ $\forall t \in (0, T_{max}]$, the agents are trapped in the Nash equilibrium.

Figure 4 shows EFE trajectories in scenarios where agents converge to the optimal outcome (above) and where agents get trapped in the Nash equilibrium (below). Convergence to the Nash equilibrium occurs in the absence of any trial where the relative value of cooperation reaches 0 simultaneously for both agents. Instead, the relative values of cooperation slowly converge to different and nonoverlapping limits[4]. If the intersection of the condition in Eq. (A.31) does occur, this guarantees that the agents will begin the oscillatory period which will eventually lead them to convergence to CC (while there may be some instances of CD and DC in the oscillatory period, this will not prevent eventual cooperation). In general, when learning rates are close together, the likelihood of convergence to CC is more likely; however, the actual pattern is more complicated than this. Figure 5 demonstrates the complex pattern of instances in which the agents converge to the cooperative steady state given different learning rate combinations, with both the deterministic and stochastic sampling.

3.2 Stochastic Sampling

Here, we introduce noise in the action selection such that agents sample actions with some probability proportional to their (negative) EFE. Action stochasticity can be controlled with an inverse temperature parameter α according to Eq. (A.22). In general, all of the principles outlined in Sect. 2 remain; however, now the agents will sometimes perform the suboptimal action. This enables agents to experience the entire state space (different combinations of defection and cooperation) and therefore estimate transitions between all the combinations of states.

[4] Note that even after the agents reach a cooperative steady state, the difference in expected free energy takes time to flatten because the entropy is still decreasing as beliefs become more precise, via learning.

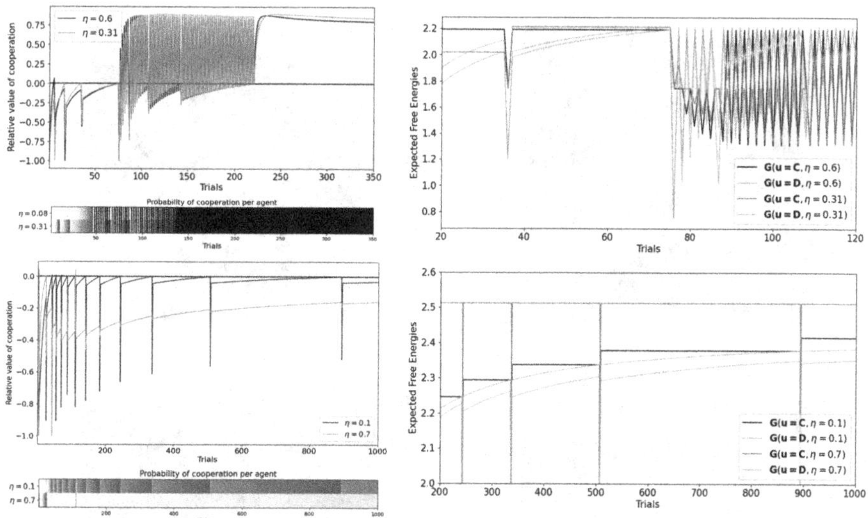

Fig. 4. Relative value of cooperation under different η parameterisations. Above: Agents are configured with ηs along the tendrils of Fig. 5. On the left, the relative values of cooperation, calculated as $\mathbf{G}(\mathbf{u} = C) - \mathbf{G}(\mathbf{u} = D)$, reach zero several times and converging around 0.75 at the optimal outcome. On the right: the fluctuations in the individual EFEs. There are periods before τ_1 and between τ_1 and τ_2 in which one player will cooperate and the opponent defects; this creates the spikes in the distribution, as one agent is punished and the other is rewarded. **Below:** Agents with ηs that are not on the tendrils in Fig. 5, meaning that they do not converge to the cooperative steady state. We can see on the left how $\mathbf{G}_t(\mathbf{u} = D)$ is converging to something less than $\mathbf{G}_t(\mathbf{u} = C)$.

We can see from Fig. 5 that, on average, endowing the agents with stochasticity enables them to converge to the cooperative steady state for a larger number of combinations of different learning rates. This makes sense, because it increases the likelihood of 'escaping' the pattern of continuous defection, and therefore learning about the advantages of cooperation. In terms of the reward, the agents that have most similar learning rates will behave most similarly and therefore accumulate more reward (along the diagonal).

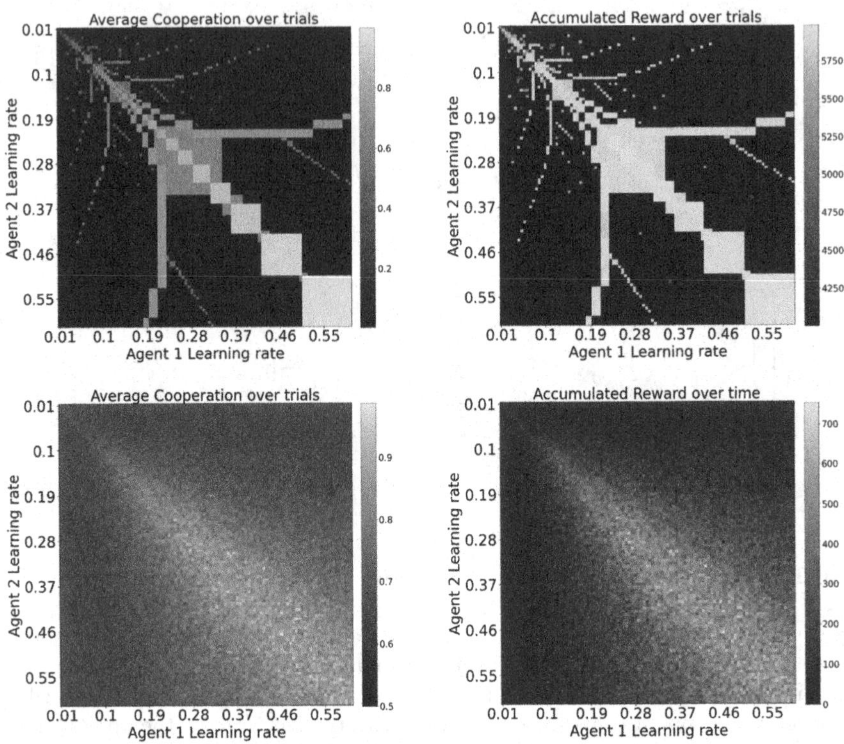

Fig. 5. Parameter sweeps over η. Top row: Agents sample actions deterministically. Wherever the average cooperation is nonzero, agents converged to the cooperative steady state—yellow cells indicate faster cooperation, which is generally associated with higher overall reward. **Bottom row:** Agents sample actions stochastically. Cooperation still occurs most often along the diagonal, tapering off as learning rates become more different.

4 Conclusion

Iterated Prisoners' Dilemma games have long been the test bed for new developments in behavioural science and game theory. Because of the relative simplicity of the game's structure—and its, at times, surprising experimental results—researchers often use it to develop mathematical frameworks for understanding decision making in social or multi-agent contexts. In this paper, we demonstrated how active inference can be used to model the IPD transparently, such that in a simple set-up, we can derive a solution to the evolution of the agents' beliefs about the game dynamics, i.e., the transition probabilities. This allows us to quantitatively reason about why the agents converge to their chosen optimal strategy and how behaviour changes as a function of different learning rates and stochastic action selection. While the simple case of similarly-configured agents resulted in both agents exhibiting the Pavlov strategy, once we introduce asymmetry in the generative models, and/or stochasticity in action sampling, then upon testing, agents are able to learn a variety of different strategies, including the Pavlov strategy, Unconditional Defection, Unconditional Cooperation, and Tit for Tat—or some variation of Tit for Tat [15,32].

This finding is a starting point for future work, in which such a model could be extended to multiple agents interacting towards a common goal, and investigating the various strategies that emerge from acting in a network order to minimise free energy. The current model did not incorporate the information-seeking components that are often leveraged in action-selection under active inference [10]. In our case, the ambiguity term of the expected free energy was zero by construction (due to zero observation uncertainty), but future work could explore the role of parameter information gain (resolving uncertainty about B) and how that changes the multi-agent dynamics in IPD. Overall, in this work we demonstrated that AIF can offer game theory a novel analytic transparency and simplicity for accounting for multi-agent dynamics using a first-principles, Bayesian account.

Acknowledgements. The authors thank Wolfram Barfuss and Christoph Riedl for valuable feedback and comments that substantially improved the quality of the manuscript. *Funding information:* DD, CH, & BK acknowledge the support of a grant from the John Templeton Foundation (61780). The opinions expressed in this publication are those of the authors and do not necessarily reflect the views of the John Templeton Foundation.

A Appendix

A.1 Generative Model

In this section, we describe the Prisoner's Dilemma game as a two-agent active inference system and determine the conditions under which the agents reach the optimal state of constant cooperative play, avoiding the Nash equilibrium. To enable active inference agents to reach the cooperative steady state, we invoke the notion of parameter learning; specifically, the ability of agents to infer likely sequences of game states by updating posterior beliefs about transition probabilities. These transition probabilities parameterise a likelihood model that describes transitions between game states (e.g., the transition from the state of 'cooperate-cooperate' to 'cooperate-defect'). Under active inference, this parameter learning is cast as a problem of inferring generative model parameters. Usually, parameter inference unfolds on a slow timescale (hence the term 'learning') relative to 'fast' inference of hidden states [9] (See Table 2 for full description of model parameters).

The agent's generative model is a Markov Decision Process [27] that encodes a joint distribution over sequences of hidden states $\mathbf{s}_{1:T}$ observations $\mathbf{o}_{1:T}$, actions $\mathbf{u}_{1:T}$, and model parameters A, B, D [13]. Markov Decision Processes assume that the dynamics are shallow, with single-timestep dependency $P(s_{t+1}|s_t, u_t; B)$; this Markov property means we can write the generative model as a product of time-dependent distributions:

$$P(\mathbf{o}_{1:T}, \mathbf{s}_{1:T}, \mathbf{u}_{1:T}, A, B) = P(\mathbf{s}_1; D)P(\pi)P(A)P(B)P(D) \prod_{t=1}^{T-1} P(\mathbf{o}_{t+1}|\mathbf{s}_{t+1}; A)P(\mathbf{s}_{t+1}|\mathbf{s}_t, \mathbf{u}_t; B) \tag{A.1}$$

multiplied by initial priors over hidden states, policies, and parameters.

The hidden states $\mathbf{s}$ consist of a single factor with four possible states or levels, corresponding to the game states (the four combinations of possible two-player choices): CC, CD, DC, and DD. This game state factor comprises the primary random variable in each agent's model.

In our notation, the first letter of each game state corresponds to the focal agent's choice, and the second letter corresponds to that of its opponent. In our formulation, agents have precise knowledge of the current game state, which they technically infer through (unambiguous) observation of their and their opponent's action. Uncertainty comes into the game insofar as agents must *predict* the subsequent game state and then act based on their predictions and their desires to maximise utility.

There is one observation modality with four observations, which again correspond directly to the four game states. Therefore, the four observations are CC, CD, DC, and DD. Note that the agents will only observe the game state after-the-fact, i.e., each observation corresponds to the game state in the previous round of iterative play. This is because in the Prisoner's Dilemma, the agents perform their actions at any given trial without knowing what their opponent will do in that trial, but in iterative play, the agents can build a strategy over time by observing the resulting game states after each trial ends.

Observation Likelihood. The observation model $P(\mathbf{o}_t|\mathbf{s}_t, A)$ is a conditional distribution encoding the agent's beliefs about the relationship between the current (hidden) game state and its concurrent observation. Also known as the likelihood model, the agent uses this distribution to infer the most likely game state, given an observation thereof.

In the simulations presented here, we assume that agents are equipped with a deterministic, unambiguous observation model, i.e., observations are deterministic indicators of the game state. In the discrete state space models common in active inference, likelihoods like $P(\mathbf{o}_t|\mathbf{s}_t, A)$ are often represented as multidimensional arrays (e.g., matrices) whose values are populated by parameters; in the case of the observation model, we represent this likelihood directly as a matrix $\mathbf{A}$ whose entries are given by the likelihood parameters A. Hereafter we use boldface $\mathbf{X}$ to indicate a representation of Categorical parameters in terms of vectors and matrices, and use the standard italic notation X to indicate the random variable in the generative model (e.g. $P(A)$). When we have an unambiguous or precise likelihood mapping, this matrix is the identity matrix, representing the mapping from hidden states (columns) to observations (rows):

$$P(\mathbf{o}_t = i|\mathbf{s}_t = j, [A]_{ij}) = \delta(i - j) \tag{A.2}$$

$$\mathbf{A} = \begin{bmatrix} 1 & 0 & 0 & 0 \\ 0 & 1 & 0 & 0 \\ 0 & 0 & 1 & 0 \\ 0 & 0 & 0 & 1 \end{bmatrix} = \mathbf{I} \tag{A.3}$$

An agent with such a precise likelihood model will infer the game state in the previous round of iterative play entirely based on the observed game state.

However, one can imagine introducing uncertainty into an agent's beliefs by adding off-diagonal, positive values into the A matrix – this would correspond to the agent believing that game state observations are ambiguous with respect to the true game state. Concretely, we could imagine that one agent might receive a misleading signal indicating that its opponent defected when they actually cooperated. A simple way to parameterise this uncertainty is through an inverse temperature parameter ψ, which makes the A matrix totally uninformative (maximum entropy columns) in the limit of $\psi \to 0$, and infinitely precise in the limits of $\psi \to \infty$:

$$\mathbf{A} = \frac{\mathbf{I}^{\psi}}{\sum \mathbf{I}^{\psi}} \tag{A.4}$$

Finally, it is worth mentioning that we assume $P(A)$ is infinitely precise and not subject to learning. Therefore, we emit any parameterisation of the priors over this likelihood, while we keep them for the transition likelihood parameters B, as we will update these in learning.

Reward. Different game states are assigned different rewards or desirabilities under the Prisoner's Dilemma problem formulation. Active inference converts the notion of 'reward' into prior probability by equipping agents with biased prior beliefs about future states or observations [8]. In the context of planning actions, this biased prior serves the role of a "goal-vector" or reward function [13]. We denote this as a biased prior over states in our agent's model $\tilde{P}(\mathbf{s}; \mathbf{C})$[5]. This special 'goal prior' is parameterised by a vector of Categorical parameters $\mathbf{C}$. Reward and prior probability can be straightforwardly related via the relation $\tilde{P}(\mathbf{s}) \propto \exp(r)$ [22]; therefore, we typically parameterise $\mathbf{C}$ using relative log probabilities or nats, i.e., $\mathbf{C} = \ln \tilde{P}(\mathbf{s}) + Z$. Following from Table 1, the most desirable observation is $s^{\mathbf{DC}}$ (the agent defects and the opponent cooperates), followed by $s^{\mathbf{CC}}$ (both players cooperate), then $s^{\mathbf{DD}}$ (both players defect), and finally $s^{\mathbf{CD}}$ (the agent cooperates and the opponent defects). Therefore, our $\mathbf{C}$ vector is $\mathbf{C} = [3, 1, 4, 2]$.

Note that the values of these numbers have an effect on the desirability of the observations and therefore will impact the agents action-planning such that they plan actions that they infer will result in the observation of the most

[5] Note that in many formulations of active inference this is formulated as a prior over observations $\tilde{P}(o; \mathbf{C})$.

desirable state. Changing the values of these rewards will change the incentive and behaviour of the agents.

Different Reward Parameterizations. We can parameterise the reward function $\mathbf{C}$ in terms of a single precision that makes a single ordered reward function with the constraints $r_{\text{CD}} < r_{\text{DD}} < r_{\text{CC}} < r_{\text{DC}}$ more or less shallow/steep. We do this using the softmax (normalised exponential transformation):

$$\mathbf{C} = \sigma\left(\begin{bmatrix} r_{\text{CC}} \\ r_{\text{CD}} \\ r_{\text{DC}} \\ r_{\text{DD}} \end{bmatrix}, \beta\right), \quad \text{where } \mathbf{C}_{\text{CC}} = \frac{\exp(\beta r_{\text{CC}})}{\sum_i \exp(\beta r_i)}$$

$$\ln \mathbf{C}_{\text{CC}} = \beta r_{\text{CC}} - \ln\left(\sum_i \exp(\beta r_i)\right)$$

$$\implies \ln \mathbf{C} \propto \beta \begin{bmatrix} r_{\text{CC}} \\ r_{\text{CD}} \\ r_{\text{DC}} \\ r_{\text{DD}} \end{bmatrix} \tag{A.5}$$

Policies. A policy π is comprised of individual actions, or control states, $\pi = \{\mathbf{u}_1, \mathbf{u}_2, ...\mathbf{u}_H\}$. At each trial of iterative play, the agents can either defect or cooperate. This means that the policy space consists of two control states, namely $u^{\mathbf{C}}$ and $u^{\mathbf{D}}$. Once the action is inferred, the intersection of both agents' actions will result in the realised game state.

Transition Likelihood. The transition matrix encodes the beliefs that the agent holds about how game states will evolve given previous trials and their actions. Because action selection under active inference depends on model-based planning, this transition model also directly determines the agent's strategy. Although in this work we focus on how agents can automatically learn the game's dynamics and thus their strategies through experience, we nevertheless begin by constraining what agents can learn by initialising agents' beliefs about transition dynamics, so that they assume that two game state transitions are always impossible. Agents believe that when they cooperate, there is zero probability that the next state will be DC or DD, and conversely, when they defect, they believe there is zero probability that the next state will be CD or CC. Therefore, the transition matrix encodes the agent's assumptions about whether the other will cooperate or defect in the next trial, given the outcome of the current trial and the agent's own action.

We use existing formulations of parameter learning under active inference to allow our agents to update their beliefs about transition model over time based on experience. Technically, the agents are updating a Dirichlet posterior belief over the Categorical parameters B that characterise its transition model (a transition probability matrix, mapping from past to current game states, further conditioned on action). They update this matrix of posterior Dirichlet parameters at the end of each trial, based on that trial's outcome.

At the beginning of iterative play, the agent will be initialised with no prior opinion or knowledge about which of the possible transitions are more likely given its actions (aside from the zero constraints laid out above). These uniform initial transition distributions are shown in Eq. (A.6) and Eq. (A.7).

$$P(\mathbf{s}_{t+1}|\mathbf{s}_t, \mathbf{u}_t = C) = \begin{bmatrix} 0.5 & 0.5 & 0.5 & 0.5 \\ 0.5 & 0.5 & 0.5 & 0.5 \\ 0 & 0 & 0 & 0 \\ 0 & 0 & 0 & 0 \end{bmatrix} \tag{A.6}$$

$$P(\mathbf{s}_{t+1}|\mathbf{s}_t, \mathbf{u}_t = D) = \begin{bmatrix} 0 & 0 & 0 & 0 \\ 0 & 0 & 0 & 0 \\ 0.5 & 0.5 & 0.5 & 0.5 \\ 0.5 & 0.5 & 0.5 & 0.5 \end{bmatrix} \tag{A.7}$$

At the conclusion of each trial during a session of iterative play, a given agent observes the game state of the previous trial and updates its beliefs about transitions based on the realised states and its actions. As these transition dynamics are learned, the agent is simultaneously learning a strategy based on planning the most optimal action (cooperate or defect), given its evolving beliefs.

A.2 Inference

State Inference. At each trial of iterative play, the agents first infer the game state by inverting their Markovian (POMDP) generative model using ongoing observations $\mathbf{o}_t$.

The agent's hidden state inference involves optimising a variational posterior over hidden states and policies $Q(\mathbf{s}_{1:T}, \pi)$ as a categorical distribution with parameters $\tilde{\boldsymbol{\phi}}$ that are factorised 'mean-field'-style across timesteps [4]:

$$Q(\mathbf{s}_{1:T}, \pi; \tilde{\boldsymbol{\phi}}) = Q(\pi; \phi_\pi) \prod_{1:T} Q(\mathbf{s}_t; \boldsymbol{\phi}_{\mathbf{s},t})$$

where the variational parameters $\tilde{\boldsymbol{\phi}} = \{\phi_\pi, \boldsymbol{\phi}_{\mathbf{s}_{1:T}}\}$ are themselves segregated into policy-specific parameters ϕ_π and hidden-state-specific parameters $\boldsymbol{\phi}_{\mathbf{s}_{1:T}}$.

At each timestep t, the agent performs inference by optimising the posterior parameters $\tilde{\boldsymbol{\phi}}$ to minimise the timestep-specific variational free energy $\mathcal{F}_t$, which due to the Markovian factorisation of the generative model and mean-field factorisation of the posterior, can be expressed in terms of only the generative model of the current timestep $P(\mathbf{o}_t, \mathbf{s}_t, \pi, \mathbf{A}, \mathbf{B}, \mathbf{C})$:

$$\mathcal{F}_t = \mathbb{E}_{Q(\mathbf{s}_t, \pi; \tilde{\boldsymbol{\phi}}_t)} \left[\ln Q(\mathbf{s}_t, \pi; \tilde{\boldsymbol{\phi}}_t) - \ln P(\mathbf{o}_t, \mathbf{s}_t, \pi, \mathbf{A}, \mathbf{B}, \mathbf{C}) \right] \tag{A.8}$$

The optimal posterior parameters $\tilde{\boldsymbol{\phi}}^*$ are those that minimise the free energy in Eq. (A.8) and can be found by solving exactly for the fixed points of $\mathcal{F}_t$. We begin by solving for the parameters of the variational beliefs about hidden states $\boldsymbol{\phi}_{\mathbf{s}_t}$:

$$\frac{\partial \mathcal{F}_t}{\partial \boldsymbol{\phi}_{\mathbf{s}_t}} = 0$$
$$\implies \boldsymbol{\phi}^*_{\mathbf{s}_t} = \sigma\left(\ln \mathbf{A}^T \mathbf{o}_t + \ln(\mathbf{B}_{\mathbf{u}_{t-1}} \cdot \boldsymbol{\phi}^*_{\mathbf{s}_{t-1}})\right) \tag{A.9}$$

where σ represents the softmax (or normalised exponential) transform of a vector. The ith entry of the softmaxed output is given by:

$$\sigma(x)_i \triangleq \frac{\exp(x_i)}{\sum_j \exp(x_j)} \tag{A.10}$$

The initial matrix-vector product in the last line of Eq. (A.9) $\ln \mathbf{A}^T \mathbf{o}_t$ represents the contribution of sensory evidence to inference, and can be thought of as picking out the row of the $\mathbf{A}$ matrix that corresponds to the observation at timestep t. The second matrix vector product $\ln(\mathbf{B}_{\mathbf{u}_{t-1}} \cdot \boldsymbol{\phi}^*_{\mathbf{s}_{t-1}})$ represents the contribution of prior information to inference. This simple form is a consequence of the mean-field factorisation of the variational parameters $\boldsymbol{\phi}_{\mathbf{s}_{1:T}}$ across timesteps and an 'empirical prior' assumption, where the prior term of the generative model $P(\mathbf{s}_t) = \mathbb{E}_{P(\mathbf{s}_{t-1})}[P(\mathbf{s}_t|\mathbf{s}_{t-1}, \mathbf{u}_{t-1}, \mathbf{B})]$ is evaluated at the parameters of the previous timestep's variational posterior, in a manner reminiscent of a belief propagation step or empirical Bayes:

$$\begin{aligned} P(\mathbf{s}_t) &= \mathbb{E}_{P(\mathbf{s}_{t-1})}\left[P(\mathbf{s}_t|\mathbf{s}_{t-1}, \mathbf{u}_{t-1}, \mathbf{B})\right] \\ &\approx \mathbb{E}_{Q(\mathbf{s}_{t-1};\boldsymbol{\phi}_{\mathbf{s}_{t-1}})}\left[P(\mathbf{s}_t|\mathbf{s}_{t-1}, \mathbf{u}_{t-1}, \mathbf{B})\right] \end{aligned} \tag{A.11}$$

We can simplify the expression for the parameters of the variational beliefs due to the unambiguous form of the observation likelihood with infinite precision in Eq. (A.3), $\mathbf{A} = \frac{\mathbf{I}^\psi}{\sum \mathbf{I}^\psi}$, as well as the fact that the agents are taking identical actions at every trial, thus limiting the state space to {CC, DD} which implies that inference can be solved for exactly for any trial $t > 0$ as

$$\boldsymbol{\phi}^*_{\mathbf{s}_t} = \sigma\left(\ln\left(\frac{\mathbf{I}^\psi}{\sum \mathbf{I}^\psi}\right)^T \mathbf{o}_t + \ln\left(\mathbf{B}_{\mathbf{u}_{t-1}} \cdot \boldsymbol{\phi}^*_{\mathbf{s}_{t-1}}\right)\right) \tag{A.12}$$

$$= \lim_{\psi \to \infty} \sigma\left(\psi \ln\left(\mathbf{I}^T \mathbf{o}_t\right) + \ln\left(\mathbf{B}_{\mathbf{u}_{t-1}} \cdot \boldsymbol{\phi}^*_{\mathbf{s}_t}\right) - \ln \sum \mathbf{I}^\psi\right) \tag{A.13}$$

$$= \sigma\left(\ln\left(\mathbf{I}^T \mathbf{o}_1\right)\right) = \mathbf{I}^T \mathbf{o}_1 \tag{A.14}$$

Policy Inference. Under active inference, action selection and planning are cast as an inference problem, where policies are treated as a latent variable to be inferred. This has deep homology to contemporary approaches to model-based planning in reinforcement learning, such as planning as inference and control as inference [1,2,5,22]. In particular, active inference agents optimise a variational posterior over policies $Q(\pi)$. However, because policies inherently require estimation of future, unobserved states, we use an augmented, 'predictive' generative model to perform this policy inference. This predictive generative model is importantly augmented with the biased prior distribution over states $\tilde{P}(\mathbf{s}; C)$. Beliefs about policies, similar to those about hidden states, are optimised by minimising a free energy functional of beliefs about the consequences of action under the predictive generative model. This functional is known as the *expected free energy* and exhibits many desirable properties such as a natural balance between information-seeking ('exploration') and goal-directedness ('exploitation') [21]. The approximate posterior over policies $Q(\pi)$ is also a Categorical distribution with parameters $\boldsymbol{\phi}_{\mathbf{u}}$; the optimal setting of these parameters $\boldsymbol{\phi}^*_{\mathbf{u}}$ minimises the expected free energy, leading to the relationship:

$$
\begin{aligned}
Q(\pi; \boldsymbol{\phi}_{\mathbf{u}}) &= \sigma(-\mathbf{G}(\pi)) \\
\mathbf{G}(\pi) &= \sum_{\tau=1}^{H} \mathbf{G}_{t+\tau}(\mathbf{u}_{t+\tau-1})
\end{aligned} \tag{A.15}
$$

The second line shows that the expected free energy of a policy is the sum of the expected free energies that accrue for each action that comprises the policy: $\pi = \{\mathbf{u}_1, \mathbf{u}_2, ...\mathbf{u}_H\}$. For the present purposes we only consider 1-step ahead policies ($H = 1$). This means that the expected free energy of a policy is simply the expected free energy computed one timestep into the future $\mathbf{G}_{t+1}(\mathbf{u}_t)$.

The expected free energy can be decomposed into expected ambiguity and risk terms:

$$
\mathbf{G}_{t+1}(\mathbf{u}_t) = \mathbf{E}_{Q(\mathbf{s}_{t+1}|u_t)} \left[\mathbf{H}\left[P(\mathbf{o}_{t+1}|\mathbf{s}_{t+1})\right]\right] + \mathrm{D}_{KL}\left(Q(\mathbf{s}_{t+1}|\mathbf{u}_t) \parallel \ln P(\mathbf{s}_{t+1}|C)\right) \tag{A.16}
$$

We can write this general expression in terms of sufficient statistics of the variational distribution over hidden states $\boldsymbol{\phi}^*_{\mathbf{s}_t}$. The ambiguity term of the expected free energy vanishes because the agent's likelihood matrix is the identity:

$$
\mathbf{A}\mathbf{B}_t \cdot \boldsymbol{\phi}^*_{\mathbf{s}_t} \cdot \left(\ln(\mathbf{A}\mathbf{B}_t \cdot \boldsymbol{\phi}^*_{\mathbf{s}_t}) - \ln \mathbf{C}\right) \tag{A.17}
$$

$$
= \mathbf{B}_t \cdot \boldsymbol{\phi}^*_{\mathbf{s}_t} \left(\ln \mathbf{B}_t \cdot \boldsymbol{\phi}^*_{\mathbf{s}_t} - \ln \mathbf{C}\right) - \underbrace{(\mathbf{A} \ln \mathbf{A}) \cdot \boldsymbol{\phi}^*_{\mathbf{s}_t}}_{=0} \tag{A.18}
$$

Action Selection. Having optimised a posterior over policies (which in this context simply reduce to control states), action selection simply consists of sampling the action at trial t that minimises the expected free energy, i.e., sampling an action from the posterior marginal over actions.

$$
\boldsymbol{\phi}_{\mathbf{u}} = \sigma(-\mathbf{G}) \tag{A.19}
$$

$$
u_{t+1} \sim Q(\mathbf{u}_{t+1}; \boldsymbol{\phi}_{\mathbf{u}}) \tag{A.20}
$$

This can be done either deterministically by selecting the most probable control state at every timestep:

$$u_{t+1} = \arg\max_{u} Q(\mathbf{u}_{t+1}; \boldsymbol{\phi}_{\mathbf{u}}) \tag{A.21}$$

Or, this can be done stochastically by sampling from the posterior over actions. The stochasticity of this sampling can be further tuned by sampling from a transformed action posterior scaled by a temperature parameter α.

$$u_{t+1} \sim Q(\mathbf{u}_{t+1}; \boldsymbol{\phi}, \alpha) \tag{A.22}$$

B Matrix Learning. After every trial of iterative play, each agent updates its posterior beliefs about the transition model B by optimizing Dirichlet parameters $\boldsymbol{\phi}_{\mathbf{b}}$, which are the sufficient statistics of a Dirichlet parameterization of the posterior $Q(B; \boldsymbol{\phi}_{\mathbf{b}})$. This is also known as 'learning' in the active inference literature, and analogised to neuronal processes such as synaptic plasticity, which typically occurs on a slower timescale than hidden state inference (analogised to rapid dynamics of neural firing rates) [9]. Dirichlet distributions are used as the parameterizations of discrete Categorical likelihood matrices, due to their natural role as conjugate priors for the Categorical distribution.

We supplement the generative model with an additional prior over the parameters of the transition model, the Dirichlet distribution $P(B; \mathbf{b})$ parameterised by a vector of positive real hyperparameters $\mathbf{b}$, that can also be interpreted as 'pseudocounts', i.e., how many times has the agent seen this particular transition occur, before the simulation starts. Alongside this prior we introduce a variational posterior over B that is also a Dirichlet distribution $Q(B; \boldsymbol{\phi}_{\mathbf{b}})$. This leads to a new expression for the variational free energy at a given time point, which includes an additional Kullback-Leibler divergence between the variational and generative model Dirichlet distributions over B [13]:

$$\begin{aligned}\mathcal{F}_t &= \mathbb{E}_{Q(\mathbf{s}_t, \mathbf{u}_t, B; \tilde{\phi})}\left[\ln Q\left(\mathbf{s}_t, \mathbf{u}_t, B; \tilde{\phi}\right) - \ln P\left(\mathbf{o}_t, \mathbf{s}_t, \mathbf{u}_t, A, B, C; \mathbf{A}, \mathbf{b}, \mathbf{C}\right)\right] \\ &= \mathbb{E}_{Q\mathbf{s}_t, \mathbf{u}_t; \boldsymbol{\phi}_{\mathbf{s},\mathbf{u}}}\left[\ln Q\left(\mathbf{s}_t, \mathbf{u}_t; \boldsymbol{\phi}_{\mathbf{s},\mathbf{u}}\right) - \ln P\left(\mathbf{o}_t, \mathbf{s}_t, \mathbf{u}_t, A, C; \mathbf{A}, \mathbf{C}\right)\right] + \mathrm{D}_{KL}\left(Q(B; \boldsymbol{\phi}_{\mathbf{b}}) \parallel P(B; \mathbf{b})\right)\end{aligned} \tag{A.23}$$

This new expression means that when we minimise $\mathcal{F}_t$ with respect to the variational (Dirichlet) parameters $\boldsymbol{\phi}_{\mathbf{b}}$, we get a closed-form expression for the variational beliefs over $\mathbf{B}$, which can be expressed in terms of the Dirichlet prior parameters $\mathbf{b}$ and the variational posterior over hidden states at current and previous timesteps $\boldsymbol{\phi}_{\mathbf{s}_t}$ and $\boldsymbol{\phi}_{\mathbf{s}_{t-1}}$.

$$\mathbf{B}_{t+1} = \frac{\boldsymbol{\phi}^*_{\mathbf{b}}}{\boldsymbol{\phi}_{\mathbf{b},0}} \tag{A.24}$$

$$\boldsymbol{\phi}^*_{\mathbf{b}} = \mathbf{b} + \eta(\boldsymbol{\phi}_{\mathbf{s}_t} \otimes \boldsymbol{\phi}_{\mathbf{s}_{t-1}}) \tag{A.25}$$

where Eq. (A.24) represents the update to the Dirichlet prior for the transition distribution during learning. This is updated with respect to the learning rate η and the transition probabilities given the previously performed action a_{t-1}. It is this normalised updated Dirichlet prior that then becomes the new transition probability distribution for the following trial.

The updates to the transition model are governed by the sequence of game states. We can imagine a fictive 1-turn sequence (two trials) to imagine how a particular sequence influences learning. If at one trial, the agents both cooperated, then they will infer that the game state was CC. Given this belief, they will infer which action to take. If they choose to defect, hoping that the opponent will cooperate again, the resulting inferred state will be that the optimal action is $\mathbf{u}_t = u^{\mathbf{D}}$, and after the trial they will observe the resulting state, DD. At this point, the agents will update their beliefs about likely transitions (encoded in the B matrix parameters), such that there will be a small incremental increase in the conditional probability of DD, given a past state of CC and a past action of u^D, i.e., $P(\mathbf{s}_{t+1} = \mathrm{DD}|\mathbf{s}_t = \mathrm{CC}, \mathbf{u}_t = u^{\mathbf{D}})$. The size of this update is determined by a learning rate parameter η.

A.3 Deriving the Analytic Form of the Transition Function

When two deterministic agents have the same learning rate, they will perform the same action at every timestep. This has the consequence that the two-agent system will only ever explore two out of four states, namely CC and DD.

The posterior belief can be represented as a vector of its parameters, and in the solution of two identical agents, it can take two possible values, which we denote as $s^{\mathbf{CC}}$ and $s^{\mathbf{DD}}$. Because the likelihood distribution is the identity matrix, these will be maximally precise vectors:

$$s^{\mathbf{CC}} = \begin{bmatrix} 1 \\ 0 \\ 0 \\ 0 \end{bmatrix} \qquad s^{\mathbf{DD}} = \begin{bmatrix} 0 \\ 0 \\ 0 \\ 1 \end{bmatrix} \tag{A.26}$$

The initial Dirichlet parameters of the prior distribution over the transition model are, for the cooperate and defect-conditioned transitions, respectively,

$$\mathbf{b}_0^{\mathbf{C}} = \begin{bmatrix} 0.5 & 0.5 & 0.5 & 0.5 \\ 0.5 & 0.5 & 0.5 & 0.5 \\ 0 & 0 & 0 & 0 \\ 0 & 0 & 0 & 0 \end{bmatrix} \qquad \mathbf{b}_0^{\mathbf{D}} = \begin{bmatrix} 0 & 0 & 0 & 0 \\ 0 & 0 & 0 & 0 \\ 0.5 & 0.5 & 0.5 & 0.5 \\ 0.5 & 0.5 & 0.5 & 0.5 \end{bmatrix} \tag{A.27}$$

This means that at each timestep, there are four possible updates to the parameters of each agent's variational posterior over the transition model $\phi_{\mathbf{b}}$, given the two variational beliefs a given agent might have ($\mathbf{s}^{\mathbf{CC}}$ and $\mathbf{s}^{\mathbf{DD}}$):

$$\phi^*_{\mathbf{b}_{t+1}} = \begin{cases} \mathbf{b}_0^{\mathbf{C}} + \eta \cdot (s^{\mathbf{CC}} \otimes s^{\mathbf{CC}})t \\ \mathbf{b}_0^{\mathbf{D}} + \eta \cdot (s^{\mathbf{DD}} \otimes s^{\mathbf{CC}})t \\ \mathbf{b}_0^{\mathbf{C}} + \eta \cdot (s^{\mathbf{CC}} \otimes s^{\mathbf{DD}})t \\ \mathbf{b}_0^{\mathbf{D}} + \eta \cdot (s^{\mathbf{DD}} \otimes s^{\mathbf{DD}})t \end{cases} \tag{A.28}$$

When the agents are both defecting (e.g., in the first timestep when the most likely action is defect), then the update rule for the weights of the Dirichlet parameters of the transition matrix is governed by:

$$\phi^*_{\mathbf{b}_{t<\tau_1}} = \mathbf{b}_0^{\mathbf{D}} + \eta \left(s^{\mathbf{DD}} \otimes s^{\mathbf{DD}}\right) t \tag{A.29}$$

$$\mathbf{B}_{t+1<\tau_1} = \frac{\phi^*_{\mathbf{b}_{t<\tau_1}}}{\phi^*_{\mathbf{b}_{t<\tau_1},0}} \tag{A.30}$$

At some critical time τ_1 the probability of cooperation exceeds that of defection, due to the change in the expected free energies of the two actions $\mathbf{G}_{\tau_1}(u = \text{C}) < \mathbf{G}_{\tau_1}(u = \text{D})$. This triggers the beginning of the so-called "oscillation period" (see Sect. 2 in the main text), where agents periodically oscillate between cooperating and defecting with the same phase. We can expand this condition according to Eq. (A.18) into the following form:

$$\mathbf{B}_0^{\mathbf{C}} \cdot s_{\tau_1}^{\mathbf{DD}} \cdot \left(\ln \mathbf{B}_0^{\mathbf{C}} \cdot s_{\tau_1}^{\mathbf{DD}} - \ln \mathbf{C}\right) = \mathbf{B}_{\tau_1}^{\mathbf{D}} \cdot s_{\tau_1}^{\mathbf{DD}} \cdot \left(\ln \mathbf{B^D}_{\tau_1} \cdot s_{\tau_1}^{\mathbf{DD}} - \ln \mathbf{C}\right) \tag{A.31}$$

As shown in Sect. A.4, the equality in Eq. (A.31) can be written in terms of η, $\mathbf{C}$ and τ_1:

$$\frac{1}{(2+\eta 2\tau_1)}\left[\ln \frac{1}{2(1+\eta\tau_1)\mathbf{C}_3} + (1+2\eta\tau_1)\ln \frac{1+2\eta\tau_1}{2(1+\eta\tau_1)\mathbf{C}_4}\right] = \frac{1}{2}\ln\left(\frac{1}{4\mathbf{C}_1\mathbf{C}_2}\right), \tag{A.32}$$

We now let $y = \frac{1}{2+2\eta\tau_1}$, which will always be between 0 and 1. We can now rewrite Eq. (A.32) as

$$y\ln y - y\ln \mathbf{C}_3 + (1-y)\ln(1-y) - (1-y)\ln \mathbf{C}_4 = \frac{1}{2}\ln\left(\frac{1}{4\mathbf{C}_1\mathbf{C}_2}\right) \tag{A.33}$$

To derive τ_1 in terms of η, we must make an approximation. We use the fact that when y is between 0 and 1, it can be approximated by $y \approx Ay^b(y-1)$. This gives us the following expression as an approximation for (35)

$$Ay^b(y-1) - y\ln \mathbf{C}_3 - A(1-y)^b y - (1-y)\ln \mathbf{C}_4 = \frac{1}{2}\ln\left(\frac{1}{4\mathbf{C}_1\mathbf{C}_2}\right) \tag{A.34}$$

The optimal values for the approximation are $A = \frac{4774}{4563}$ and $b = \frac{3}{5}$, however, for simplicity, we let $A = 1$ and $b = 1$ and then the desired root of Eq. (A.34) can be solved as:

$$y = \frac{1}{4}\Big(\ln \frac{\mathbf{C}_3}{\mathbf{C}_4} + 2 - \sqrt{(\ln \frac{\mathbf{C}_4}{\mathbf{C}_3} - 2)^2 - 8(-\ln \frac{\mathbf{C}_4}{2\sqrt{\mathbf{C}_1\mathbf{C}_2}} - \frac{1}{5}))} \tag{A.35}$$

Therefore, since $y = \frac{1}{2+2\eta\tau_1}$, we have that

$$\tau_1 \approx \frac{R_1}{\eta} \tag{A.36}$$

where

$$R_1 = \frac{2}{\ln\frac{\mathbf{C}_3}{\mathbf{C}_4} + 2 - \sqrt{(\ln\frac{\mathbf{C}_4}{\mathbf{C}_3} - 2)^2 - 8(-\ln\frac{\mathbf{C}_4}{2\sqrt{\mathbf{C}_1\mathbf{C}_2}} - \frac{1}{5})}} - 1 \tag{A.37}$$

We now have an approximation for τ_1 in terms of η and a constant R_1, which depends on the reward $\mathbf{C}$ which can be parameterised by β according to Eq. (A.5).

$$\tau_1 = \frac{R_1(\beta)}{\eta} \tag{A.38}$$

for some precision β. We can plot this equation for different values of β to see how the values in the reward function influence τ_1 (see Fig. 6).

For $\tau_1 < t < \tau_2$ (i.e., during the period of oscillation dynamics shown in Fig. 2), the update rules then become:

$$\phi^{\mathbf{D}}_{\mathbf{b}_{\tau_1<t<\tau_2}} = \mathbf{b}^{\mathbf{D}}_{\tau_1} + \frac{1}{2}\eta(s^{\mathbf{DD}} \otimes s^{\mathbf{CC}})(t - \tau_1) \tag{A.39}$$

$$\phi^{\mathbf{C}}_{\mathbf{b}_{\tau_1<t<\tau_2}} = \mathbf{b}^{\mathbf{C}}_{0} + \frac{1}{2}\eta(s^{\mathbf{CC}} \otimes s^{\mathbf{DD}})(t - \tau_1) \tag{A.40}$$

The update rule changes from Eq. (A.39) to Eq. (A.40) at every other trial, from conditioning on the previous action being D, to being C. The oscillation period persists until some time τ_2. At τ_2 we will have that, for the first time, $\mathbf{G}_0(u = \mathrm{C}, \phi^{\mathrm{C}}) < \mathbf{G}_0(u = \mathrm{C}, \phi^{\mathrm{D}})$. Again, we can expand this according to Eq. (A.18) as:

$$\mathbf{B}^{\mathbf{C}}_{\tau_2} \cdot s^{\mathbf{CC}} \cdot \left(\ln \mathbf{B}^{\mathbf{C}}_{\tau_2} \cdot s^{\mathbf{CC}} - \ln \mathbf{C}\right) = \mathbf{B}^{\mathbf{D}}_{\tau_2} \cdot s^{\mathbf{CC}} \cdot \left(\ln \mathbf{B^D}_{\tau_1} \cdot s^{\mathbf{CC}} - \ln \mathbf{C}\right) \tag{A.41}$$

Rewriting this equation in terms of η, τ_1, τ_2, and $\mathbf{C}$ leads to the following inequality (for full derivation, see Sect. A.5):

$$\frac{1}{2+\eta(\tau_2-\tau_1)}\left[\ln[\frac{1}{\mathbf{C}_3}(\frac{1}{2+\eta(\tau_2-\tau_1)})] + (1+\eta(\tau_2-\tau_1))\ln\frac{1+\eta(\tau_2-\tau_1)}{\mathbf{C}_4(2+\eta(\tau_2-\tau_1))}\right] = -\frac{1}{2}\ln(4\mathbf{C}_1\mathbf{C}_2) \tag{A.42}$$

This time, we let $y = \frac{1}{2+\eta(\tau_2-\tau_1)}$ and we have:

$$(y-1)\ln y - y\ln \mathbf{C}_3 + (1-y)\ln(1-y) - (1-y)\ln \mathbf{C}_4 = -\frac{1}{2}\ln(4\mathbf{C}_1\mathbf{C}_2) \tag{A.43}$$

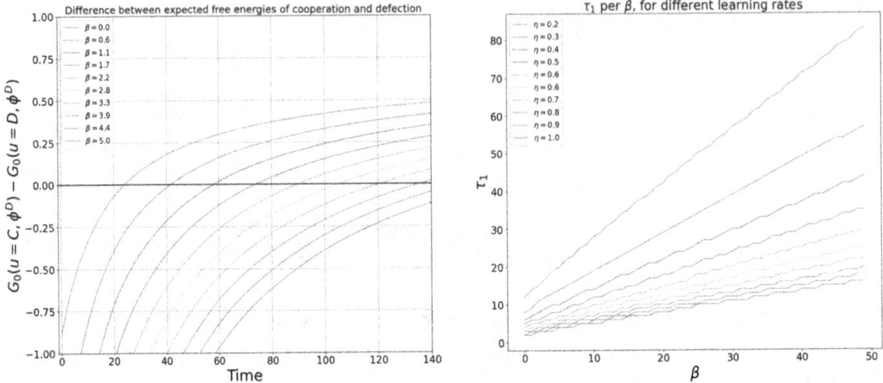

Fig. 6. Dynamics of the expected free energy. Left: The difference of EFE for cooperation and defection (vertical axis). The roots of this equation are the values of τ_1 for different values of β, parameterizing the values in the reward function $\mathbf{C}$ as per Eq. (A.42), with $\eta = 0.2$. It is clear that with a higher value of β, it will take agents longer to cooperate, i.e. τ_1 will be larger, demonstrated by the horizontal translations of the curves as β increases. **Right:** Values of τ_1 for different values of β parameterizing the reward function, at different learning rates. Again, we see that as β increases, τ_1 increases. We can also see that larger η competes with higher β to decrease τ_1, as the agents update their transition probability distributions at a higher frequency.

Now, notice that this is the exact same equation as Eq. (A.33) above, which we know we can approximate as Eq. (A.34). We can then write our solution in terms of R_1:

$$\tau_2 \approx \frac{1}{\eta}(\frac{1}{y} - 2) + \tau_1 = \frac{1}{\eta}(\frac{3}{2}R_1) \tag{A.44}$$

The resulting equation is obtained in terms of R_2, where $R_2 = \frac{3}{2}R_1$.

$$\tau_2 \approx \frac{R_2(\beta)}{\eta} \tag{A.45}$$

After τ_2, agents will cooperate indefinitely according to the final steady state update rule:

$$\phi^*_{\mathbf{b}_{t>\tau_2}} = \mathbf{b}^{\mathbf{C}}_{\tau_2} + \eta\left(s^{\mathbf{CC}} \otimes s^{\mathbf{CC}}\right) t \tag{A.46}$$

$$\mathbf{B}_{t+1>\tau_2} = \frac{\phi^*_{\mathbf{b}_{t>\tau_2}}}{\phi^*_{\mathbf{b}_{t>\tau_2},0}} \tag{A.47}$$

A.4 Full Derivation of τ_1

Here we derive τ_1 for the following equality from Eq. (A.31):

$$\mathbf{B}^{\mathbf{C}}_0 \cdot s^{\mathbf{DD}}_{\tau_1} \cdot \left(\ln \mathbf{B}^{\mathbf{C}}_0 \cdot s^{\mathbf{DD}}_{\tau_1} - \ln \mathbf{C}\right) = \mathbf{B}^{\mathbf{D}}_{\tau_1} \cdot s^{\mathbf{DD}}_{\tau_1} \cdot \left(\ln \mathbf{B^D}_{\tau_1} \cdot s^{\mathbf{DD}}_{\tau_1} - \ln \mathbf{C}\right) \tag{A.48}$$

Using the following:

$$\mathbf{B}_t = \frac{\phi_{\mathbf{b}_t}}{\phi_{\mathbf{b}_t,0}} \tag{A.49}$$

$$\phi^{\mathbf{D}}_{\mathbf{b}_{t<\tau_1}} = \mathbf{b}^{\mathbf{D}}_0 + \eta(s^{\mathbf{DD}} \otimes s^{\mathbf{DD}})t \tag{A.50}$$

$$\phi^{\mathbf{C}}_{\mathbf{b}_{t<\tau_1}} = \mathbf{b}^{\mathbf{C}}_0 \tag{A.51}$$

$$\mathbf{b}^{\mathbf{C}}_0 = \begin{bmatrix} 0.5 & 0.5 & 0.5 & 0.5 \\ 0.5 & 0.5 & 0.5 & 0.5 \\ 0 & 0 & 0 & 0 \\ 0 & 0 & 0 & 0 \end{bmatrix} \qquad \mathbf{b}^{\mathbf{D}}_0 = \begin{bmatrix} 0 & 0 & 0 & 0 \\ 0 & 0 & 0 & 0 \\ 0.5 & 0.5 & 0.5 & 0.5 \\ 0.5 & 0.5 & 0.5 & 0.5 \end{bmatrix} \tag{A.52}$$

$$s^{\mathbf{DD}} = \mathbf{e}_4, \tag{A.53}$$

we have:

$$\frac{\phi^{\mathbf{C}}_{\mathbf{b}_0}}{\phi^{\mathbf{C}}_{\mathbf{b}_0,0}} \cdot s^{\mathbf{DD}} \cdot \left(\ln \frac{\phi^{\mathbf{C}}_{\mathbf{b}_0}}{\phi^{\mathbf{C}}_{\mathbf{b}_0,0}} \cdot s^{\mathbf{DD}} - \ln \mathbf{C}\right) = \frac{\phi^{\mathbf{D}}_{\mathbf{b}_{\tau_1}}}{\phi^{\mathbf{D}}_{\mathbf{b}_{\tau_1},0}} \cdot s^{\mathbf{DD}} \cdot \left(\ln \frac{\phi^{\mathbf{D}}_{\mathbf{b}_{\tau_1}}}{\phi^{\mathbf{D}}_{\mathbf{b}_{\tau_1},0}} \cdot s^{\mathbf{DD}} - \ln \mathbf{C}\right) \tag{A.54}$$

On the LHS:

$$\frac{\phi^{\mathbf{C}}_{\mathbf{b}_0}}{\phi^{\mathbf{C}}_{\mathbf{b}_0,0}} \cdot s^{\mathbf{DD}} \cdot \left(\ln \frac{\phi^{\mathbf{C}}_{\mathbf{b}_0}}{\phi^{\mathbf{C}}_{\mathbf{b}_0,0}} \cdot s^{\mathbf{DD}} - \ln \mathbf{C}\right) = (\mathbf{b}^{\mathbf{C}}_0 \cdot \mathbf{e}_4) \cdot \ln \frac{\mathbf{b}^{\mathbf{C}}_0 \cdot \mathbf{e}_4}{\mathbf{C}} = -\frac{1}{2}\ln(4\mathbf{C}_1\mathbf{C}_2) \tag{A.55}$$

On the RHS:

$$\frac{\phi^{\mathbf{D}}_{\mathbf{b}_{\tau_1}}}{\phi^{\mathbf{D}}_{\mathbf{b}_{\tau_1},0}} \cdot s^{\mathbf{DD}} \cdot \left(\ln \frac{\phi^{\mathbf{D}}_{\mathbf{b}_{\tau_1}}}{\phi^{\mathbf{D}}_{\mathbf{b}_{\tau_1},0}} \cdot s^{\mathbf{DD}} - \ln \mathbf{C}\right) = \frac{\phi_{\mathbf{b}^{\mathbf{D}}_{\tau_1},j=4}}{\phi^{\mathbf{D}}_{\mathbf{b}_{\tau_1},0}} \cdot \left(\ln \frac{\phi_{\mathbf{b}^{\mathbf{D}}_{\tau_1},j=4}}{\phi^{\mathbf{D}}_{\mathbf{b}_{\tau_1},0}} - \ln \mathbf{C}\right) \tag{A.56}$$

$$= \frac{1}{2}\frac{1}{1+\eta\tau_1}\ln(\frac{1}{2(1+\eta\tau_1)\mathbf{C}_3}) + \frac{1}{2}\frac{1+2\eta\tau_1}{1+\eta\tau_1}\ln(\frac{1+\eta\tau_1}{2(1+\eta\tau_1)\mathbf{C}_4}) \tag{A.57}$$

Our equality is therefore:

$$\frac{1}{(2+2\eta\tau_1)}\left[\ln \frac{1}{2(1+\eta\tau_1)\mathbf{C}_3} + (1+2\eta\tau_1)\ln \frac{1+2\eta\tau_1}{2(1+\eta\tau_1)\mathbf{C}_4}\right] = -\frac{1}{2}\ln(4\mathbf{C}_1\mathbf{C}_2) \tag{A.58}$$

A.5 Full Derivation of τ_2

Our condition for deriving τ_2 in terms of the expected free energies is

$$\mathbf{B}^{\mathbf{C}}_{\tau_2} \cdot s^{\mathbf{CC}} \cdot (\ln \mathbf{B}^{\mathbf{C}}_{\tau_2} \cdot s^{\mathbf{CC}} - \ln \mathbf{C}) = \mathbf{B}^{\mathbf{D}}_{\tau_2} \cdot s^{\mathbf{CC}} \cdot (\ln \mathbf{B^D}_{\tau_2} \cdot s^{\mathbf{CC}} - \ln \mathbf{C}) \tag{A.59}$$

Here our ϕs between trials τ_1 and τ_2 are:

$$\phi^{\mathbf{D}}_{\mathbf{b}_{\tau_1<t<\tau_2}} = \phi^{\mathbf{D}}_{\mathbf{b}_{t<\tau_1}} + \frac{1}{2}\eta(s^{\mathbf{DD}} \otimes s^{\mathbf{CC}})(t-\tau_1) \tag{A.60}$$

$$\phi^{\mathbf{C}}_{\mathbf{b}_{\tau_1<t<\tau_2}} = \mathbf{b}^{\mathbf{C}}_0 + \frac{1}{2}\eta(s^{\mathbf{CC}} \otimes s^{\mathbf{DD}})(t-\tau_1) \tag{A.61}$$

And to solve for τ_2 our inequality is

$$\frac{\phi^{\mathbf{C}}_{\mathbf{b}_{\tau_2}}}{\phi^{\mathbf{C}}_{\mathbf{b}_{\tau_2},0}} \cdot s^{\mathbf{CC}} \cdot \left(\ln \frac{\phi^{\mathbf{C}}_{\mathbf{b}_{\tau_2}}}{\phi^{\mathbf{C}}_{\mathbf{b}_{\tau_2},0}} \cdot s^{\mathbf{CC}} - \ln \mathbf{C}\right) = \frac{\phi^{\mathbf{D}}_{\mathbf{b}_{\tau_2}}}{\phi^{\mathbf{D}}_{\mathbf{b}_{\tau_2},0}} \cdot s^{\mathbf{DD}} \cdot \left(\ln \frac{\phi^{\mathbf{D}}_{\mathbf{b}_{\tau_2}}}{\phi^{\mathbf{D}}_{\mathbf{b}_{\tau_2},0}} \cdot s^{\mathbf{DD}} - \ln \mathbf{C}\right) \tag{A.62}$$

On the LHS we have:

$$\frac{\phi^{\mathbf{C}}_{\mathbf{b}_{\tau_2}}}{\phi^{\mathbf{C}}_{\mathbf{b}_{\tau_2},0}} \cdot s^{\mathbf{CC}} \cdot \left(\ln \frac{\phi^{\mathbf{C}}_{\mathbf{b}_{\tau_2}}}{\phi^{\mathbf{C}}_{\mathbf{b}_{\tau_2},0}} \cdot s^{\mathbf{CC}} - \ln \mathbf{C}\right) = -\frac{1}{2}\ln(4\mathbf{C}_1\mathbf{C}_2) \tag{A.63}$$

On the RHS:

$$\frac{\phi^{\mathbf{C}}_{\mathbf{b}_{\tau_2}}}{\phi^{\mathbf{C}}_{\mathbf{b}_{\tau_2},0}} \cdot s^{\mathbf{DD}} = \frac{1}{2+\eta(\tau_2-\tau_1)}\begin{bmatrix} 0 \\ 0 \\ 1 \\ 1+\eta(\tau_2-\tau_1)\end{bmatrix} \tag{A.64}$$

$$\frac{1}{2+\eta(\tau_2-\tau_1)}\left[\ln\left[\frac{1}{\mathbf{C}_3}(\frac{1}{2+\eta(\tau_2-\tau_1)})\right] + (1+\eta(\tau_2-\tau_1))\ln\frac{1+\eta(\tau_2-\tau_1)}{\mathbf{C}_4(2+\eta(\tau_2-\tau_1))}\right] \tag{A.65}$$

$$\frac{1}{2+\eta(\tau_2-\tau_1)}\ln\left[\frac{\mathbf{C}_4}{\mathbf{C}_3(1+\eta(\tau_2-\tau_1)})\right] + \ln\frac{1+\eta(\tau_2-\tau_1)}{\mathbf{C}_4(2+\eta(\tau_2-\tau_1))} \tag{A.66}$$

Finally, our inequality is:

$$\frac{1}{2+\eta(\tau_2-\tau_1)}\left[\ln\left[\frac{1}{\mathbf{C}_3}\left(\frac{1}{2+\eta(\tau_2-\tau_1)}\right)\right] + (1+\eta(\tau_2-\tau_1))\ln\frac{1+\eta(\tau_2-\tau_1)}{\mathbf{C}_4(2+\eta(\tau_2-\tau_1))}\right] = -\frac{1}{2}\ln(4\mathbf{C}_1\mathbf{C}_2) \tag{A.67}$$

References

1. Abdolmaleki, A., Springenberg, J.T., Tassa, Y., Munos, R., Heess, N., Riedmiller, M.: Maximum a posteriori policy optimisation. In: International Conference on Learning Representations (2018). https://openreview.net/forum?id=S1ANxQW0b

2. Attias, H.: Planning by probabilistic inference. In: Bishop, C.M., Frey, B.J. (eds.) Proceedings of the Ninth International Workshop on Artificial Intelligence and Statistics. Proceedings of Machine Learning Research, R4, pp. 9–16 (2003). https://proceedings.mlr.press/r4/attias03a.html
3. Axelrod, R., Hamilton, W.D.: The evolution of cooperation. Science **211**(4489), 1390–1396 (1981). https://doi.org/10.1126/science.7466396
4. Blei, D.M., Kucukelbir, A., McAuliffe, J.D.: Variational inference: a review for statisticians. J. Am. Stat. Assoc. **112**(518), 859–877 (2017). https://doi.org/10.1080/01621459.2017.1285773
5. Botvinick, M., Toussaint, M.: Planning as inference. Trends Cogn. Sci. **16**(10), 485–488 (2012). https://doi.org/10.1016/j.tics.2012.08.006
6. Farooqui, A.D., Niazi, M.A.: Game theory models for communication between agents: a review. Complex Adapt. Syst. Model. **4**(1), 1–31 (2016). https://doi.org/10.1186/s40294-016-0026-7
7. Fountas, Z., Sajid, N., Mediano, P.A., Friston, K.J.: Deep active inference agents using Monte-Carlo methods. In: Proceedings of the 34th International Conference on Neural Information Processing Systems, NIPS 2020. Curran Associates Inc. (2020). https://doi.org/10.5555/3495724.3496702
8. Friston, K.J., Daunizeau, J., Kiebel, S.J.: Reinforcement learning or active inference? PLoS ONE **4**(7), e6421 (2009). https://doi.org/10.1371/journal.pone.0006421
9. Friston, K.J., FitzGerald, T., Rigoli, F., Schwartenbeck, P., O'Doherty, J., Pezzulo, G.: Active inference and learning. Neurosci. Biobehav. Rev. **68**, 862–879 (2016). https://doi.org/10.1016/j.neubiorev.2016.06.022
10. Friston, K.J., Rigoli, F., Ognibene, D., Mathys, C., Fitzgerald, T., Pezzulo, G.: Active inference and epistemic value. Cogn. Neurosci. **6**(4), 187–214 (2015). https://doi.org/10.1080/17588928.2015.1020053
11. Heide, J.B., Miner, A.S.: The shadow of the future: effects of anticipated interaction and frequency of contact on buyer-seller cooperation. Acad. Manag. J. **35**(2), 265–291 (1992). https://www.jstor.org/stable/256374
12. Heins, C., Klein, B., Demekas, D., Aguilera, M., Buckley, C.L.: Spin glass systems as collective active inference. In: Buckley, C.L., et al. (eds.) IWAI 2022. CCIS, vol. 1721, pp. 75–98. Springer, Cham (2023). https://doi.org/10.1007/978-3-031-28719-0_6
13. Heins, C., et al.: pymdp: a Python library for active inference indiscrete state spaces. J. Open Source Softw. **7**(73), 4098 (2022). https://doi.org/10.21105/joss.04098
14. Holt, C.A., Roth, A.E.: The Nash equilibrium: a perspective. Proc. Natl. Acad. Sci. **101**(12), 3999–4002 (2004). https://doi.org/10.1073/pnas.0308738101
15. Imhof, L.A., Fudenberg, D., Nowak, M.A.: Tit-for-tat or win-stay, lose-shift? J. Theor. Biol. **247**(3), 574–580 (2007). https://doi.org/10.1016/j.jtbi.2007.03.027
16. Kuhn, S.: Prisoner's Dilemma. In: Zalta, E.N. (ed.) The Stanford Encyclopedia of Philosophy. Metaphysics Research Lab, Stanford University, Winter 2019 edn. (2019). https://plato.stanford.edu/archives/win2019/entries/prisoner-dilemma/
17. Kuhn, S.: Strategies for the Iterated Prisoner's Dilemma. In: Zalta, E.N. (ed.) The Stanford Encyclopedia of Philosophy. Metaphysics Research Lab, Stanford University, Winter 2019 edn. (2019). https://plato.stanford.edu/entries/prisoner-dilemma/strategy-table.html
18. Lin, B., Bouneffouf, D., Cecchi, G.: Online learning in Iterated Prisoner's Dilemma to mimic human behavior. In: Khanna, S., Cao, J., Bai, Q., Xu, G. (eds.) PRICAI

2022. LNCS, vol. 13631, pp. 134–147. Springer, Cham (2022). https://doi.org/10.1007/978-3-031-20868-3_10
19. Marković, D., Stojić, H., Schwöbel, S., Kiebel, S.J.: An empirical evaluation of active inference in multi-armed bandits. Neural Netw. **144**, 229–246 (2021). https://doi.org/10.1016/j.neunet.2021.08.018
20. Dyer, M., Mohanaraj, V.: The Iterated Prisoner's Dilemma on a cycle. arXiv (2018). https://doi.org/10.48550/arXiv.1102.3822
21. Millidge, B., Tschantz, A., Buckley, C.L.: Whence the expected free energy? Neural Comput. **33**(2), 447–482 (2021). https://doi.org/10.1162/neco_a_01354
22. Millidge, B., Tschantz, A., Seth, A.K., Buckley, C.L.: On the relationship between active inference and control as inference. In: Verbelen, T., Lanillos, P., Buckley, C.L., De Boom, C. (eds.) IWAI 2020. CCIS, vol. 1326, pp. 3–11. Springer, Cham (2020). https://doi.org/10.1007/978-3-030-64919-7_1
23. Nowak, M., Sigmund, K.: A strategy of win-stay, lose-shift that outperforms tit-for-tat in the Prisoner's Dilemma game. Nature **364**(6432), 56–58 (1993). https://doi.org/10.1038/364056a0
24. Nowak, M.A.: Five rules for the evolution of cooperation. Science **314**(5805), 1560–1563 (2006). https://doi.org/10.1126/science.1133755
25. Parr, T., Pezzulo, G., Friston, K.J.: Active Inference: The Free Energy Principle in Mind, Brain, and Behavior. MIT Press, Cambridge (2022)
26. Press, W.H., Dyson, F.J.: Iterated Prisoner's Dilemma contains strategies that dominate any evolutionary opponent. Proc. Natl. Acad. Sci. **109**(26), 10409–10413 (2012). https://doi.org/10.1073/pnas.1206569109
27. Puterman, M.L.: Markov decision processes. In: Handbooks in Operations Research and Management Science, Stochastic Models, vol. 2, pp. 331–434. Elsevier (1990). https://doi.org/10.1016/S0927-0507(05)80172-0
28. Ramstead, M.J., et al.: On Bayesian mechanics: a physics of and by beliefs. Interface Focus **13**(3), 20220029 (2023). https://doi.org/10.1098/rsfs.2022.0029
29. Sandholm, T.W., Crites, R.H.: Multiagent reinforcement learning in the Iterated Prisoner's Dilemma. Biosystems **37**(1), 147–166 (1996). https://doi.org/10.1016/0303-2647(95)01551-5
30. Simon, H.A.: Bounded rationality. In: Eatwell, J., Milgate, M., Newman, P. (eds.) Utility and Probability, pp. 15–18. The New Palgrave, Palgrave Macmillan UK, London (1990)
31. Vukov, J., Szabó, G., Szolnoki, A.: Cooperation in the noisy case: Prisoner's Dilemma game on two types of regular random graphs. Phys. Rev. E **73**, 067103 (2006). https://doi.org/10.1103/PhysRevE.73.067103
32. Wedekind, C., Milinski, M.: Human cooperation in the simultaneous and the alternating Prisoner's Dilemma: Pavlov versus Generous Tit-for-Tat. Proc. Natl. Acad. Sci. **93**(7), 2686–2689 (1996). https://doi.org/10.1073/pnas.93.7.2686

Toward Design of Synthetic Active Inference Agents by Mere Mortals

Bert de Vries(✉)

Eindhoven University of Technology, Eindhoven, The Netherlands
bert.de.vries@tue.nl

Abstract. The theoretical properties of active inference agents are impressive, but how do we realize effective agents in working hardware and software on edge devices? This is an interesting problem because the computational load for policy exploration explodes exponentially, while the computational resources are very limited for edge devices. In this paper, we discuss the necessary features for a software toolbox that supports a competent non-expert engineer to develop working active inference agents. We introduce a toolbox-in-progress that aims to accelerate the democratization of active inference agents in a similar way as TensorFlow propelled applications of deep learning technology.

Keywords: active inference agent · factor graphs · free energy minimization · reactive message passing · rxinfer · structural adaptation

1 Introduction

This position paper aims to complement a recent white paper on designing future intelligent ecosystems where autonomous Active InFerence (AIF) agents learn purposeful behavior through situated interactions with other AIF agents [11]. The white paper states that these agents "... can be realized via (variational) message passing or belief propagation on a factor graph" [11, abstract]. Here, we discuss the computational requirements for a factor graph software toolbox that supports this vision. Noting that the steep rise of commercialization opportunities for deep learning systems was greatly facilitated by the availability of professional-level toolboxes such as TensorFlow and successors, we claim that a high-quality AIF software toolbox is needed to realize the proposition in [11]. Therefore, in this paper, we ask the question: what properties should a factor graph toolbox possess that enable a competent engineer to develop relevant AIF agents? The question is important since the number of applications for autonomous AIF agents is expected to vastly outgrow the number of world-class experts in AIF and robotics.

As an illustrating example, consider an engineer (Sarah) who needs to design a quad-legged robot that is tasked to enter a building and switch off a valve. We assume that Sarah is a competent engineer with an MS degree and a few

C. L. Buckley et al. (Eds.): IWAI 2023, CCIS 1915, pp. 173–185, 2024.
https://doi.org/10.1007/978-3-031-47958-8_11

years of experience in coding and control systems. She has some knowledge of probabilistic modeling but is not a top expert in those fields.

In order to relieve Sarah from designing every detail of the robot, we expect that the robot possesses some "intelligent" adaptation capabilities. Firstly, the robot should be able to define sub-tasks and solve these tasks autonomously. Secondly, since we do not know a-priori the inside terrain of the building, the robot should be capable of adapting its walking and other locomotive skills under situated conditions. Thirdly, we expect that the robot performs robustly, in real-time, and cleverly manages the consumption of its computational resources.

All these robot properties should be supported seamlessly by Sarah's AIF software toolbox. For instance, she should not need to know the specifics of how to implement robustness in her algorithms or how many time steps the robot needs to look ahead in any given situation for effective planning purposes. We want a toolbox that enables competent engineers to develop effective AIF agents, not a toolbox for a select group of world-class machine learning experts. We do expect that Sarah is capable of describing her beliefs about desired robot behavior through the high-level specification of a probabilistic (world or generative) model or, at least, the prior preferences or constraints that underwrite behavior.

After reviewing some motivating agent properties that follow immediately from committing to free energy minimization (Sect. 2), we proceed to discuss why message passing in a factor graph is the befitting framework for implementing AIF agents (Sect. 3.1). More specifically, we argue that a reactive programming-based implementation of message passing will be the standard in professional-level AIF tools (Sect. 3.2). In comparison to the usual procedural coding style, reactive message passing leads to increased robustness (Sect. 3.3), lower power consumption (Sect. 3.5), hard real-time processing (Sect. 3.4), and support for continual model structure adaptation (Sect. 4). In Sect. 5.3 we introduce `RxInfer`, a toolbox-in-progress for developing AIF agents that robustly minimize free energy in real-time by reactive message passing.

2 The Free Energy Principle and Active Inference

2.1 FEP for Synthetic AIF Agents

The Free Energy Principle (FEP) describes self-organizing behavior in persistent natural agents (such as a brain) as the minimization of an information-theoretic functional that is known as the variational Free Energy (FE).[1] Essentially, the FEP is a commitment to describing adaptive behavior by Hamilton's Principle of Least Action [14]. The process of executing FE minimization in an agent that interacts with its environment through both active and sensory states is called *Active Inference* (AIF). Crucially, the FEP claims that, in natural agents,

[1] For reference, we use the following abbreviations in this paper: Active Inference (AIF), Constrained Bethe Free Energy (CBFE), Expected Free Energy (EFE), (variational) Free Energy (FE), Free Energy Principle (FEP), Free Energy Minimization (FEM), Message Passing (MP), Reactive Message Passing (RMP).

FE minimization is *all that is going on*. While engineering fields such as signal processing, control, and machine learning are considered different disciplines, in nature these fields all relate to the same computational mechanism, namely FE minimization.

For an engineer, this is good news. If we wish to design a synthetic AIF agent that learns purposeful behavior solely through self-directed environmental interactions, we can focus on two tasks:

1. Specification of the agent's model and inference constraints. This is equivalent to the specification of a (constrained) FE functional.
2. A recipe to continually minimize the FE in that model under situated conditions, driven by environmental interactions.

We are interested in the development of an engineering toolbox to support these two tasks.

2.2 FEM for Simultaneous Refinement of Problem Representation and Solution Proposal

An important quality of the robot will be to define tasks for itself and solve these tasks autonomously. Here, we shortly discuss how the FEP supports this objective.

Consider a generative model $p(x, s, u)$, where x are observed sensory inputs, u are latent control signals and s are latent internal states. For notational ease, we collect the latent variables by $z = \{s, u\}$. The variational FE for model $p(x, z)$ and variational posterior $q(z)$ is then given by

$$F[q,p] = \underbrace{-\log p(x)}_{\text{surprise}} + \underbrace{\sum_z q(z) \log \frac{q(z)}{p(z|x)}}_{\text{bound}} \tag{1a}$$

$$= \underbrace{\sum_z q(z) \log \frac{q(z)}{p(z)}}_{\text{complexity}} - \underbrace{\sum_z q(z) \log p(x|z)}_{\text{accuracy}} . \tag{1b}$$

The FE functional in (1a) can be interpreted as the sum of surprise (negative log-evidence) and a non-negative bound that is the Kullback-Leibler divergence between the variational and the optimal (Bayesian) posterior. The first term, surprise, can be interpreted as a performance score for the problem representation in the model. This term is completely independent of any inference performance issues. The second term (the bound) scores how well actual solutions are inferred, relative to optimal (Bayesian) inference solutions. In other words, the FE functional is a universal cost function that can be interpreted as the sum of problem representation and solution proposal costs. FE minimization leads toward improving both the problem representation and solving the problem through inference over latent variables. In particular, FE minimization over a particular model structure p should lead to nested sub-models that reflect the causal structure of the sensory data. Sub-tasks are solved by FE minimization in these sub-models. Hence, both creation of subtasks and solving these subtasks are driven solely by FE minimization.

In conclusion, a high-end toolbox should be capable to minimize FE both over (beliefs over) latent variables through adaptation of $q(z)$ (leading to better solution proposals for the current model p), and over the model structure p (leading to a better problem representation).

As an aside, an interesting consequence of the FE decomposition into problem plus solution costs is that a relatively poor problem representation with a superior inference process may be preferred (evidenced by lower FE), over a model with a good problem representation (high Bayesian evidence) where inference costs are high. The notion that the model with the largest Bayesian evidence may not be the most useful model in a practical application, casts an interesting light on the common interpretation of FE as a mere upper bound on Bayesian evidence. We argue here that FE is actually a more principled performance score for a model, since in addition to Bayesian model evidence, FE also scores the performance loss in a model due to an inaccurate inference process.

2.3 AIF for Smart Data Sets and Resource Management

If we want the robot to cope with unknown physical terrain conditions, it is not sufficient to pre-train the robot offline on a large set of relevant examples. The robot must be able to acquire relevant new data and update its model under real-world conditions.

FE minimization in the generative model's roll-out to the future results in the minimization of a cost functional known as the Expected Free Energy (EFE). It can be shown that the EFE decomposes into a sum of pragmatic (goal-driven, exploitation) and epistemic (information-seeking, exploration) costs [9]. As a result, inferred actions balance the need to acquire informative data (to learn a better predictive model) with the goal to reach desired future behavior.

In contrast to the current AI direction towards training larger models on larger data sets, an active inference process elicits an optimally informative, small ("smart") data set for training of just "good-enough" models to achieve a desired behavior. AIF agents adapt enough to accomplish the task at hand while minimizing the consumption of resources such as energy, data, and time. The trade-off between data accuracy and resource consumption is driven by the decomposition in (1b) of FE as a measure of complexity minus accuracy. According to this decomposition, more accurate models are only pursued if the increase in accuracy outweighs the resource consumption costs.

In short, AIF agents that are driven solely by FE minimization will inherently manage their computational resources. These agents automatically infer actions that elicit appropriately informative data to upgrade their skills toward good-enough performance levels. Since both the agent and environment mutually affect each other in a real-time information processing loop, it would not be possible

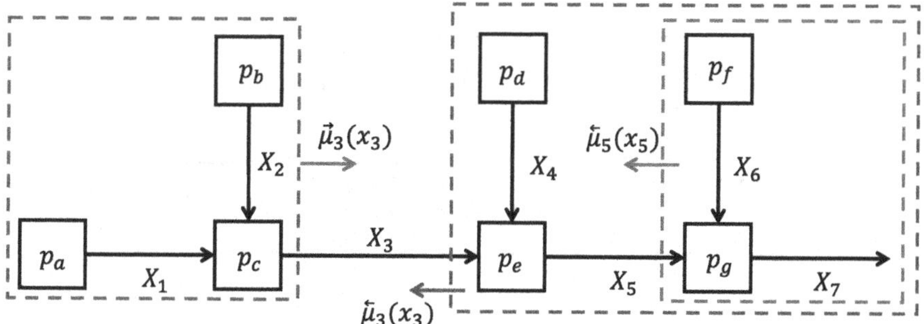

Fig. 1. Forney-style Factor Graph representation of the factorization (2).

to acquire the same data set through the sampling of the environment without the agent's participation.

3 FE Minimization by Reactive Message Passing

3.1 Why Message Passing-Based Inference?

Up to this point, our arguments strongly supported AIF as an information processing engine for the robot. Unfortunately, the computational demands for simulating a non-trivial synthetic AIF agent are extreme. For comparison, consider the human brain that minimizes in real-time, for less than 20 W, a highly time-varying FE functional (visual data rate about of about a million bits per second) over about 100 trillion latent variables (synapses). It has been estimated that the human brain consumes about a million times less energy than a high-tech silicon computer on quantitatively comparable information processing tasks [17].

Clearly, the human brain minimizes FE in a very different way than is available in standard optimization toolboxes. In this section, we will argue for developing a FE minimization toolbox based on reactive message passing in a factor graph.

First, we shortly recapitulate why message passing in factor graphs is an effective inference method for large models. Consider a factorized multivariate function

$$\begin{aligned} p(x_1,&x_2,\ldots,x_7) \\ &= f_a(x_1)f_b(x_2)f_c(x_1,x_2,x_3)f_d(x_4)f_e(x_3,x_4,x_5)f_f(x_6)f_g(x_5,x_6,x_7) \end{aligned} \tag{2}$$

Assume that we are interested in inferring (the so-called marginal distribution)

$$p(x_3) = \sum_{x_1}\sum_{x_2}\sum_{x_4}\sum_{x_5}\sum_{x_6}\sum_{x_7} p(x_1,x_2,\ldots,x_7) \tag{3}$$

If each variable x_i in (3) has about 10 possible values, then the sum contains about 1 million terms. However, making use of the factorization (2) and the distributive law [7], we can rewrite this sum as

$$p(x_3) = \Bigg(\overbrace{\sum_{x_1} \sum_{x_2} f_a(x_1) f_b(x_2) f_c(x_1, x_2, x_3)}^{\overrightarrow{\mu}_3(x_3)} \Bigg) \cdot$$
$$\cdot \underbrace{\Bigg(\sum_{x_4} \sum_{x_5} f_d(x_4) f_e(x_3, x_4, x_5) \big(\overbrace{\sum_{x_6} \sum_{x_7} f_f(x_6) f_g(x_5, x_6, x_7)}^{\overleftarrow{\mu}_5(x_5)} \big) \Bigg)}_{\overleftarrow{\mu}_3(x_3)} \tag{4}$$

The computation in (4), which requires only a few hundred summations and multiplications, is clearly preferred from a computational load viewpoint. To execute (4), we need to compute intermediate results $\overrightarrow{\mu}_i(x_i)$ and $\overleftarrow{\mu}_i(x_i)$ that afford an interpretation of local messages in a Forney-style Factor Graph (FFG) representation of the model, see Fig. 1.

Variational FE minimization can also be executed by message passing in a factor graph. In fact, nearly all known effective variational inference methods on factorized models can be interpreted as minimization of a so-called "constrained Bethe Free Energy" (CBFE) functional [16]. In this formulation, posterior variational beliefs are factorized into beliefs over both the nodes and the edges of the graph. It is possible to add constraints to these local beliefs such as requiring that a particular variational posterior is expressed by a Gaussian distribution. In general, CBFE minimization by message passing in a factor graph supports local adaptation of a plethora of constraints to optimize accuracy vs resource consumption [1,16].

Useful dynamic models for real-time processing of data streams with a large number of latent variables are necessarily sparsely connected because otherwise, real-time inference would not be tractable. In sparse models, the computational complexity of inference can be vastly reduced by message passing in a factor graph representation of the model. In particular, automated CBFE minimization by message passing in a factor graph supports refined optimization of the accuracy vs resource consumption balance.

3.2 Reactive vs Procedural Coding Style

Next, we discuss a key technological component for a synthetic AIF agent, namely the requirement to execute FE minimization through a *reactive* programming paradigm.

A crucial feature of all MP-based inference is that the inference process consists entirely of a (parallelizable) series of small steps (messages) that individually and independently contribute to FE minimization. As a result, a message passing-based FE minimization process can be interrupted *at any time* without loss of important intermediate computational results.

In a practical setting, it is very important that an ongoing inference process can be robustly (without crashing) interrupted at any time with a result. These intermediate inference results can only be reliably retrieved if the inference process iteratively updates its beliefs in small steps, or, in other words, by message passing. Moreover, the inference process should not be subject to a prescribed control flow that contains for-loops. Rather, if we were to write code for an anytime-interruptable inference process in a programming language, we should use a *reactive* rather than the more common *procedural* programming style. In a reactively coded inference engine, there is no code for control flow, such as "do first this, then that", but instead only a description of how a processing module (a factor graph node) should react to changes in incoming messages. We will call this process *Reactive Message Passing* (RMP) [2]. In an RMP inference process, there is no prescribed schedule for passing messages such as the Viterbi or Bellman algorithm. Rather, an RMP inference process just *reacts* by FE minimization whenever FE increases due to new observations.

In Fig. 2, we display the consequences of choosing a reactive programming style for an application engineer like Sarah. The procedural programming style in Algorithm-1 requires Sarah to provide the control flow (the "procedure") for the inference process. Sarah needs to write code for when to collect observations, when to update states, etc. The specific control flow in Algorithm-1 is just an example and there exists literature that aims to improve the efficiency of the control flow [5,10]. In order to write an efficient inference control flow recipe for a complex AIF agent, Sarah needs to be an absolute expert in this field.

Consider in contrast the code for reactive inference in Algorithm-2. In a reactive programming paradigm, there is no control flow. Rather, the only inference instruction is for the agent to react to any opportunity to minimize FE. When FE minimization is executed by a reactive message passing toolbox, the application engineer only needs to specify the model.

Aside from lowering the competence bar for application engineers to design effective AIF agents, the procedural style of implementing FE minimization is fundamentally inappropriate. The control flow in Algorithm-1 necessarily contains many design choices that only become known during deployment. For instance, how far should the agent roll out its model to the future for computing the EFE? This kind of information is highly contextual and not available to the application engineer. In contrast, the application engineer's code for reactive inference ("react to any FEM opportunity") works for any model in any context. In a reactive inference setting, the appropriate planning horizon is going to be continually updated (inferred) with contextual information. In other words, it is the reactive FEM process itself that leads to optimizing the inference control flow.

3.3 RMP for Robustness

Since an AIF agent executes under situated conditions, it must perform the FE minimization process robustly in real-time. Consider an agent whose computational resources are represented by a graph and FE minimization results from executing MP-based inference on that graph. Any MP schedule that visits the

Algorithm 1 Procedural AIF

Specify model $p(x, s, u, \theta)$
for $t = 1, 2, \ldots$ **do** ▷ Deploy
  Collect new observation x_t
  Update state $q(s_t | x_{1:t})$
  Update desired future $\tilde{p}(x_{>t})$
  Upd. candidate policies $\{\pi^{(i)}\}$
  for all $\pi^{(i)}$ **do**
    Predict future $p(x_{>t} | s_t, \pi^{(i)})$
    Compute EFE $G(\pi^{(i)})$
  end for
  Select $\pi^* = \arg\min_{\pi \in \{\pi^{(i)}\}} G(\pi)$
end for

Algorithm 2 Reactive AIF

Specify model $p(x, s, u, \theta)$
while true **do** ▷ Deploy
  React to any FEM opportunity
end while

Fig. 2. Pseudo-code for procedural and reactive coding styles for AIF agents.

nodes in the graph in a prescribed fixed order (as would be the case in a procedural approach to FE minimization) is vulnerable to malfunction in any of the nodes in the schedule. In principle, the FE minimization process needs to stop after such a malfunction and proceed to compute a new MP schedule. Since FE minimization is the *only* ongoing computational process, the robot basically moves blindfolded after a reset. Clearly, for robustness, we need a system that continues to minimize FE, even after parts of the graph break down over time. In a reactive inference framework, collapse of a component is simply a switch to an alternative model structure. The new model may perform better or worse at FE minimization, but there is no reason to stop processing.

3.4 RMP for Real-Time, Situated Processing

An ongoing RMP process can always be interrupted when computational resources have run out on a given platform. In this way, by trading computational complexity (i.e., the number of messages) for accuracy, any RMP-based inference process can be scaled down to a real-time processing procedure, where of course a prediction accuracy price may have to be paid, depending on the available computational resources. In short, FE minimization in any model can be executed in real-time on any computational platform if we implement inference by RMP in a factor graph.

3.5 RMP for Low Power Consumption

Similarly, an ongoing RMP process can always be terminated if the expected improvement in accuracy does not outweigh the expected computational load that additional messages would incur.[2] Note that, since FE decomposes as computational complexity minus accuracy, interrupting an RMP-based inference process for this reason is fully consistent with the goal of FE minimization.

Interrupting an ongoing MP process by any of the above-mentioned reasons (e.g., node malfunction, running out of computational resources, expected processing costs outweighing expected accuracy gains, etc.), in principle always leads to sacrificing some prediction accuracy in favor of saving computational costs. Crucially, these interrupts will not cause a system-wide crash in a reactive system.

4 Model Structure Adaptation

In Sect. 2.2, we touched upon the notion that FE minimization should ideally drive the generative model p to evolve to structurally segregated but communicating sub-models that reflect the causal structure of the environment. Technically, this is due to the drive for a lower surprise ($-\log p(x)$).

There is another reason why online structural adaptation is important. Free energy minimization over the structure of p should also lead to a model structure for which inference costs $D_{\mathrm{KL}}[q(z)||p(z|x)]$ are lower by moving $p(z|x)$ closer to $q(z)$. Consider again the procedural and reactive inference code in Fig. 2. The control flow in the procedural code aims to cleverly steer the inference process toward maximal inference accuracy for minimal computational costs. In contrast, the reactive code just declares that the system should react (by message passing) to any FE minimization opportunity. In the reactive framework, *clever* inference is learned over time by continual minimization over all movable parts of the CBFE, i.e., by FEM over states, parameters, structure (adaptation of p), and constraints (adaptation of the structure of q). To learn the most effective paths for inference, the toolbox should support structural adaptation over both p and q.

Unfortunately, online structural adaptation during the deployment of the robot is still an ongoing research issue, e.g., [3,8,15]. One technical difficulty is that an efficient inference control flow (which states are updated at what time, etc.) may change if the structure of the generative model changes. In a procedural programming style, we would need to reset the system and reprogram the inference code in Algorithm-1 (in Fig. 2). This is incompatible with the demand that the agent adapts during deployment. As discussed above, a reactive programming style solves this issue since the application inference code (Algorithm-2 in Fig. 2) is independent of the model structure.

[2] The computational load and complexity can only be equated in the absence of a Von Neumann bottleneck (i.e., with mortal computation or in-memory processing). This is because energy and time are 'wasted' by reading and writing to memory.

5 Discussion

5.1 Review of Arguments

We shortly summarize our view on a professional-level supporting software toolbox for the design of relevant AIF agents, see also Table 1. In Sect. 2, we discussed a few extraordinary features that follow straightaway from committing to free energy minimization as the sole computational mechanism for a future AI ecosystem as proposed in Friston et al. [11]. First, the FE functional in an AIF agent can be interpreted as a universal performance criterion that applies in principle to all problems. If FEM can be extended to structural model adaptation, then an AIF agent is naturally able to create and solve sub-problems. Moreover, by virtue of the decomposition of EFE into a sum of information- and goal-seeking costs, AIF agents naturally seek out small "smart" data sets.

In terms of FEM implementation, we asserted that useful models are highly factorized and sparse. Efficient inference in factorized models can always be described as message passing in a factor graph. In particular, nearly all known variants of highly efficient message passing algorithms for FEM can be formulated in a single framework as minimizing a Constrained Bethe Free Energy (CBFE).

We then claimed that a *reactive* rather than procedural processing strategy is essential. Reactive message passing-based (RMP) inference is always interruptible with an inference result, thus supporting guaranteed real-time processing, which is a hard requirement for AIF agents in the real world. In comparison to the more common procedural programming approach to FEM, reactive processing also improves robustness, resource consumption, and the capability to make structural changes without the need for resetting the inference process.

This latter feature, support for online structural adaptation is also a vital feature of a high-quality AIF toolbox. Online structural adaptation leads to both continual problem representation refinement (by lowering surprise) and to a more efficient inference process.

Table 1. Summary of benefits for supporting reactive message passing and structural adaptation in an AIF agent.

	realization technology	benefits
1	FEP, AIF	one solution approach; smart data
2	reactive message passing	low power; robustness; real-time
3	structural adaptation	problem refinement; clever inference

5.2 Review of Existing Tools

Currently, there exists a small but vibrant research community on the development of open-source tools for simulating synthetic AIF agents. In this community, a few supporting packages have been released, including SPM [12], PyMDP [13] and `ForneyLab` [6]. The SPM toolbox was originally written by Karl Friston and colleagues, and has developed into a very large set of tools and demonstrations for experimental validation of the scientific output of the UCL team and collaborators. PyMDP is a more recent Python package for simulating discrete-state POMDP models by Conor Heins, Alexander Tschantz and a team of collaborators. `ForneyLab.jl` is a Julia package from BIASlab (http://biaslab.org) for simulating FE minimization by message passing in Forney-style factor graphs. Unfortunately, none of the above-mentioned tools support *reactive* message passing-based inference. Therefore, we believe that these tools will serve the community well as AIF prototyping and validation tools, but they will not scale to support real-time, robust simulation of AIF agents with commercializable value.

5.3 Reactive Message Passing with `RxInfer`

More recently, BIASlab has released the open-source Julia package `RxInfer` (http://rxinfer.ml) to support an engineer at Sarah's level to develop commercially relevant AIF agents that minimize FE by automated reactive message passing in a factor graph [2]. Julia is a modern open-source scientific programming language with roughly the syntax of MATLAB and out-of-the-box speed of C [4].

The development process of `RxInfer` focuses on the following priorities:

1. model space coverage
   - `RxInfer` aims to support reactive message passing-based FEM for a very large set of freely definable relevant probabilistic models.
2. user experience
   - `RxInfer` aims to support a busy, competent researcher or developer who understands probabilistic modeling (but doesn't know Julia) to design and deploy an AIF agent into the world. In particular, a user-friendly specification of nested AIF agents should be supported.
3. adaptation
   - `RxInfer` aims to support continual adaptation by automated FEM over all movable parts of the CBFE functional, including states, parameters, structure, and variational constraints.
4. real-time
   - `RxInfer` aims to process data streams in "hard" real-time, under situated conditions, even for large models. Larger models may lead to less accurate inference (in terms of KL-divergence between variational and Bayesian posteriors), but no crashes.

5. low-power
   - `RxInfer` aims to process data streams on any, possibly time-varying, power budget. Lower power budgets may lead to less accurate inference but no crashes.

At the time of writing this paper, `RxInfer` supports fast and robust automated CBFE minimization by reactive message passing for states and parameters in a large set of freely definable models. `RxInfer` processes streaming data very fast, but not yet guaranteed in hard real-time. User-friendly specifications of AIF agents will be released this summer. Model structure adaptation is supported by NUV priors (normal priors with unknown variance) [15], but not yet by online Bayesian model reduction [3,8]. `RxInfer` comes with a large set of examples and is slated to support the above priority list in the future.

6 Conclusions

Supported by `RxInfer` or a similar toolbox, future AI engineers will no longer design end-product algorithms, but will instead design the designers (AIF agents) of production algorithms in short and easy-readable code scripts. Along with [11], we think that the potential benefits of shared intelligence in ecosystems of communicating AIF agents are hard to overstate. As we have argued in this position paper, the required underlying technology for realizing this vision is very demanding and currently not yet available. Still, we also think it is not out of reach and is one of the most exciting ongoing research threads in the AI field.

Acknowledgments. I would like to acknowledge my colleagues at BIASlab (http://biaslab.org) for the stimulating work environment and the anonymous reviewers for excellent feedback on the draft version. Some wording in this document, such as footnote (see footnote 2), comes straight from a reviewer.

References

1. Akbayrak, S., Bocharov, I., de Vries, B.: Extended variational message passing for automated approximate Bayesian inference. Entropy **23**(7), 815 (2021). ISSN: 1099-4300. https://doi.org/10.3390/e23070815. https://www.mdpi.com/1099-4300/23/7/815. Accessed 26 May 2023
2. Bagaev, D., de Vries, B.: Reactive message passing for scalable bayesian inference. Sci. Program. **2023** (2023). ISSN: 1058-9244. https://doi.org/10.1155/2023/6601690. https://www.hindawi.com/journals/sp/2023/6601690/. Accessed 28 May 2023
3. Beckers, J., et al.: Principled Pruning of Bayesian Neural Networks through Variational Free Energy Minimization (2022). https://doi.org/10.48550/arXiv.2210.09134. arXiv: 2210.09134. http://arxiv.org/abs/2210.09134. Accessed 26 May 2023
4. Bezanson, J., et al.: Julia: a fresh approach to numerical computing. SIAM Rev. **59**(1), 65–98 (2017). ISSN: 0036-1445. https://doi.org/10.1137/141000671. https://epubs.siam.org/doi/10.1137/141000671. Accessed 03 Feb 2022

5. Champion, T., Grze, M., Bowman, H.: Realizing active inference in variational message passing: the outcome-blind certainty seeker. Neural Comput. **33**(10), 2762–2826 (2021). ISSN: 0899-7667. https://doi.org/10.1162/neco_a_01422. Accessed 26 May 2023
6. Cox, M., van de Laar, T., de Vries, B.: A factor graph approach to automated design of Bayesian signal processing algorithms. Int. J. Approx. Reason. **104**, 185–204 (2019). ISSN: 0888-613X. https://doi.org/10.1016/j.ijar.2018.11.002. http://www.sciencedirect.com/science/article/pii/S0888613X18304298. Accessed 16 Nov 2018
7. Distributive property. Wikipedia. Page Version ID: 1124679546 (2022). https://en.wikipedia.org/w/index.php?title=Distributive_property&oldid=1124679546. Accessed 26 May 2023
8. Friston, K., Parr, T., Zeidman, P.: Bayesian model reduction. arXiv:1805.07092 (2018). http://arxiv.org/abs/1805.07092. Accessed 28 May 2018
9. Friston, K., et al.: Active inference and epistemic value. Cogn. Neurosci. (2015). ISSN: 1758-8928. https://doi.org/10.1080/17588928.2015.1020053. Accessed 22 Feb 2015
10. Friston, K., et al.: Sophisticated inference. Neural Comput. **33**(3), 713–763 (2021). ISSN 0899-7667. https://doi.org/10.1162/neco_a_01351. Accessed 14 Feb 2022
11. Friston, K.J., et al.: Designing Ecosystems of Intelligence from First Principles (2022). https://doi.org/10.48550/arXiv.2212.01354. arXiv:2212.01354. http://arxiv.org/abs/2212.01354. Accessed 08 Dec 2022
12. Friston, K.J., et al.: SPM12 toolbox (2014). http://www.fil.ion.ucl.ac.uk/spm/software/
13. Heins, C., et al.: pymdp: a Python library for active inference in discrete state spaces. arXiv:2201.03904 (2022). http://arxiv.org/abs/2201.03904. Accessed 03 Feb 2022
14. Lanczos, C.: The Variational Principles of Mechanics, 4th Revised edition, 464 p. Dover Publications, New York (1986). ISBN 978-0-486-65067-8
15. Loeliger, H.-A., et al.: On sparsity by NUV-EM, Gaussian message passing, and Kalman smoothing. In: 2016 Information Theory and Applications Workshop (ITA). 2016 Information Theory and Applications (ITA), La Jolla, CA, USA, pp. 1–10. IEEE (2016). ISBN: 978-1-5090-2529-9. https://doi.org/10.1109/ITA.2016.7888168. http://ieeexplore.ieee.org/document/7888168/. Accessed 21 July 2021
16. Senöz, I., et al.: Variational message passing and local constraint manipulation in factor graphs. Entropy **23**(7), 807 (2021). ISSN: 1099-4300. https://doi.org/10.3390/e23070807
17. Smirnova, L., et al.: Organoid intelligence (OI): the new frontier in biocomputing and intelligence-in-a-dish. Front. Sci. (2023)

Learning Representations for Active Inference

Exploring Action-Centric Representations Through the Lens of Rate-Distortion Theory

Miguel De Llanza Varona[1,2(✉)], Christopher Buckley[1,2], and Beren Millidge[3]

[1] School of Engineering and Informatics, University of Sussex, Brighton, UK
{M.De-Llanza-Varona,C.L.Buckley}@sussex.ac.uk
[2] VERSES Research Lab, Los Angeles, CA, USA
[3] MRC Brain Networks Dynamics Unit, University of Oxford, Oxford, UK
beren@millidge.name

Abstract. Organisms have to keep track of the information in the environment that is relevant for adaptive behaviour. Transmitting information in an economical and efficient way becomes crucial for limited-resourced agents living in high-dimensional environments. The efficient coding hypothesis claims that organisms seek to maximize the information about the sensory input in an efficient manner. Under Bayesian inference, this means that the role of the brain is to efficiently allocate resources in order to make predictions about the hidden states that cause sensory data. However, neither of those frameworks accounts for how that information is exploited downstream, leaving aside the action-oriented role of the perceptual system. Rate-distortion theory, which defines optimal lossy compression under constraints, has gained attention as a formal framework to explore goal-oriented efficient coding. In this work, we explore action-centric representations in the context of rate-distortion theory. We also provide a mathematical definition of abstractions and we argue that, as a summary of the relevant details, they can be used to fix the content of action-centric representations. We model action-centric representations using VAEs and we find that such representations i) are efficient lossy compressions of the data; ii) capture the task-dependent invariances necessary to achieve successful behaviour; and iii) are not in service of reconstructing the data. Thus, we conclude that full reconstruction of the data is rarely needed to achieve optimal behaviour, consistent with a teleological approach to perception.

Keywords: Rate-distortion theory · Action-centric representations · Efficient coding · Bayesian Inference

1 Introduction

Embodied agents have to focus on the relevant information from their environment to achieve adaptive behaviour. Their resource-limited cognition and the

C. L. Buckley et al. (Eds.): IWAI 2023, CCIS 1915, pp. 189–203, 2024.
https://doi.org/10.1007/978-3-031-47958-8_12

high-complexity structure inherent to the environment force them to economize the transmission of information. Thus, the goal of the perceptual system is to generate representations that are useful for successful behaviour while at the same being encoded in the most efficient manner.

A well-known hypothesis in theoretical neuroscience called the efficient coding hypothesis proposes that the neural coding in the brain is optimized to maximize sensory information under metabolic and capacity constraints [3,13,23]. In particular, this hypothesis suggests that neurons are tuned to the statistical properties of the environment, which allows them to efficiently allocate signaling resources to generate compressed low-dimensional representations of the environment. In this theoretical framework, it is commonly assumed that the function of neurons is to maximize their capacity to account for all the variability in the sensory input. In information theory terms, this means that the brain seeks to maximize the mutual information between stimuli and neurons' response to reduce as much as possible the uncertainty about the environment, which is defined by its entropy. While this hypothesis answers the question about information processing under biological constraints, it leaves aside the utilitarian aspect of perception [11,14,15,18,20].

Cognition can't be fully understood without its ecological context, as agents are coupled with their environment forming a perception-action feedback loop [22]. In this sense, the functional role of perceptual processing has to be in service of achieving behavioural objectives, and to do that, perceptual representations must efficiently encode the relevant information needed by the motor system to guide future actions. Thus, a key component of the perceptual system is to summarize relevant sensory information to generate action-centric representations.

The teleological essence of the perceptual system imposes a normativity on representations: a perceptual representation is accurate if it captures the relevant information needed downstream and discards the irrelevant details. Thus, we need an extra ingredient to account for the goodness of representations under constraints. This is where the rate-distortion theory comes into play [19]. This subfield of information theory defines the optimal trade-off between channel capacity and expected communication error. When error-free communication is not necessary to guide behaviour, the optimal encoding is a lossy compression of the input.

Interestingly, rate-distortion theory can be seen as a way to perform Bayesian inference under constraints. Under a Bayesian approach to cognition, the brain performs inference to compute an optimal posterior distribution over hidden environmental states given sensory data [8]. As computing the true posterior is usually intractable, the brain approximates the true posterior by optimizing the variational free energy [4,9,10]. The main conceptual contribution of rate-distortion theory is to define the "goodness" of that approximation, as computing the true posterior is not always necessary to act optimally. In the context of active inference, it has been shown that action-oriented models learn parsimonious representations of the environment by capturing relevant information for behaviour [21]. In the same spirit, we investigate the information-theoretic properties of action-centric representations and their relation to the formal definition of abstractions we propose.

In this work, we explore action-centric representations under the lens of rate-distortion theory to account for the teleological aspect of perception. To do that, we provide a mathematical definition of abstraction that allows us to specify the task-relevant information that should carry an action-centric representation. Given the tight connection between Bayesian Inference and rate-distortion theory, we use a Variational Autoencoder (VAE) framework to model action-centric representations as optimal lossy compression. Our results show that action-centric representations are optimal lossy compressions of the data; can be successfully used in downstream tasks; and crucially, they achieve that without being in service of reconstructing the data.

2 Efficient Coding and Rate-Distortion Theory

2.1 Efficient Coding

The efficient coding hypothesis states that neurons are optimized to maximize the information they carry about sensory states. In doing so, neurons have to generate minimal redundancy codes to economically use limited resources. In particular, neurons seek to maximize the ratio between information about sensory inputs, defined by the mutual information $I(X;Z)$ between sensory data X and neural responses Z, and the channel capacity C: $\frac{I(X;Z)}{C}$. The maximum mutual information is upper bounded by the channel capacity

$$C \geq I(X;Z) \tag{1}$$

so the best efficient coding satisfies

$$I(X;Z) = C \tag{2}$$

where neuronal encoding exploits the whole bandwidth of the channel.

2.2 Rate-Distortion Theory as Goal-Oriented Efficient Coding

Under the classical conception of efficient coding, the exploitation of information downstream is ignored. When not all sensory information is needed to guide behaviour, error-free communication is not expected. This is precisely what is addressed by the rate-distortion theory, which provides the theoretical foundations for optimal lossy data compression. Formally, the rate-distortion function defines an optimal lossy compression Z of some data X as the minimization of their mutual information $I(X;Z)$ given some expected distortion D associated with reconstructing X from its lossy compression Z. It is defined as [6]

$$R(D) = \min_{q(z|x): D_{q(x,z)} \leq D} I(X;Z) \tag{3}$$

where q is the optimal distribution of z given x that satisfies the expected distortion constraint and the rate R is an upper bound on the mutual information:

$$R \geq I(X; Z) \tag{4}$$

The expected distortion D is defined by some arbitrary loss function (e.g., mean-squared error) that quantifies the faithfulness of information transmission (i.e., how well can the data be recovered from its optimal lossy compression). Lossy compression sacrifices the capacity to represent all the information in the input in service of transmitting information that allows adaptive behaviour. Having a faithfulness criterion allows the brain to efficiently represent the world by allocating just the necessary amount of resources required to navigate the environment (Fig. 1). Thus, rate-distortion adds a teleological perspective to efficient coding that shifts the focus from efficient information maximization to efficient transmission of action-oriented information.

In the lossy regime of the rate-distortion (i.e., all points such that $D > 0$), the obtained representations can be understood as abstractions of the data, as their function is to summarize the relevant properties of the data needed downstream. In the next section, we provide a mathematical definition of abstractions based on the intuition that are entities that convey the necessary information to answer a set of queries about the data. The mathematical formulation of abstractions is crucial to determine the content of action-centric representations.

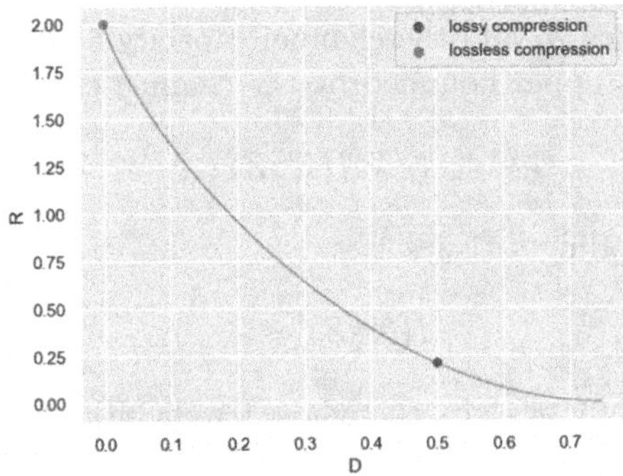

Fig. 1. Rate-distortion function for a discrete random variable with four uniformly distributed states. Assuming that behavioral objectives are achievable even when half of the information generated at the source is missing; that is, when the expected distortion D does not exceed 0.5 (x-axis), an optimal agent with bounded rationality can rely on a lossy compression scheme and transmit information at a rate R of 0.20 bits (y-axis).

3 Abstractions and Action-Centric Representations

3.1 Mathematical Formalization of Abstractions

An abstraction is the reduction of complexity by discarding certain features while preserving others. As a relational concept, an abstraction involves two components: its object (what is being abstracted) and its content (what the abstraction is about). The content of the abstraction is a summary of the *relevant* properties of its object and the relevancy is fixed, ultimately, by the agent's needs. Thus, abstractions are intrinsically *teleological* entities; that is, their meaning or content is fixed by their function or purpose, which is to transmit information about the properties of interest for an agent.

Following [16], we address the content of abstractions as the information necessary to answer a set of queries about the data. A query captures what the agent wants to know about the data (i.e., what is relevant). Formally, given a set of queries about the data $\mathbb{Q} = \{Q_1, Q_2 \dots Q_3\}$, each query is a mapping from the data distribution to a probability distribution over a subset of elements of the data $Q : \mathbb{X} \to p(q|x)$. A good abstraction is one that fulfills its purpose; namely, one that keeps track of those properties that make it possible to answer a particular query. Thus, the 'goodness' of an abstraction z for a given query can be defined as:

$$\mathcal{L}_Q(x, z) = \mathcal{D}[Q(x)||Q(r(z))] \tag{5}$$

where $Q(x)$ is the query distribution over the true system or data, $Q(r(z))$ is the query distribution over a lossy reconstruction of the data $r(z)$ produced by the abstraction z, and $\mathcal{D}$ is an arbitrary divergence function. Without loss of generality, the 'goodness' of an abstraction given a set of queries can be defined as the weighted loss over all the queries given the abstraction:

$$\mathcal{L}(x, z) = \sum_{Q_i \in \mathbb{Q}} p(Q_i)\mathcal{L}_{Q_i}(x, z) \tag{6}$$

Ideally, the mutual information between the abstractions and the query should be the same as the information transmitted between the data and the query:

$$I(X; Q) = I(Q; Z) \tag{7}$$

The intuition is that a good abstraction Z of the data should reduce the uncertainty of the data X in the same way as the query Q does.

3.2 Abstractions as Sufficient and Non-superfluous Representations

Following [7], an abstract representation Z that captures the relevant details of the data X to answer a query Q should be sufficient ($I(X; Q|Z) = 0$) and non-superfluous ($I(X; Z|Q) = 0$):

$$\underbrace{I(X; Z|Q)}_{\text{Superfluous}} = I(X; Z) - I(X; Q; Z) \tag{8}$$

$$= I(X; Z) - I(X; Q) + \underbrace{I(X; Q|Z)}_{\text{Sufficient}} \tag{9}$$

$$= I(X; Z) - I(X; Q) \tag{10}$$

therefore

$$\big(I(X;Z|Q)=0\big)\wedge\big(I(X;Q|Z)=0\big)\iff I(X;Z)=I(X;Q) \tag{11}$$

From an information theory perspective, a good abstraction Z only carries the relevant information in the data; that is, the information necessary to answer a query. Note that this is a continuum where at one extreme the optimal compression captures all the information in the data when the query contains the same information as the data $H(Q) = H(X)$ (an ideal scenario in the efficient coding hypothesis, where the goal is to maximize mutual information):

$$I(X;Q) = H(X) - H(X|Q) = H(X) - H(X|X) = H(X) \tag{12}$$

therefore

$$I(X;Z) = I(X;Q) \tag{13}$$
$$I(X;Z) = H(X) \tag{14}$$

which corresponds to the lossless compression regime of the rate-distortion function. On the contrary, when the communication channel is closed, then we recover the other extreme of the rate-distortion curve, where the mutual information is zero. This is the case when knowing the query does not reduce the uncertainty of the data:

$$I(X;Q) = H(X) - H(X|Q) = H(X) - H(X) = 0 \tag{15}$$

therefore

$$I(X;Z) = I(X;Q) \tag{16}$$
$$I(X;Z) = 0 \tag{17}$$

Any other stage in between is a case where the query carries partial information about the data. Importantly, these information-theoretic entities are implicitly optimized in rate-distortion theory. On the one hand, sufficient information is related to predictability and, therefore, to communication fidelity, which is satisfied when the expected distortion allows for answering the query (i.e., successful behaviour). On the other hand, non-superfluous information is related to the minimization of the mutual information up to a point in which only query-relevant information is encoded in the abstraction. Thus, optimal lossy representations, whose function is to encode the relevant invariances and symmetries in the data, lie in the rate-distortion curve.

4 Variational Free Energy and Rate-Distortion Theory

As computing the rate-distortion function is intractable in high-dimensional systems [6], variational inference can be used as a proxy of the amount of information transmitted through a communication channel. In variational inference, a quantity called variational free energy sets an upper bound on the sensory surprisal (i.e., the entropy of sensory states), and by minimizing it is reduced the uncertainty about the sensory data allowing for predictability of future states and adaptive behaviour. One common variational free energy decomposition is the ELBO, which involves two terms, accuracy and complexity, and is formally defined as [17]

$$F = \int q(z|x) \ln \frac{q(z|x)}{p(x,z)} \tag{18}$$

$$F = \int q(z|x) \ln \frac{q(z|x)}{p(z)} - \int q(z|x) \ln p(x|z) \tag{19}$$

$$F = \underbrace{D_{KL}[q(z|x)||p(z)]}_{\text{Complexity}} - \underbrace{\mathop{\mathbb{E}}_{z \sim q}[\ln p(x|z)]}_{\text{Accuracy}} \tag{20}$$

To model lossy representations of the data, we use VAEs due to its close relation to variational inference and rate-distortion theory. VAEs is an unsupervised learning framework that captures the underlying data distribution by using i) an encoder that learns a latent representation of the data; and ii) a decoder that generates data-like samples from the latent representation. The objective function optimized by VAEs is the ELBO, where the complexity term can be seen as a regularizer applied to the latent space, and the accuracy term as the faithfulness of the decoder's reconstruction.

As has been recently shown [1,12], the ELBO is implicitly optimizing the rate-distortion function. On the one hand, the expected complexity is an upper bound on the mutual information $I(X;Z)$ (see Appendix C for full derivation):

$$\mathop{\mathbb{E}}_{p(x)}\left[D_{KL}[q(z|x)||p(z)]\right] \geq I(X;Z) \tag{21}$$

just as the rate R is in rate-distortion theory. On the other hand, the expected distortion can be measured using any loss function that captures how faithful the reconstruction of the decoder resembles the input data (e.g., hamming distance). In this case, the negative log-likelihood used in VAEs can be used as a distortion measure between the input and its reconstruction, so D can be defined as:

$$D = - \mathop{\mathbb{E}}_{z \sim q_\phi(z|x)} \left[\log p_\theta(x|z)\right] \tag{22}$$

Thus, variational inference can be understood through the lens of rate-distortion is characterized as

$$F = \underbrace{- \mathop{\mathbb{E}}_{z \sim q_\phi(z|x)} \left[\log p_\theta(x|z)\right]}_{\text{Distortion}} + \underbrace{\mathcal{D}_{KL}\left[q_\phi(z|x)||p(z)\right]}_{\text{Rate}} \tag{23}$$

5 Methods

5.1 Model

Inspired by the utilitarian perspective on the efficient coding hypothesis and the mathematical foundations of abstractions, we present a modified VAEs to model action-centric representations (Fig. 2). The main novelty of the VAEs presented here lies in the accuracy term of the free energy (Eq. (20)). Contrary to vanilla VAEs, where the goal is to learn latent representations of the data to reconstruct it as faithfully as possible, here we are interested in learning action-centric representations that convey sufficient and non-superfluous information about a query. In this model, full reconstruction of the data is not expected. The final form of the objective function for our action-centric VAEs is:

$$F = -\mathcal{D}[Q(x)||Q(r(z))] + \beta\mathcal{D}_{KL}\left[q_\phi(z|x)||p(z)\right] \tag{24}$$

where β is the gradient of the rate with respect to the distortion $\frac{\partial R}{\partial D} = \beta$ and here it's used to target specific regimes of the rate-distortion plane [5]. The accuracy is modified to account for the goodness of the abstraction. The training pipeline is as follows. First, we define a query to be the discrimination of the ten classes of the FASHION-MNIST dataset. We first trained a classifier, using a CNN, on the task specified by the query (i.e., multiclass classification). Once the discriminator is trained we trained both the vanilla VAEs and our action-centric VAEs. Importantly, both VAEs have the same channel capacity, as they share the same architecture, so the maximum achievable rate in both models is the same. The crucial difference is that our VAEs is not trained to fully reconstruct the data but to generate reconstructions that can be well-classified by the discriminator.

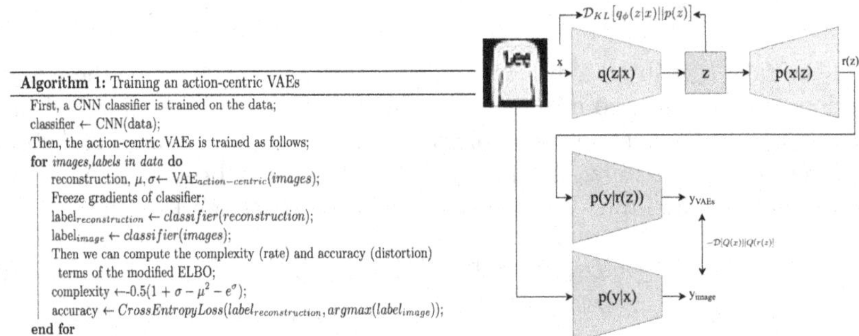

(a) Algorithmic details of the action-centric VAEs

(b) Architecture of the action-centric VAEs

Fig. 2. Action-centric VAEs. Left: algorithmic level of the action-centric VAEs. Right: a schematic overview of the architecture. The novel component is in the accuracy term of the ELBO. Instead of measuring the faithfulness of the reconstruction, it is measured how good is the reconstruction for a specific task, which in this case is an image classifier.

By doing this divergence measure, we can evaluate the goodness of the abstract representations for the given query. To compute the rate-distortion function, we trained several VAEs using different β to study the rate-distortion trade-off in different regimes and the potential differences between vanilla VAEs and our model.

Using this model we can investigate whether the latent space can efficiently encode just the relevant invariances and symmetries required for the downstream task without the need to generate faithful reconstructions of the data. If that is the case, full reconstruction no longer becomes a necessary condition for goal-oriented representations. In the next section, we present the main results and their connection to the theoretical framework presented previously.

5.2 Results

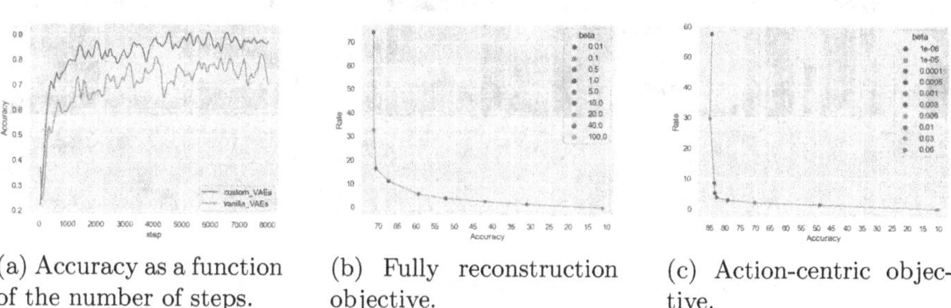

(a) Accuracy as a function of the number of steps.

(b) Fully reconstruction objective.

(c) Action-centric objective.

Fig. 3. Rate-distortion curve made up of different VAEs. Note that as a distortion measure here we use the accuracy. Left figure: vanilla VAEs that try to maximize the mutual information given the channel constraints. Right figure: lossy compression VAEs whose function is to maximize utility downstream given the channel constraints.

The results regarding the transmission of information in the two different VAEs are shown in Fig. 3. In (Fig. 3a) it can be seen how action-oriented VAEs converges faster to an encoding-decoding scheme that is useful for the downstream task (measured by the accuracy), compared to the VAEs. This indicates that action-centric representations might require less exposure to data, which makes them more efficient in terms of exploiting the available information.

Figure 3b and Fig. 3c show the rate-distortion curve for both types of VAEs. It is clear how action-oriented representations require significantly less information from the data to achieve better results in the downstream task. In particular, it can be seen that transmitting at a rate of around 10 bits the action-centric VAEs reaches almost 85% of accuracy, compared to the 67% achieved by the vanilla VAEs at approximately the same rate. This suggests that lossy compression leads to efficient codings and, importantly, to better behaviour.

The results so far indicate that the main function of representations might not be to fully reconstruct the data, but not capture the relevant invariances in the data exploited by optimal behaviour. We explicitly show this by investigating the reconstructions obtained by action-oriented representations. Figure 4 shows a sample of the reconstructions obtained by the vanilla and action-centric VAEs, respectively. While the vanilla VAEs generates relatively faithful reconstructions of the data, the action-centric VAEs generates meaningless and uninterpretable images. Interestingly, these action-oriented reconstructions are classified with approximately 85% of accuracy, which suggests that the underlying structure of these reconstructions is preserving some important invariances and symmetries of the data. On the contrary, the full reconstruction might carry irrelevant information that is non-task specific, which could explain why they are more difficult to classify.

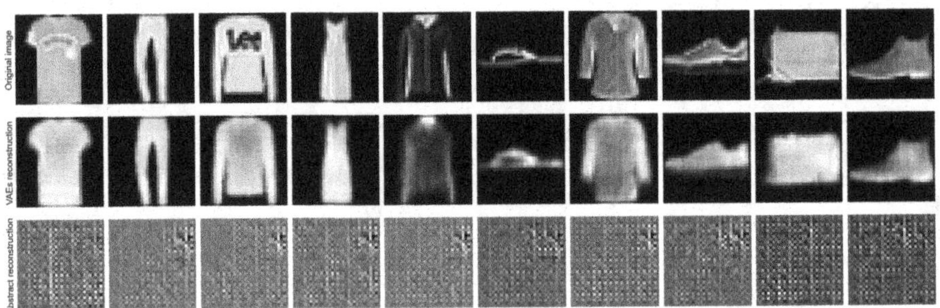

Fig. 4. Data reconstruction by VAEs and action-centric VAEs.

6 Discussion

Agents need to navigate complex environments with limited biological information processing. Under this circumstance, an optimal perceptual system has to efficiently allocate cognitive resources to transmit the relevant sensory information to achieve successful behaviour. Thus, the goal of perception is not to generate faithful reconstructions of the sensory input, but abstract representations that are useful downstream.

A common approach to representations in Artificial Intelligence and Neuroscience is that they should be in service of fully reconstructing the data. However, such representations will carry irrelevant information for downstream tasks that only depend on the exploitation of specific invariances and symmetries of the data.

In this work, we explore useful efficient coding within the framework of rate-distortion to explore optimal information processing for task-dependent contexts. We have provided a formal definition of abstractions that can be used to learn action-centric representations whose main function is to capture the task-dependant invariances in the data. Such lossy compressions of the data lie near optimal points of the rate-distortion curve. Crucially, we show that action-centric representations i) are efficient lossy compressions of the data; ii) capture the task-dependent invariances necessary to achieve adaptive behaviour; and iii) are not in service of reconstructing the data. This could shed some light on how organisms are not optimized to reconstruct their environment; instead, their representational system is tuned to convey action-relevant information.

Interestingly, our work resonates with recent research on multimodal learning such as the joint embedding predictive architecture and multiview systems [2, 7]. The main objective of these models is to obtain representations that are useful downstream but from which it's not possible to reconstruct the data. These representations learn the relevant invariances by maximizing only the information shared across different views or modalities of the data. We argue that action-centric representations operate in a similar way, as shared information across views is an implicit way to define a query (see Appendix D).

An interesting line of research is to explore faithful reconstruction in the context of fine-grained queries such as pixel predictability. We hypothesize that, as the number of pixel-specific queries approaches the pixel space of the image, the abstract representation might allow for faithful reconstruction of the data. Although that could be to the detriment of worse performance on downstream tasks.

In conclusion, this work sets a promising line of research in the field of representational theory by understanding representations not as faithful reconstructions of the data but as action-driven entities.

A Model Details

The classifier used to implement the query is a deep convolutional network (CNN) with three convolutional layers. The number of filters for the first layer is 16, and it is doubled in each layer. The kernel size is 3 in all layers, and padding is set to 1, also in all layers. Stride is 1 in the first two layers, and 2 in the third one. In addition, batch normalization is applied in each layer; 16 for the first one, and doubled in each layer. The activation function in each layer is ReLU, and max pooling is applied in the first two layers, both with a kernel size of 2, and stride of 2 in the first and 1 in the second. Between the first two fully connected layers it is used a dropout of 0.2. The number of neurons for the fully connected layers is 512, 128, and 10. We use the Adam optimizer with a learning rate of 0.001. We trained the classifier for 15 epochs with a batch size of 64.

Regarding the VAEs, the encoder is a CNN of 4 layers with the same parameters as the CNN. Every VAEs trained has 8 latent dimensions and are trained for 20 epochs using a batch size of 64. In the case of the vanilla VAEs, the β used to draw the rate-distortion curve are 100, 40, 20, 10, 5, 1, 0.5, 0.1, and 0.01. For the custom VAEs, the β values are 6e−2, 3e−2, 1e−2, 6e−3, 3e−3, 1e−3, 5e−4, 1e−4, 1e−5, 1e−6.

B Latent Space of VAEs

PCA to explore and show the latent space of the vanilla and action-centric VAEs that achieve a good performance downstream (Fig. 5):

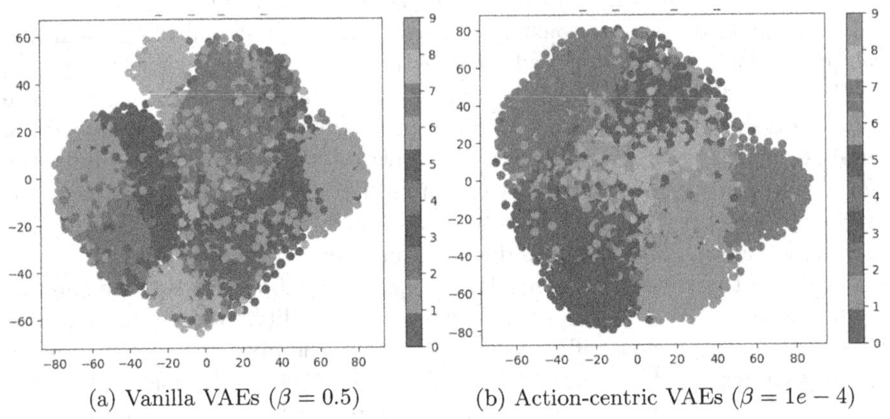

(a) Vanilla VAEs ($\beta = 0.5$)

(b) Action-centric VAEs ($\beta = 1e-4$)

Fig. 5. Latent spaces of the two types of VAEs.

As can be seen, our VAEs achieves a compact meaningful encoding of the data, with an apparent better separability among classes than the vanilla VAEs.

C ELBO and RDT

One way to derive the upper bound on mutual information from the complexity term of the ELBO is:

$$\begin{aligned}
\mathop{\mathbb{E}}_{p(x)}\left[D_{KL}[q(z|x)||p(z)]\right] &= \mathop{\mathbb{E}}_{p(x)}\left[\int q(z|x)\ln\frac{q(z|x)}{p(z)}dxdz\right] && (25)\\
&= \mathop{\mathbb{E}}_{p(x)}\left[\int q(z|x)\ln\frac{q(z|x)q(z)}{p(z)q(z)}dxdz\right] && (26)\\
&= \mathop{\mathbb{E}}_{p(x)}\left[\int q(z|x)\ln\frac{q(z|x)}{q(z)}dxdz + \int q(z|x)\ln\frac{q(z)}{p(z)}dxdz\right] && (27)\\
&= \int q(z|x)q(x)\ln\frac{q(z|x)}{q(z)}dxdz + \int q(z|x)q(x)\ln\frac{q(z)}{p(z)}dxdz && (28)\\
&= \int q(x,z)\ln\frac{q(x,z)}{q(x)q(z)}dxdz + \int q(z)\ln\frac{q(z)}{p(z)}dz && (29)\\
&= I(X;Z) + D_{KL}[q(z)||p(z)] && (30)\\
&\geq I(X;Z) && (31)
\end{aligned}$$

Another way to derive this upper bound is by splitting the expected complexity into conditional entropy and entropy terms:

$$\underset{p(x)}{\mathbb{E}}\left[D_{KL}[q(z|x)||p(z)]\right] = \underset{p(x)}{\mathbb{E}}\left[\int q(z|x)\ln\frac{q(z|x)}{p(z)}dxdz\right] \tag{32}$$

$$= \int q(z|x)q(x)\ln q(z|x)dxdz - \int q(z|x)q(x)\ln p(z)dxdz \tag{33}$$

$$= \int q(x,z)\ln q(z|x)dxdz - \int q(z)\ln p(z)dz \tag{34}$$

In the last equation, we can see that the first term is the negative conditional entropy $-H(Z|X)$ which is one of the two terms in which the mutual information is decomposed: $I(Z;X) = H(Z) - H(Z|X)$. To get the entropy $H(Z)$ we need to replace $p(z)$ by an approximate distribution $q(z)$. By Jensen's inequality, we know that $D_{KL}[q(z)||p(z)] \geq 0$, therefore, we know that:

$$\int q(z)\ln q(z) - \int q(z)\ln p(z) \geq 0 \tag{35}$$

$$\int q(z)\ln q(z) \geq \int q(z)\ln p(z) \tag{36}$$

Replacing that term in the previous expression (34) we get:

$$\int q(x,z)\ln q(z|x)dxdz - \int q(z)\ln p(z)dz \geq \int q(x,z)\ln q(z|x)dxdz - \int q(z)\ln q(z)dz \tag{37}$$

$$\int q(x,z)\ln q(z|x)dxdz - \int q(z,x)\ln q(z)dxdz \geq \int q(x,z)\ln\frac{q(x,z)}{q(x)q(z)}dxdz = I(X;Z) \tag{38}$$

D Multiview Architectures and Queries

Given a query $Q(X)$ over X in a multiview scenario it can be understood as the subset of information contained in the intersection of X and $t(X)$ such that:

$$Q(X) \in X \cap t(X) \tag{39}$$

as the transformation t only preserves those symmetries relevant for the query (i.e., relevant to solve a set of tasks that only depend on those invariances). Therefore, the relevant query in a multiview scenario can be defined as:

$$Q(X) = p(X, t(X)) \tag{40}$$

Mutual information between X and Z and between $Q(X)$ and Z is (assuming that X, X' and Z form a dag where Z only depends on X):

$$I(X;Z) = \int p(x,z) \ln \frac{p(x,z)}{p(x)p(z)} dxdz \tag{41}$$

$$I(Q(X);Z) = \int p(q,z) \ln \frac{p(q,z)}{p(q)p(z)} dqdz \tag{42}$$

$$= \int p(x,x',z) \ln \frac{p(x,x',z)}{p(x,x')p(z)} dxdx'dz \tag{43}$$

$$= \int p(x,x',z) \ln \frac{p(x')p(x|x')p(z|x)}{p(x')p(x|x')p(z)} dxdx'dz \tag{44}$$

$$= \int p(x,x',z) \ln \frac{p(z|x)}{p(z)} dxdx'dz \tag{45}$$

$$= \int p(x,x')p(z|x) \ln \frac{p(x,z)}{p(x)p(z)} dxdx'dz \tag{46}$$

$$= \int p(x)\frac{p(x,z)}{p(x)} \ln \frac{p(x,z)}{p(x)p(z)} dxdz \tag{47}$$

$$= \int p(x,z) \ln \frac{p(x,z)}{p(x)p(z)} dxdz \tag{48}$$

$$= I(X;Z) = I(X;X') \tag{49}$$

The mutual information between the latent Z and one of the views X is equal to the mutual information between the query distribution $Q(X)$ and the latent Z. As the mutual information between an optimal lossy representation and its corresponding view is equal to the mutual information between views, then, the information conveyed by the query is the one shared by the views. This shows that the multiview architecture is essentially a query-oriented system where the transformations applied to the data keep specific invariances with respect to a set of implicit queries of interest.

References

1. Alemi, A., Poole, B., Fischer, I., Dillon, J., Saurous, R.A., Murphy, K.: Fixing a broken ELBO. In: International Conference on Machine Learning, pp. 159–168 (2018)
2. Bardes, A., Ponce, J., LeCun, Y.: VICReg: variance-invariance-covariance regularization for self-supervised learning. arXiv preprint arXiv:2105.04906 (2021)
3. Barlow, H.B.: Possible principles underlying the transformation of sensory messages. Sensory Commun. **1**(01), 217–233 (1961)
4. Buckley, C.L., Kim, C.S., McGregor, S., Seth, A.K.: The free energy principle for action and perception: a mathematical review. J. Math. Psychol. **81**, 55–79 (2017)
5. Burgess, C.P., et al.: Understanding disentangling in β-VAE. arXiv preprint arXiv:1804.03599 (2018)
6. Cover, T., Thomas, J.: Elements of Information Theory. Wiley, New York (2006)

7. Federici, M., Dutta, A., Forré, P., Kushman, N., Akata, Z.: Learning robust representations via multi-view information bottleneck. arXiv preprint arXiv:2002.07017 (2020)
8. Friston, K.: The free-energy principle: a rough guide to the brain? Trends Cogn. Sci. **13**(07), 293–301 (2009)
9. Friston, K.: The free-energy principle: a unified brain theory? Nat. Rev. Neurosci. **11**(2), 127–138 (2010)
10. Friston, K.: A free energy principle for biological systems. Entropy **14**(11), 2100–2121 (2012)
11. Genewein, T., Leibfried, F., Grau-Moya, J., Braun, D.A.: Bounded rationality, abstraction, and hierarchical decision-making: an information-theoretic optimality principle. Front. Robot. AI **2**(27) (2015)
12. Hoffman, M.D., Johnson, M.J.: ELBO surgery: yet another way to carve up the variational evidence lower bound. In: Workshop in Advances in Approximate Bayesian Inference, NIPS, vol. 1, no. 2 (2016)
13. Laughlin, S.: A simple coding procedure enhances a neuron's information capacity. Zeitschrift für Naturforschung c **36**(9–10), 910–912 (1981)
14. Lieder, F., Griffiths, T.L.: Resource rational analysis: understanding human cognition as the optimal use of limited computational resources. Behav. Brain Sci. **47** (2020)
15. Manookin, M.B., Rieke, F.: Two sides of the same coin: efficient and predictive neural coding. Ann. Rev. Vis. Sci. (9) (2023)
16. Millidge, B.: Towards a mathematical theory of abstraction. arXiv preprint arXiv:2106.01826 (2021)
17. Millidge, B., Seth, A., Buckley, C.L.: Predictive coding: a theoretical and experimental review. arXiv preprint arXiv:2107.12979 (2021)
18. Park, I.M., Pillow, J.W.: Bayesian efficient coding. BioRxiv 178418 (2017)
19. Shannon, C.E.: Coding theorems for a discrete source with a fidelity criterion. IRE Nat. Conv. **4**(142–163) (1959)
20. Sims, C.R.: Rate-distortion theory and human perception. Cognition **152**(46), 181–198 (2016)
21. Tschantz, A., Seth, A.K., Buckley, C.L.: Learning action-oriented models through active inference. PLoS Comput. Biol. **16**(4) (2009)
22. de Wit, M.M., de Vries, S., van der Kamp, J., Withagen, R.: Affordances and neuroscience: steps towards a successful marriage. Neurosci. Biobehav. Rev. **80**, 622–629 (2017)
23. Zhou, D., et al.: Efficient coding in the economics of human brain connectomics. Netw. Neurosci. **6**(1), 234–274 (2022)

Integrating Cognitive Map Learning and Active Inference for Planning in Ambiguous Environments

Toon Van de Maele[1(✉)], Bart Dhoedt[1], Tim Verbelen[2], and Giovanni Pezzulo[3]

[1] IDLab, Department of Information Technology, Ghent University - imec, Ghent, Belgium
toon.vandemaele@ugent.be
[2] VERSES Research Lab, Los Angeles, USA
[3] Institute of Cognitive Sciences and Technologies, National Research Council, Rome, Italy

Abstract. Living organisms need to acquire both cognitive maps for learning the structure of the world and planning mechanisms able to deal with the challenges of navigating ambiguous environments. Although significant progress has been made in each of these areas independently, the best way to integrate them is an open research question. In this paper, we propose the integration of a statistical model of cognitive map formation within an active inference agent that supports planning under uncertainty. Specifically, we examine the clone-structured cognitive graph (CSCG) model of cognitive map formation and compare a naive clone graph agent with an active inference-driven clone graph agent, in three spatial navigation scenarios. Our findings demonstrate that while both agents are effective in simple scenarios, the active inference agent is more effective when planning in challenging scenarios, in which sensory observations provide ambiguous information about location.

Keywords: Cognitive map · Active inference · Navigation · Planning

1 Introduction

Cognitive maps [1] are mental representations of spatial and conceptual relationships. They are considered essential components for intelligent reasoning and planning, as they are often associated with navigation in humans and rodents [2]. For this reason, a lot of recent developments in both neuroscience and computer science have been building computational models of cognitive maps [3].

These advances in the field [4,5] are very impressive in learning abstract representations and even show that biological patterns such as grid cells [4], or

T. Verbelen and G. Pezzulo—Equal Contribution.

C. L. Buckley et al. (Eds.): IWAI 2023, CCIS 1915, pp. 204–217, 2024.
https://doi.org/10.1007/978-3-031-47958-8_13

splitter cells [5] can emerge from learning. However, these works typically do not focus on complex planning tasks and only consider naive or greedy strategies.

In this paper, we investigate the potential of active inference as a planning mechanism for these cognitive maps. Active inference is a corollary of the free energy principle which states that intelligent agents infer actions that minimize their expected free energy. This is a proxy or bound on expected surprise, yielding a natural trade-off between exploration and goal-driven exploitation [6,7]. We aim to investigate the impact of active inference as a planning mechanism on the performance of cognitive maps in spatial navigation strategies, especially in terms of disambiguating the "mental position" and decision-making efficiency.

In particular, we look at the clone-structured cognitive graph (CSCG) [5]: a unifying model for two essential properties of cognitive maps. First, flexible planning behavior, i.e. if observations are not consistent with the expected observation in the plan, the plan can be adapted. Second, the model is able to disambiguate aliased observations depending on the context in which it is encountered, e.g. in spatial alternation tasks at the same location different decisions are made depending on context [8]. Given the CSCG's inherent mechanism for disambiguating aliased observations, we hypothesize that coupling it with active inference as a planning system will enable the identification of the optimal sequence that accurately represents the agent's location.

To investigate this hypothesized benefit of active inference, we compare both a naive clone graph and an active inference-driven clone graph for navigating toward goals on two separate metrics: the number of steps it takes for an agent to reach the goal and the overall success rate. We design three distinct spatial navigation scenarios, each with a different complexity. First, we consider a slightly ambiguous (open room) environment described by [5] where we evaluate the structure learning mechanism and planning algorithms for both models. We then increase the level of ambiguity in a maze described in [9] where we believe that information-seeking behavior will be crucial for self-localization. Finally, we evaluate the performance in the T-maze, where an agent is punished for making the wrong choice by ending the episode. To summarize, the contributions of this paper are: (i) we show how to use the learned structure of a CSCG as the generative model within the active inference framework, (ii) we show that active inference agents are significantly faster in disambiguating the state in highly ambiguous environments than greedy planning agents, and (iii) we show that active inference agents make more careful decisions by first gathering evidence, yielding higher success rates for finding the reward in the T-maze environment.

2 Methods

In this section, we first describe the mechanisms driving standard clone-structured cognitive graphs for structure learning. Then we provide a brief summary of the active inference framework and how the action is driven through Bayesian inference. Finally, we conclude this section by showing how the CSCG can be used as a generative model within the active inference framework.

2.1 Clone-Structured Cognitive Graphs

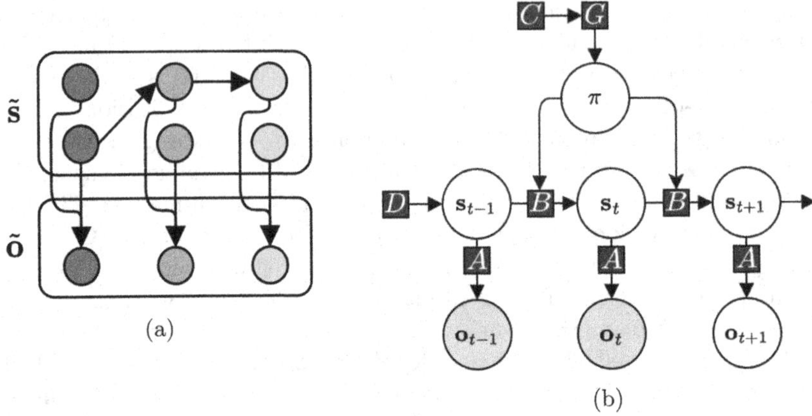

Fig. 1. (a) A mapping of a sequence of observations to distinct clone states in the clone-structured cognitive graph. The color indicates clones belonging to a specific observation, i.e. for each colored observation there are two clones states from which it can transition into either clone state belonging to the next observation. (b) The factor graph describing an active inference driven partially observable Markov decision process (POMDP). π denotes the policy, which is sampled according to the expected free energy G, dependent on the preference matrix C. The hidden states of the agent $\mathbf{s}_t$ are initialized using the prior matrix D. These states are then transitioned according to the B matrix, conditioned on the selected policy. Finally, the observed outcome variables are generated through the likelihood factor (A matrix). Observed variables are denoted in light blue circles, while unobserved variables are denoted in white circles. The factors describing the generative model are denoted in a dark blue square. (Color figure online)

Clone-structured cognitive graphs (CSCG) [5] are a computational implementation of a cognitive map that models the joint probability of a sequence of action and observation pairs. They are a variation of the action-augmented hidden Markov model, where the next state and action are conditioned on the current state and action. The crucial difference is that these clone-structured cognitive graphs are able to disambiguate aliased observations based on the context (e.g. the previously visited trajectory), which is a property that is also observed in hippocampal splitter cells.

In order for a CSCG to be able to disambiguate observations, it needs distinct states for each observation based on its context - in this case, the previous observations and actions. All states corresponding to a single observation are called the clones of this observation, and by design, each state deterministically maps to a single observation. In essence, a CSCG is a hidden Markov model in which multiple different values of the hidden state predict identical observations (i.e. their corresponding columns in the transition matrix are non-identical). A pair of the clone states in a CSCG is therefore a set of two values that a hidden state might take which share identical likelihood contingencies, but differ in their transition probabilities. A depiction of the clone graph, as described in [5] is shown in Fig. 1a.

The CSCGs are optimized by minimizing the variational free energy over a sequence of observation-action pairs using the Baum-Welch algorithm [10], an expectation-maximization scheme for hidden Markov models. Through this optimization and random initialization, the model will converge to use distinct clone states for different sequences in the data. This distinction between clones is further improved by optimizing the learned model parameters through a Viterbi decoding step, only keeping the states necessary for the maximum likelihood paths in the learned model.

2.2 Clone Graph Agent

We define a clone graph agent that uses a greedy planning approach to select the actions. Planning using the clone-structured cognitive graphs is done by setting a fixed target state (or states), and forward propagating the messages starting from the current state. When one of the target states is assigned a non-zero probability, a path is found and the maximum likelihood states are backward propagated to retrieve the corresponding action sequence, or policy. The probability of each policy is computed as the belief over the current state $Q(\mathbf{s}|\tilde{\mathbf{o}}, \tilde{\mathbf{a}})$. Once the agent's belief over state collapses to a single state, the planning mechanism falls back to the one described in [5], where the current state is known.

2.3 Active Inference Agent

Actionable agents, whether biological or artificial, are separated by their environment through sensory inputs (perception) and action. The agent's observations are indirectly observed through its different sensory modalities, while the world state is also only indirectly affected by the agent's actions. This separation between the hidden variables (action, observation, agent state, and world state) is commonly referred to as the Markov blanket.

The free energy principle proposes that an agent possesses a generative model that describes how outcomes are generated from the world state and how the world state is affected by the agent's actions. The principle states that the agents will minimize their surprise, bounded by the variational free energy by updating the parameters of the generative model (learning) or inferring the hidden state (perception). Active inference agents can infer the action that minimizes the "expected free energy (G)" (or in other words, the free energy of the future courses of actions) [6].

Active inference assumes that actions are inferred through the minimization of the expected free energy G. This means that the posterior over a policy is proportional to the expected free energy G, which can be computed for each policy. More specifically, approximate posterior over policy $Q(\pi)$ is computed as the softmax (σ) over the categorical over all the policies with a value of the respective expected free energy G, γ is a temperature variable:

$$Q(\pi) = \sigma(-\gamma G(\pi)),$$

Where the expected free energy G of this model, for a fixed time horizon T, is defined as in [11]:

$$G(\pi) = \sum_{\tau=t+1}^{T} G(\pi, \tau)$$

$$G(\pi, \tau) \geq -\underbrace{\mathbb{E}_{Q(\mathbf{o}_\tau|\pi)}\big[D_{KL}[Q(\mathbf{s}_\tau|\mathbf{o}_\tau, \pi)||Q(\mathbf{s}_\tau|\pi)]\big]}_{\text{Epistemic value}} - \underbrace{\mathbb{E}_{Q(\mathbf{o}_\tau|\pi)}\big[\log P(\mathbf{o})\big]}_{\text{Pragmatic Value}}$$

This equation decomposes in two distinct terms: an epistemic value computing the information gain term over the belief over the state, and a pragmatic value (or utility) term with respect to a preferred distribution over the observation $P(\mathbf{o})$. In active inference, the goal of an agent is encoded in this prior belief as a preference. In a CSCG, planning is done by setting a preferred state, whereas in active inference this is typically done by setting the preferred observation. In order to make both approaches comparable, here we always plan by setting preferred states (and assume an identity mapping between the state and observation).

Evaluating the expected free energy G for all the considered policies is exponential w.r.t. the time horizon T. This limits the tree depth to low values for which this is practically computable. To mitigate this limitation, we set the preference for each state proportional to the distance toward the goal state (in the cognitive map). While this system simplifies computing the utility to be sufficient for a depth of one, the planning mechanism still requires larger depths for achieving (non-greedy) long-term information-seeking behavior.

CSCG as the Generative Model for Active Inference. We consider active inference in the discrete state space formulation [12], as shown in the factor graph in Fig. 1b. The generative model is therefore described by a set of four specific matrices: the A matrix defines the likelihood model, or how observations are generated from states: $P(\mathbf{o}|\mathbf{s})$, the B matrix defines the transition model, or how the belief over state changes conditioned on an action $\mathbf{a}_t$: $P(\mathbf{s}_{t+1}|\mathbf{s}_t, \mathbf{a}_t)$. The C matrix describes the preference of the agent $P(\mathbf{s})$, and finally, the D matrix describes the prior belief over the initial state $P(\mathbf{s})$.

First, we learn the world structure using a CSCG through the minimization of the evidence lower bound with respect to the model parameters as described in [5]. We then map the parameters of the learned hidden Markov model to the four matrices describing the active inference model.

First, we reduce the model by only considering the states for which the transition probability marginalized over action and next state $\sum_{\mathbf{s}} \sum_{\mathbf{a}} p(s_t|\mathbf{s}, \mathbf{a})$, assuming a uniform distribution over $\mathbf{s}$ and $\mathbf{a}$, is larger than the threshold of

0.0001. The A matrix can be directly constructed by setting $P(\mathbf{o}_i|\mathbf{s}_j) = 1$ for all remaining clones $\mathbf{s}_j$ of observation $\mathbf{o}_i$.

To construct the B matrix, the transition matrix from the trained CSCG can be taken directly. A crucial difference between the POMPD in discrete time active inference and the CSCG is that the actions are state-conditioned in the latter. This means that starting in some states, an action can not be taken. In the learned transition matrix, the following condition does not always hold: $\sum_{\mathbf{s}_{t+1}} P(\mathbf{s}_{t+1}|\mathbf{s}_t, \mathbf{a}_t) = 1$. We convert this transition matrix to proper probabilities by adding a novel dispreferred state $\mathbf{s}_d$, for which we set the transition probability to 1 in these illegal cases, and for which this state transitions to itself for each possible action. We then normalize the transition matrix such that probabilities sum to 1. We also add a $P(\mathbf{o}_d|\mathbf{s}_d) = 1$ mapping in the A matrix.

The preference of the agent, or C matrix, is not present in the standard formulation of the CSCG. However, the agent is able to plan toward a goal that is set in state space. We model this by setting a preference over this state, or set of states in case of an observation-space preference or multiple target goals. Additionally, for the newly added state $\mathbf{s}_d$ to which the illegal actions are mapped, we set a very low value (as if it would drive you to a state that is farther away from the goal than the maximum distance) in order to drive the agent to avoid these actions when planning according to its expected free energy.

The prior distribution over the initial state, matrix D, is initialized as a uniform prior over all the states. The agent thus starts with no knowledge about the state it is in and has to gather evidence to change this belief.

3 Results

In this work, we compare the behavior of two agents that select their actions using a CSCG: the former ("clone graph", Sect. 2.2) agent plans using a greedy approach, whereas the latter ("active inference", Sect. 2.3) agent uses active inference and expected free energy to plan ahead. We also compare these two agents with a random ("random") agent baseline. In particular, we look at goal-driven behavior in three distinct environments each requiring a different level of information-seeking behavior. First, we consider an open room as proposed in [5] in which the agent has to reach a uniquely defined corner, for which the goal is provided as a goal observation. Second, we consider a more ambiguous environment in which the agent has to reach the uniquely defined center of a room, but it first needs to localize itself within the room. Finally, we evaluate the approach on the T-maze, where the agent should first observe a cue, as a wrong decision is "fatal".

In each experiment, we first train the generative models as CSCGs and then convert them to discrete state space matrices for active inference within the PyMDP framework [11].

3.1 Navigating in an Open Room Environment

In this first experiment, we investigate the performance of all agents in a simple environment where we hypothesize that there is no immediate gain in using the active inference framework for information-seeking behavior. As the clone graph agent is still able to integrate observations to improve its belief over its current state, we expect both agents to gather enough evidence to accurately plan toward the goal.

For this maze, we consider an open room environment based on the one described in [5]. We recreate the environment within the Minigrid [13] framework. The room is defined by a four-by-four grid in which the agent can freely navigate by selecting actions like "turn left", "turn right" or "move forward". The agent observes a three-by-three patch around its current position, as shown in Fig. 2b. Each corner of the environment is uniquely defined by an observable colored patch, as shown in Fig. 2a and Fig. 2b. Each observed patch is mapped to a unique index as observation. In this environment, this corresponds to 21 observations.

We learn the structure of the room by first training a CSCG, initialized with 20 clones for each observation, as described in Sect. 2. The model parameters were learned using a random-walk sequence consisting of 100k observation-action pairs. We then set the preference of the agent to the two observations reaching the corner, e.g. for the bottom right corner this is the observation of reaching it from the left and from the top. As described in Sect. 2, we select the clone states for which the likelihood of this observation is 1 and set the preference for all these states for both the clone graph and active inference planning schemes.

We run an experiment for all three agents where the agent starts in a random (ambiguous, i.e. looking at the center) pose and has to reach a randomly selected corner as the goal. We run this for 400 separate trials, where each trial was seeded with the same random seed, ensuring that the different agents start with the same starting position and goal. We provide the agents with 25 timesteps to reach the goal and report the success rate and episode length for each of the agents. Qualitatively, in Fig. 2a, we observe that the behavior between the clone graph agent and the active inference agent is very similar; it first picks a corner which is either the goal and the episode ends or an informative landmark, and then the agent moves towards the goal.

Quantitatively, we observe the duration of the episode and see that the average episode length shown in Fig. 2c is significantly larger for the random agent with respect to both the clone graph agent (2-sample independent t-test, p-value $= 7.6 \cdot 10^{-6}$) and the active inference agent (2-sample independent t-test, p-value $= 3.6 \cdot 10^{-5}$), illustrating that the model has learned the structure of the world and is not moving randomly. Secondly, we observe that the average episode length of the clone graph agent does not significantly differ from the active inference agent (2-sample independent t-test, p-value $= 0.237$), illustrating that for this environment the information-seeking behavior does not benefit performance. This is further evidenced by the success rate shown in Fig. 2d, where the performance of both agents does not significantly differ as they are identical at a 100% success rate.

From this experiment, we conclude that in an environment where the agent can quickly find an unambiguous landmark such as the corners in the open room, both agents have similar performance.

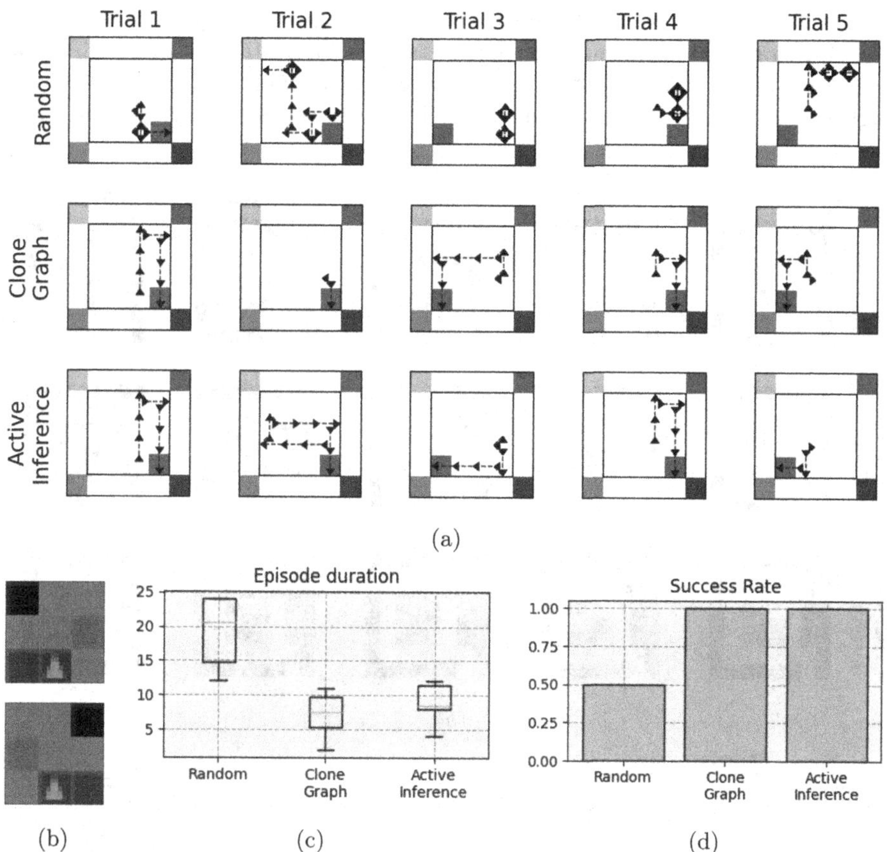

Fig. 2. (a) Qualitative results of navigating the open room maze for the different agents with different random seeds. The agent is tasked with reaching a particular corner in the maze. The trajectory of the agent is marked, and the arrow points the direction in which the agent is looking. (b) The two three-by-three observations defining a goal in a corner of the open room maze. (c) A box plot representing the statistics of the amount of time until the goal is reached (only the success scenarios are considered) over 400 trials. (d) The success rate of the agent in reaching the goal observation (computed over 400 trials).

3.2 Self-localization in an Ambiguous Maze

In the previous environment, the agent was able to quickly self-localize as random actions would easily disambiguate where in the environment they are. In this experiment, we increase the level of ambiguity and evaluate whether the active inference agent is able to self-localize faster than the clone graph agent.

For this experiment, we consider the highly ambiguous maze from Friston et al. [9] shown in Fig. 3a. In this environment, the agent is only able to observe the one-by-one tile the agent is currently standing on, i.e. if it is a red, white, or green tile. While the red and white tiles are highly ambiguous, there is only a single green tile at the center of the maze. The agent is able to navigate the maze

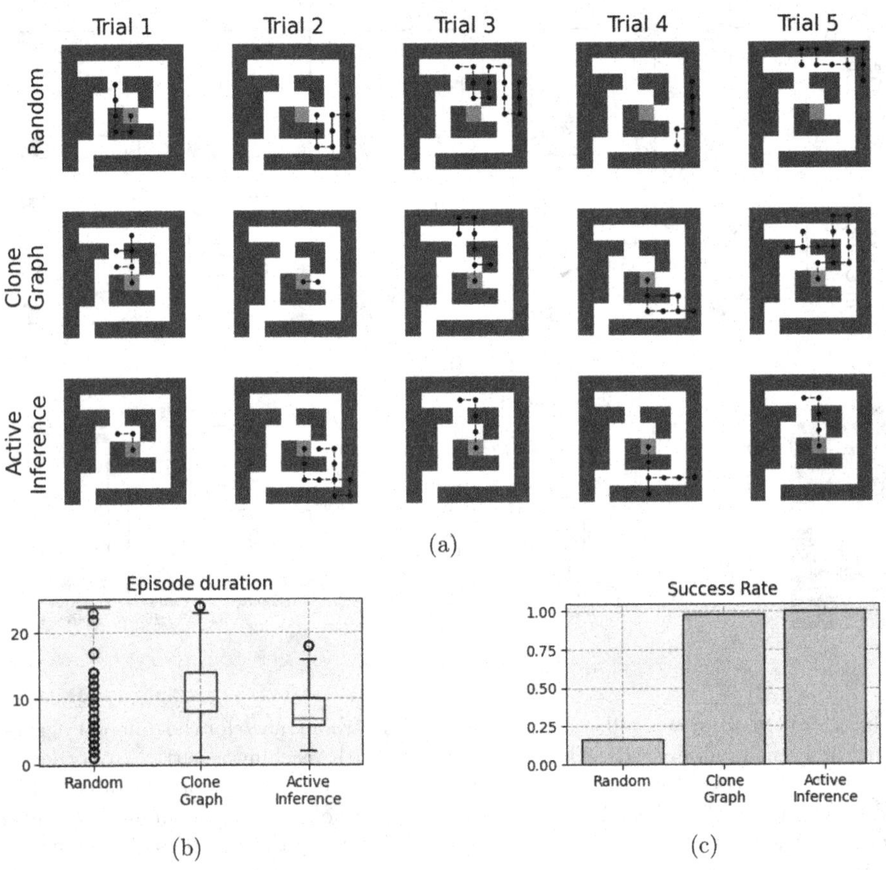

Fig. 3. (a) Qualitative results of navigating the ambiguous maze with the three different agents. The green square marks the goal observation, the trajectory of the agent is marked in black. In this maze, the agent can only observe the current tile, and the color of the tile represents the observation the agent receives. (b) Shows the amount of steps needed for reaching the target, only measured for the success cases. (c) Shows the success rate, computed over 400 trials for the three agents. (Color figure online)

through actions like "up", "down", "left" or "right", and is only limited by a wall around the maze. Unique observation tiles are again mapped to categorical indices.

We construct a CSCG with 40 clones per observation and optimize it over a sequence of 10k steps in the environment until convergence. We then set the preference for this environment as the green tile, in a similar fashion as we did in the experiment in Sect. 3.1 for both the clone graph agent and the active inference agent.

In this environment, the agent's goal is always to go to the green tile in the center of the room. However, the agent starts at a random position on a white tile. We again run this experiment for 400 trials for each agent, seeded over trials such that the starting position is the same for each agent. Each episode has a max duration of 25 steps, and we record the episode length and the success rate of the agents. Qualitatively, we can see the trajectories taken by the clone and active inference agents in Fig. 3a. We observe that both agents are able to solve the task, seemingly moving randomly in the maze. However, we also observe the random agent navigating in the maze, which typically does not reach the goal. Quantitatively, we again measure that the clone graph agent (2-sample independent t-test, p-value $= 1 \cdot 10^{-99}$) and active inference agent (2-sample independent t-test, p-value $= 1 \cdot 10^{-168}$) significantly differ from the random agent, showing goal-directed behavior. However, we now observe that the clone graph agent with a mean episode duration of 10.92 steps is significantly slower than the active inference agent with a mean episode duration of 7.92 steps (2-sample independent t-test, p-value $= 3.46 \cdot 10^{-22}$) even though their success rate is similar with 98.5% for the clone graph agent and 100% for the active inference agent.

From this experiment, we conclude that in highly ambiguous environments, agents using active inference for goal-driven behavior disambiguate their location and reach the goal faster than agents who do not.

3.3 Solving the T-Maze

In this final experiment, we consider an environment where making informative decisions is crucial. We compare the performance of the agents in the quintessential active inference environment: the T-maze [14]. In this environment, the agent must make a choice to go either in the left or the right corridor without being able to observe the location of the reward (we hide it behind a door), and the episode ends when it makes a decision. The agent is, however, able to disambiguate the location of the reward by observing a colored cue behind itself.

We create the environment again in the Minigrid environment [13], and the agent has three-by-three patches as observations and can act by either "turning left", "turning right" or "moving forward". The agent always starts in an upwards-looking position, looking away from the cue. Additionally, when the agent wants to walk through a door, it immediately goes to the tile behind the door, ending the episode either in reward or not.

We train a CSCG with 5 clones per observation on 500 distinct episodes with a maximum length of 50 steps, however, these episodes are typically shorter as the agent goes through a door. Similar to the open room environment, we map each three-by-three observation patch to a unique index and additionally, we also map the reward to a separate observation. This yields 17 unique observations the agent can observe. We then set the preference to the rewarding observation for both the clone and active inference agents, and depending on context, the agent should be able to infer a different path towards the goal.

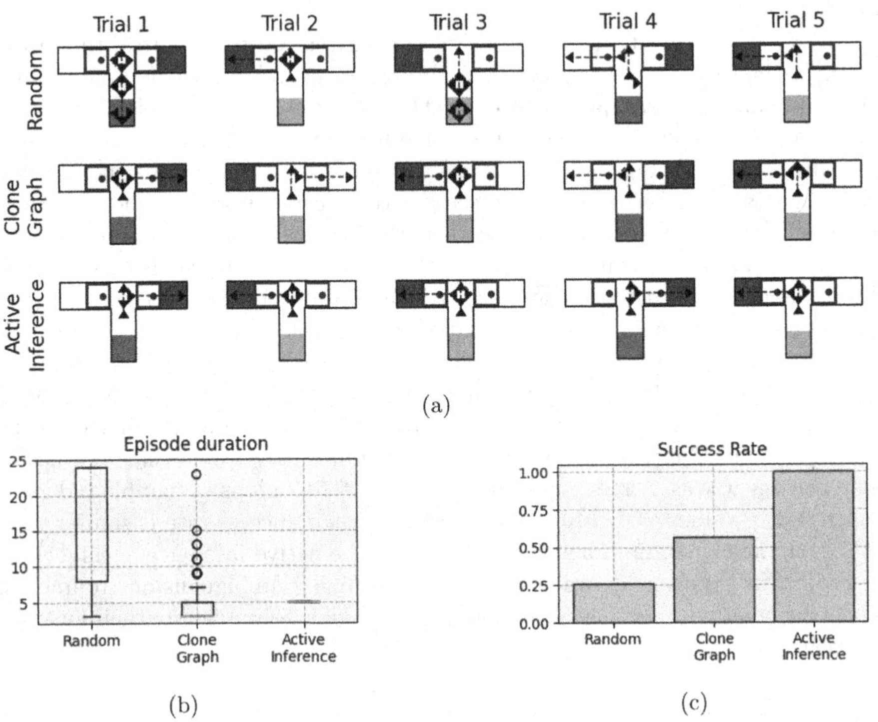

Fig. 4. (a) Qualitative results of navigating the T-maze with the three different agents. The green square marks the goal observation, and the black arrows the trajectory followed by the agent. At the bottom of the T, there is a colored cue, blue marks that the goal is on the right, while red marks that the goal is on the left. (b) Shows the number of steps needed for reaching the target, only measured for the success cases. (c) Shows the success rate, computed over 400 trials for the three agents. (Color figure online)

We again conduct 400 random trials, where the seed is again fixed for each trial within an agent, ensuring that for each trial the goal location is the same. When we evaluate the behavior of the agents qualitatively (Fig. 4a), we observe that the active inference agent always moves forward, turns around and checks for the cue, and then moves towards the correct goal location. In contrast, the clone graph agent randomly picks a direction as it has not accurately inferred in which state it currently is. Interestingly, when the stochasticity of the action sampling forces the agent to turn around and it observes the cue, it chooses the correct action. This explains the 56.75% success rate, which is slightly higher than the expected 50% of selecting actions randomly. In this environment, where thoughtless decisions are punished, the active inference agent is significantly more accurate with a success rate of 100% (2-sample independent z-test for proportions, p-value $= 6.25 \cdot 10^{-50}$). Interestingly, the clone graph agent is significantly faster with an average of 4.5 steps than the active inference agent with an average of 5 steps (2-sample independent t-test, p-value $= 2.86 \cdot 10^{-5}$). This is attributed to the fact that the agent does not take the time to observe the cue and moves towards wherever it believes the goal is.

From this experiment, we conclude that in information-critical decision-making environments using active inference provides a significant benefit over greedy planning strategies.

4 Discussion

We relate our work to representation learning in complex environments. In the context of learning cognitive maps, work has been done that explicitly separates the underlying spatial structure of the environments with the specific items observed [4]. While this model does not entail a generative model, other approaches do consider the hippocampus as a generative model [15] and show that through generative processes novel plans can be created. Model-based reinforcement learning systems learn similar world models directly from pixels [16] and are able to achieve high performance on RL benchmarks. All these approaches typically treat planning as a trivial problem that can be solved through forward rollouts, or by value optimization using the Bellman equation, however, they do not consider the belief over the state as a parameter.

Within the active inference community, a lot of work has been applied to planning in different types of environments. Casting navigation as inferring the sequence of actions under the generative model using deep neural networks has been done before in [17,18], where the approximate posterior is implemented through a variational deep neural network. The active inference framework has also been successful in solving various RL benchmarks [19,20]. These approaches show that inferring action through surprise minimization is powerful in solving a wide range of tasks, although they do not explicitly deal with aliasing in observations.

We believe that the combination of both approaches can yield a promising avenue for building cognitive maps in silico that can be used to solve important real-world tasks such as navigation.

The CSCG has been shown to be a powerful model for flexible planning and disambiguating aliased observation, making it the perfect candidate for integration within the active inference framework. Through this interaction with the inherent uncertainty-resolving behavior of active inference, we have observed significant improvements in terms of success rate or episode lengths depending on the specific environment.

Another open issue that we plan to resolve in the future is the fact that the CSCG is currently learned in an offline fashion. Therefore our current approach is not benefitting from the curiosity- or novelty-based scheme of active inference [7,21], which we hypothesize to improve the training efficiency with respect to the number of required samples.

5 Conclusion

We first propose a mechanism for using the clone-structured cognitive graph within the active inference framework. This allows us to use the naturally context-dependent disambiguating of aliased observations in the generative model within the active inference framework that naturally will seek the sequence best aligned with this purpose. Through evaluation in three distinct environments, we have highlighted the advantages of active inference compared to more simplistic and greedy planning methods. We show that in naturally unambiguous environments, the active inference and clone agents perform similarly in both success rate and time to reach the goal. Additionally, we have observed that the active inference agent exhibits a significantly higher success rate in environments requiring informed decision-making. Finally, we show that in environments where an agent has to make an informed decision, the active inference agent has a significantly higher success rate. These results corroborate the benefits of using an active inference approach.

Acknowledgments. This research received funding from the Flemish Government (AI Research Program). This research was supported by a grant for a research stay abroad by the Flanders Research Foundation (FWO) grant V408223N.

References

1. O'Keefe, J., Nadel, L.: Précis of O'Keefe & Nadel's The hippocampus as a cognitive map. Behav. Brain Sci. **2**, 487–494 (1979)
2. Peer, M., Brunec, I.K., Newcombe, N.S., Epstein, R.A.: Structuring knowledge with cognitive maps and cognitive graphs. Trends Cogn. Sci. **25**, 37–54 (2021)
3. Whittington, J.C.R., McCaffary, D., Bakermans, J.J.W., Behrens, T.E.J.: How to build a cognitive map. Nat. Neurosci. **25**, 1257–1272 (2022)
4. Whittington, J.C., et al.: The Tolman-Eichenbaum machine: unifying space and relational memory through generalization in the hippocampal formation. Cell **183**, 1249–1263.e23 (2020)
5. George, D., Rikhye, R.V., Gothoskar, N., Guntupalli, J.S., Dedieu, A., Lázaro-Gredilla, M.: Clone-structured graph representations enable flexible learning and vicarious evaluation of cognitive maps. Nat. Commun. **12**, 2392 (2021)
6. Parr, T., Pezzulo, G., Friston, K.J.: Active Inference: The Free Energy Principle in Mind, Brain, and Behavior. The MIT Press, Cambridge (2022)
7. Schwartenbeck, P., Passecker, J., Hauser, T.U., FitzGerald, T.H., Kronbichler, M., Friston, K.J.: Computational mechanisms of curiosity and goal-directed exploration. eLife **8**, e41703 (2019)

8. Jadhav, S.P., Kemere, C., German, P.W., Frank, L.M.: Awake hippocampal sharp-wave ripples support spatial memory. Science **336**, 1454–1458 (2012)
9. Friston, K., Da Costa, L., Hafner, D., Hesp, C., Parr, T.: Sophisticated Inference (2020). Publisher: arXiv Version Number: 1
10. Wu, C.F.J.: On the convergence properties of the EM algorithm. Ann. Stat. **11**(1), 95–103 (1983)
11. Heins, C., et al.: pymdp: a Python library for active inference in discrete state spaces (2022). Publisher: arXiv Version Number: 2
12. Da Costa, L., Parr, T., Sajid, N., Veselic, S., Neacsu, V., Friston, K.: Active inference on discrete state-spaces: a synthesis. J. Math. Psychol. **99**, 102447 (2020)
13. Chevalier-Boisvert, M., Willems, L., Pal, S.: Minimalistic gridworld environment for gymnasium (2018)
14. Friston, K., FitzGerald, T., Rigoli, F., Schwartenbeck, P., Pezzulo, G.: Active inference: a process theory. Neural Comput. **29**, 1–49 (2017)
15. Stoianov, I., Maisto, D., Pezzulo, G.: The hippocampal formation as a hierarchical generative model supporting generative replay and continual learning. Prog. Neurobiol. **217**, 102329 (2022)
16. Hafner, D., Lillicrap, T., Ba, J., Norouzi, M.: Dream to control: learning behaviors by latent imagination (2020). arXiv:1912.01603 [cs]
17. Çatal, O., Wauthier, S., De Boom, C., Verbelen, T., Dhoedt, B.: Learning generative state space models for active inference. Front. Comput. Neurosci. **14**, 574372 (2020)
18. Çatal, O., Verbelen, T., Van De Maele, T., Dhoedt, B., Safron, A.: Robot navigation as hierarchical active inference. Neural Netw. **142**, 192–204 (2021)
19. Tschantz, A., Baltieri, M., Seth, A.K., Buckley, C.L.: Scaling active inference (2019). Publisher: arXiv Version Number: 1
20. Fountas, Z., Sajid, N., Mediano, P.A.M., Friston, K.: Deep active inference agents using Monte-Carlo methods (2020). arXiv:2006.04176 [cs, q-bio, stat]
21. Kaplan, R., Friston, K.J.: Planning and navigation as active inference. Biol. Cybern. **112**, 323–343 (2018)

Relative Representations for Cognitive Graphs

Alex B. Kiefer[1,2(✉)] and Christopher L. Buckley[1,3]

[1] VERSES Research Lab, Los Angeles, CA, USA
akiefer@gmail.com
[2] Monash University, Melbourne, Australia
[3] Sussex AI Group, Department of Informatics, University of Sussex, Brighton, UK

Abstract. Although the latent spaces learned by distinct neural networks are not generally directly comparable, even when model architecture and training data are held fixed, recent work in machine learning [13] has shown that it is possible to use the similarities and differences among latent space vectors to derive "relative representations" with comparable representational power to their "absolute" counterparts, and which are nearly identical across models trained on similar data distributions. Apart from their intrinsic interest in revealing the underlying structure of learned latent spaces, relative representations are useful to compare representations across networks as a generic proxy for convergence, and for zero-shot model stitching [13].

In this work we examine an extension of relative representations to discrete state-space models, using Clone-Structured Cognitive Graphs (CSCGs) [16] for 2D spatial localization and navigation as a test case in which such representations may be of some practical use. Our work shows that the probability vectors computed during message passing can be used to define relative representations on CSCGs, enabling effective communication across agents trained using different random initializations and training sequences, and on only partially similar spaces. In the process, we introduce a technique for zero-shot model stitching that can be applied *post hoc*, without the need for using relative representations during training. This exploratory work is intended as a proof-of-concept for the application of relative representations to the study of cognitive maps in neuroscience and AI.

Keywords: Clone-structured cognitive graphs · Relative representations · Representational similarity

1 Introduction

In this short paper we explore the application of relative representations [13] to discrete (graph-structured) models of cognition in the hippocampal-entorhinal system — specifically, Clone-Structured Cognitive Graphs (CSCGs) [16]. In the first two sections we introduce relative representations and their extension to

C. L. Buckley et al. (Eds.): IWAI 2023, CCIS 1915, pp. 218–236, 2024.
https://doi.org/10.1007/978-3-031-47958-8_14

discrete latent state spaces via continuous messages passed on graphs. We then introduce CSCGs and their use in SLAM (Simultaneous Localization And Mapping). Finally, we report preliminary experimental results using relative representations on CSCGs showing that (a) relative representations can indeed be applied successfully to model the latent space structure of discrete, graph-like representations such as CSCGs, and more generally POMDPs such as those employed in discrete active inference modeling [1,8]; (b) comparison of agents across partially disparate environments reveals important shared latent space structure; and (c) it is possible to use the messages or beliefs (probabilities over states) of one agent to reconstruct the corresponding belief distributions of another via relative representations, without requiring the use of relative representations during training. These examples illustrate an extension of existing representational analysis techniques developed within neuroscience [10], which we hope will prove applicable to the study of cognitive maps in biological agents.

2 Relative Representations

Relative representation [13] is a technique recently introduced in machine learning that allows one to map the intrinsically distinct continuous latent space representations of different models to a common shared representation identical (or nearly so) across the source models, so that latent spaces can be directly compared, even when derived from models with different architectures. The technique is conceptually simple: given anchor points $\mathcal{A} = [\mathbf{x}_1, \mathbf{x}_2, ..., \mathbf{x}_N]$ sampled from a data or observation space and some similarity function sim (e.g. cosine similarity)[1], the relative representation $\mathbf{r}_i^M$ of datapoint $\mathbf{x}_i$ with respect to model M can be defined in terms of M's latent-space embeddings $\mathbf{e}_i^M = f_{enc_M}(\mathbf{x}_i)$ as:

$$\mathbf{r}_i^M = [sim(\mathbf{e}_i^M, \mathbf{e}_{a_1}^M), sim(\mathbf{e}_i^M, \mathbf{e}_{a_2}^M), ..., sim(\mathbf{e}_i^M, \mathbf{e}_{a_N}^M)] \quad (1)$$

where $\mathbf{e}_{a_i}^M$ is the latent representation of anchor i in M.

Crucially, the anchor points $\mathcal{A}$ must be matched across models in order for their relative representations to be compatible. "Matching" is in the simplest case simply identity, but there are cases in which it is feasible to use pairs of anchors related by a map $g(x) \to y$ (see below).

In [13] it is shown that the convergence of a model M_{target} during training is well predicted by the average cosine similarity between its relative representations of datapoints and those of an independently validated reference model M_{ref}. This is to be expected, given that there is an optimal way of partitioning the data for a given downstream task, and that distinct models trained on the same objective approximate this optimal solution more or less closely, subject to variable factors like random initialization and hyperparameter selection.

While relative representations were recently introduced in machine learning, they take their inspiration in part from prior work on representational similarity analysis (RSA) in neuroscience [4,10]. Indeed, there is a formal equivalence

[1] The selection of both suitable anchor points and similarity metrics is discussed at length in [13]. We explain our choices for these hyperparameters in Sect. 5.2 below.

between relative representations and the Representational Dissimilarity Matrices (RDMs) proposed as a common format for representing disparate types of neuroscientific data (including brain imaging modalities as well as simulated neuronal activities in computational models) in [10]. Specifically, if a similarity rather than dissimilarity metric is employed[2], then each row (or, equivalently, column) of the RDM used to characterize a representational space is, simply, a relative representation of the corresponding datapoint.

Arguably the main contribution of [13] is to exhibit the usefulness of this technique in machine learning, where relative representations may be employed as a novel type of latent space in model architectures. Given a large enough sample of anchor points, relative representations bear sufficient information to play functional roles similar to those of the "absolute" representations they model, rather than simply functioning as an analytical tool (e.g. to characterize the structure of latent spaces and facilitate abstract comparisons among systems).

The most obvious practical use of relative representations is in enabling "latent space communication": Moschella et al. [13] show that the projection of embeddings from distinct models onto the same relative representation enables "zero-shot model stitching", in which for example the encoder from one trained model can be spliced to the decoder from another (with the relative representation being the initial layer supplied as input to the decoder). A limitation of this procedure is that it depends on using a relative representation layer during training, precluding its use for establishing communication between "frozen" pretrained models. Below, we make use of a parameter-free technique that allows one to map from the relative representation space back to the "absolute" representations of the input models with some degree of success.

3 Extending Relative Representations to Discrete State-Space Models

Despite the remarkable achievements of continuous state-space models in deep learning systems, discrete state spaces continue to be relevant, both in machine learning applications, where discrete "world models" are responsible for state-of-the-art results in model-based reinforcement learning [6], and in neuroscience, where there is ample evidence for discretized, graph-like representations, for example in the hippocampal-entorhinal system [16,18,25] and in models of decision-making processes such as the POMDPs (Partially Observable Markov Decision Processes) used in active inference models [19] and elsewhere.

While typical vector similarity metrics such as cosine distance behave in a somewhat degenerate way when applied to many types of discrete representations (e.g., the cosine similarity between two one-hot vectors in the same space is 1 if the vectors are identical and 0 otherwise), they can still be usefully applied in this case (see Sect. 5 below). More generally, the posterior belief distributions inferred over discrete state spaces during simulations in agent-based models may provide suitable anchor points for constructing relative representations.

[2] See [10] fn.2.

Concretely, such posterior distributions are often derived using message-passing algorithms, such as belief propagation [14] or variational message passing [27]. We pursue such a strategy for deriving relative representations of a special kind of hidden Markov model (the Clone-Structured Hidden Markov Model or (if supplemented with actions) Cognitive Graph [16]), in which it is simple to compute forward messages which at each discrete time-step give the probability of the hidden states z conditioned on a sequence of observations o (i.e. $P(z_t|o_{1:t})$). The CSCG/CHMM is particularly interesting both because of its fidelity as a model of hippocampal-entorhinal representations in the brain and because, as in the case of neural networks, distinct agents may learn superficially distinct CSCGs that nonetheless form nearly isomorphic cognitive maps, as shown below.

4 SLAM Using Clone-Structured Cognitive Graphs

An important strand of research in contemporary machine learning and computational neuroscience has focused on understanding the role of the hippocampus and entorhinal cortex in spatial navigation [16,20,23,25], a perspective that may be applicable to navigation in more abstract spaces as well [18,21]. This field of research has given rise to models like the Tolman-Eichenbaum machine [25] and Clone-Structured Cognitive Graph [5,16]. We focus on the latter model in the present study, as it is easy to implement on toy test problems and yields a suitable representation for our purposes (an explicit discrete latent space through which messages can be propagated).

The core of the CSCG is a special kind of "clone-structured" Hidden Markov Model (CHMM) [17], in which each of N possible discrete observations are mapped deterministically to only a single "column" of hidden states by the likelihood function, i.e. $p(o|z) = \begin{cases} 1 & \text{if } z \in C(o) \\ 0 & \text{if } z \notin C(o) \end{cases}$, where $C(o)$ is the set of "clones" of observation o. The clone structure encodes the inductive bias that the same observation may occur within a potentially large but effectively finite number of contexts (i.e. within many distinct sequences of observations), where each "clone" functions as a latent representation of o in a distinct context. This allows the model to efficiently encode higher-order sequences [3] by learning transition dynamics ("lateral" connections) among the clones. CSCGs supplement this architecture with a set of actions which condition transition dynamics, creating in effect a restricted form of POMDP.

The most obvious use of CSCG models (mirroring the function of the hippocampal-entorhinal system) is to allow agents capable of moving in a space to perform SLAM (Simultaneous Localization And Mapping) with no prior knowledge of the space's topology. Starting with a random transition matrix, CSCGs trained on random walks in 2D "rooms", in which each cell corresponds to an observation, are shown in [16] to be capable of learning action-conditioned transition dynamics among hidden states that exhibit a sparsity structure precisely recapitulating the spatial layout of the room (see Fig. 1).[3]

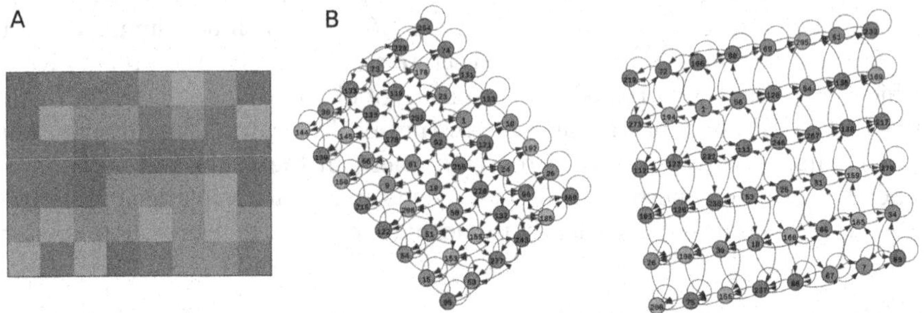

Fig. 1. Example of two cognitive graphs (B) learned by CSCG agents via distinct random walks on the same room (A). Following the convention in [16], colors indicate distinct discrete observations (in the room) or latent "clones" corresponding to those observations (in the graphs). Code for training and producing plots is provided in the supplementary materials for [16]. Note that the two graphs are obviously isomorphic upon inspection (the left graph is visually rotated about 50 degrees clockwise relative to the right one, and the node labels differ).

Given a sequence of observations, an agent can then infer states that correspond to its location in the room, with increasing certainty and accuracy as sequence length increases. Crucially, location is not an input to this model but the agent's representation of location is entirely "emergent" from the unsupervised learning of higher-order sequences of observations.

Building on the codebase provided in [16], we examined the certainty of agents' inferred beliefs about spatial location during the course of a random walk (see Fig. 2.). Though less than fully confident, such agents are able to reliably infer room location from observation sequences alone after a handful of steps. Conditioning inference as well on the equivalent of "proprioceptive" information (i.e., about which actions resulted in the relevant sequence of observations) dramatically increases the certainty of the agents' beliefs. We explored both of these regimes of (un)certainty in our experiments.

5 Experiments: Communication Across Cognitive Maps

We investigate the extent to which common structure underlying the "cognitive maps" learned by distinct CSCG agents can be exploited to enable communication across them. As in the case of neural networks trained on similar data,

[3] The training used to obtain this result is based on an efficient implementation of the Baum-Welch algorithm for E-M learning, followed by Viterbi training — please see [16] for details.

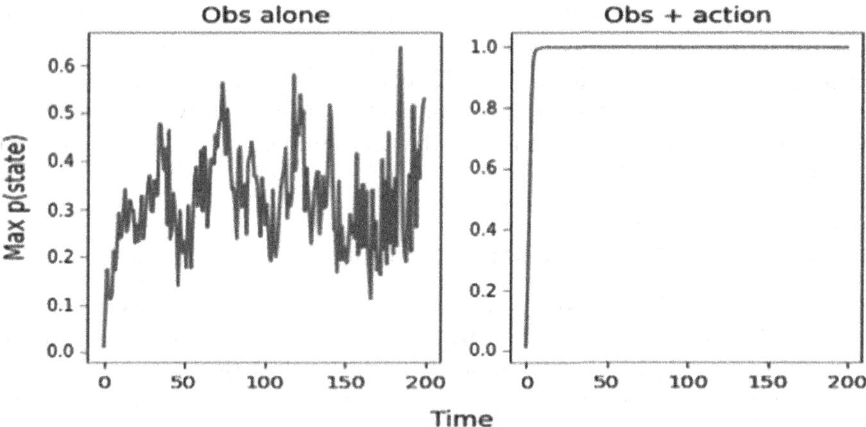

Fig. 2. Maximum probability assigned to any hidden state of a CSCG over time (during a random walk). The left panel shows confidence derived from messages inferred from observations alone, and the right panel shows the case of messages inferred from both actions and observations.

CSCG agents trained on the same room but with distinct random initializations and observation sequences learn distinct representations that are nonetheless isomorphic at one level of abstraction (i.e. when comparing the structural relationships among their elements, which relative representations make explicit — cf. Appendix B, Fig. 5).

We also explore whether partial mappings can be obtained across agents trained on somewhat dissimilar rooms. We used two metrics to evaluate the quality of cross-agent belief mappings: (1) recoverability of the maximum *a posteriori* belief of one agent at a given timestep, given those of another agent following an analogous trajectory; (2) cosine similarity between a given message and its "reconstruction" via such a mapping. The main results of these preliminary experiments are reported in Table 1.

5.1 Mapping via Permutation

We first confirmed that CSCG agents trained on distinct random walks of the same room (and with distinct random transition matrix initializations) learn functionally identical cognitive maps if trained to convergence using the procedure specified in [16]. Visualizations of the learned graphs clearly demonstrate topological isomorphism (see references as well as Fig. 1B), but in addition we found that the forward messages for a given sequence of observations are identical across agents up to a permutation (i.e., which "clones" are used to represent which observation contexts depends on the symmetry breaking induced by different random walks and initializations). It is thus possible to "translate" across such cognitive maps in a simple way. First, we obtain message sequences $\mathbf{M}$ and $\mathbf{M}'$ from the first and second CSCGs conditioned on the same observation sequence, and extract messages $\mathbf{m}$ and $\mathbf{m}'$ corresponding to some particular observation o_t. We then construct a mapping $sort\_index_{\mathbf{m}_{o_t}}(z) \rightarrow sort\_index_{\mathbf{m}'_{o_t}}(z')$ from the sort order of entries z in $\mathbf{m}$ to that of entries z' in $\mathbf{m}'$. Using this mapping, we can predict the maximum *a posteriori* beliefs in $\mathbf{M}'$ nearly perfectly given those in $\mathbf{M}$ under ideal conditions (see the "Permutation (identical)" condition in Table 1).[4]

5.2 Mapping via Relative Representations

Though it is thus relatively simple to obtain a mapping across cognitive graphs in the ideal case of CSCGs trained to convergence on identical environments, we confirm that relative representations can be used in this setting to obtain comparable results. A message $\mathbf{m}'$ from the second sequence (associated with model B) can be reconstructed from message $\mathbf{m}$ in the first (model A's) by linearly combining model B's embeddings $\mathbf{E}^B_{\mathcal{A}}$ of the anchor points, via a softmax (σ) function (with temperature T) of the relative representation $\mathbf{r}^A_{\mathbf{m}}$ of $\mathbf{m}$ derived from model A's anchor embeddings:[5]

$$\hat{\mathbf{m}'} = (\mathbf{E}^B_{\mathcal{A}})\sigma\left[\frac{\mathbf{r}^A_{\mathbf{m}}}{T}\right] \tag{2}$$

Intuitively, the softmax term scales the contribution of each vector in the set of anchor embeddings to the reconstruction $\hat{\mathbf{m}'}$ in proportion to its relative similarity to the input embedding, so that the reconstruction is a weighted superposition (convex combination) of the anchor points. The reconstruction of a sequence $\mathbf{M}'$ of m d'-dimensional messages from an analogous "source" sequence $\mathbf{M}$ of d-dimensional messages, with the "batch" relative representation operation[6] $\mathbf{R}^A_{\mathbf{M}} \in \mathbb{R}^{m\times|\mathcal{A}|}$ written out explicitly in terms of the matrix product between $\mathbf{M} \in \mathbb{R}^{m\times d}$ and anchor embeddings $\mathbf{E}^A_{\mathcal{A}} \in \mathbb{R}^{|\mathcal{A}|\times d}$, is then precisely analogous to the self-attention operation in transformers:

$$\hat{\mathbf{M}'} = \sigma\left[\frac{\mathbf{M}\left[\mathbf{E}^A_{\mathcal{A}}\right]^T}{T}\right]\mathbf{E}^B_{\mathcal{A}} \tag{3}$$

[4] This procedure does not work if the chosen message represents a state of high uncertainty, e.g. at the first step of a random walk with no informative initial state prior. The mapping also fails for many states since CSCGs, by construction, assign zero probability to all states not within the clone set of a given observation, leading to degeneracy in the mapping. We also found that accuracy of this method degrades rapidly to the extent that the learned map fails to converge to the ground truth room topology.

[5] In practice, a softmax with a low temperature worked best for reconstruction.

[6] If $\mathbf{M} = \mathcal{A}$, this term is a representational similarity matrix in the sense of [10].

Here, the source messages $\mathbf{M}$ play the role of the queries $\mathbf{Q}$, model A's anchor embeddings $\mathbf{E}_{\mathcal{A}}^{A}$ act as keys $\mathbf{K}$, and model B's anchor embeddings act as values $\mathbf{V}$ in the attention equation which computes output $\mathbf{Z} = \sigma\left[\mathbf{QK}^{T}\right]\mathbf{V}$.[7]

Since self-attention may be understood though the lens of its connection to associative memory models [12,15], this correspondence goes some way toward theoretically justifying our choice of reconstruction method. In particular, following [12], reconstruction via relative representations can be understood as implementing a form of heteroassociative memory in which model A and B's anchor embeddings are, respectively, the memory and projection matrices.

Though empirical performance against a wider range of alternative methods of latent space alignment remains to be assessed, we note a formal connection to regression-based approaches such as [22], in which a representation $\mathcal{Y}$ of the data is expressed as a mixture of "guesses" (linear projections of local embeddings) from k experts, weighted according to the fidelity of each expert's representation of the input data $\mathcal{X}$. This can be expressed as a system of linear equations $\mathcal{Y} = UL$ in which $\mathcal{Y}$, U and L play roles analogous to those of $\hat{\mathbf{M}}$, $\sigma\left[\mathbf{R}_{\mathbf{M}}^{A}\right]$ and $\mathbf{E}_{\mathcal{A}}^{B}$ above, with the "repsonsibility" terms (weights) introducing nonlinearity, as the softmax does in our approach (see Appendix C for further details).

Not surprisingly, the results of our procedure improve with the number of anchors used (see Appendix A, Fig. 4). In our experiments, we used $N = 5000$ anchors. We obtained more accurate mappings using this technique when the anchor points were sampled from the trajectory being reconstructed, which raises the probability of an exact match in the anchor set; for generality, all reported results instead sample anchor points (uniformly, without replacement) from distinct random walks. While would be possible in the present setting to use similarity metrics tailored to probability distributions to create relative representations, we found empirically that replacing cosine similarity with the negative Jensen-Shannon distance slightly adversely affected performance.

5.3 Mapping Across Dissimilar Models

As shown in [13], relative representations can reveal common structure across superficially quite different models — for example those trained on sentences in distinct natural languages — via the use of "parallel" anchor points, in which the anchors chosen for each model are related by some mapping (e.g. being translations of the same text). In the context of CSCGs, anchors (forward messages) are defined relative to an observation sequence. To sample parallel anchors across agents, we therefore require partially dissimilar rooms in which similar but distinct observation sequences can be generated.

We used four experimental manipulations to generate pairs of partially dissimilar rooms (see Fig. 3), which we now outline along with a brief discussion of our results on each.

[7] In the present setting, one might even draw a parallel between the linear projection of transformer inputs to the key, query and value matrices and the linear projection of observations and prior beliefs onto messages via likelihood and transition tensors.

Table 1. Mapping across distinct CSCG models*

Condition	Max belief recovery % accurate (±SD)	Reconstruction accuracy mean cosine similarity (±SD)
Baseline: AR† (identical)	0.01(±0.01)	0.07(±0.07)
Permutation (identical)	84.09(±28.9)	0.69(±0.01)
Permutation (shifted)	3.41(±1.48)	0.69(±0.01)
Permutation (landmark)	20.70(±19.14)	0.89(±0.003)
RR‡ (identical)	89.44(±1.84)	0.99(±0.003)
RR (isomorphic)	41.0(±3.17)	0.67(±0.02)
RR (expansion: large → small)	97.42(±3.24)	0.98(±0.02)
RR (expansion: small → large)	47.47(±2.74)	0.59(±0.02)
RR (shifted)	34.81(±3.81)	0.63(±0.03)
RR (landmark)	34.13(±6.47)	0.52(±0.06)

† Absolute Representations ‡ Relative Representations *For each condition, mean results and standard deviation over 100 trials (each run on a distinct random graph) are reported, for the more challenging case of messages conditioned only on observations. For all but the (expansion) conditions, the results of mapping in either direction were closely comparable and we report the mean.

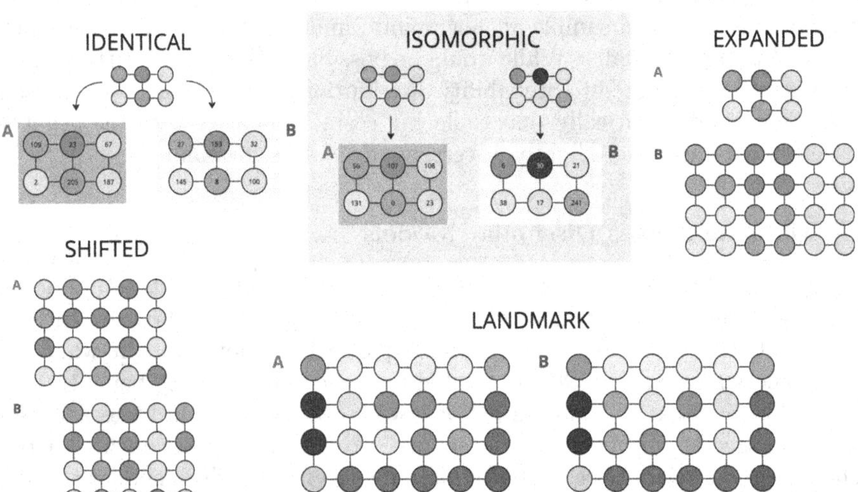

Fig. 3. Schematic illustration of experimental conditions. **A** and **B** indicate distinct rooms on which parallel models were trained, except for the "IDENTICAL" condition, where multiple models are trained on a single room. Numbers within nodes illustrate stochastic association of particular hidden state indices with positions in the learned graphs. Graph sizes depicted here do not reflect those used in the experiments.

Isomorphism. Any randomly generated grid or "room" of a given fixed size will (if CSCG training converges) yield a cognitive map with the same topology. It should thus be possible to generate parallel sequences of (action, observation) pairs — and thus parallel anchor points for defining relative representations — across two such random rooms, even if each contains a distinct set of possible observations or a different number of clones, either of which would preclude the use of a simple permutation-based mapping.

The relationships among observations will differ across such rooms, however, which matters under conditions of uncertainty, since every clone of a given observation will be partially activated when that observation is received, leading to different conditional belief distributions. This effect should be mitigated or eliminated entirely when beliefs are more or less certain, in which case "lateral" connections (transition dynamics) select just one among the possible clones corresponding to each observation. Indeed, we found that it is possible to obtain near-perfect reconstruction accuracy across models trained on random rooms with distinct observation sets, provided that messages are conditioned on both actions and observations; whereas we only obtained a <50% success rate in this scenario when conditioning on observations alone.

Expansion. In this set of experiments, we generated "expanded" versions of smaller rooms and corresponding "stretched" trajectories (paired observation and action sequences) using Kroenecker products, so that each location in the smaller room is expanded into a 2×2 block in the larger room, and each step in the smaller room corresponds to two steps in the larger one. We can then define parallel anchors across agents trained on such a pair of rooms, by taking (a) all messages in the smaller room, and (b) every other message in the larger one. In this condition, the large $\rightarrow$ small mapping can be performed much more accurately than the opposite one, since each anchor point in the smaller ("downsampled") room corresponds to four potential locations in the larger. Superior results on the (large $\rightarrow$ small) condition VS our experiments on identical rooms may be explained by the fact that the "small" room containts fewer candidate locations than the room used in the "Identical" condition.

Shifting. In a third set of experiments, we generated rooms by taking overlapping vertical slices of a wider room, such that identical sequences were observed while traversing the rooms, but within different wider contexts. In this case only the messages corresponding to overlapping locations were used as anchor points, but tests were performed on random walks across the entire room. Under conditions of certainty, mapping across these two rooms can be solved near-perfectly by using all messages as candidate anchor points, since the rooms are isomorphic. Without access to ground-truth actions, it was possible to recover the beliefs of one agent given the other's only ~35% of the time. We hypothesize that this problem is more challenging than the "Isomorphic" condition because similar patterns of observations (and thus similar messages) correspond to distinct locations across the two rooms, which should have the effect of biasing reconstructions toward the wrong locations.

Landmarks. Finally, partially following the experiments in [16] on largely featureless rooms with unique observations corresponding to unique locations (e.g. corners and walls), we define pairs of rooms with the same (unique) observations assigned to elements of the perimeter, filled by otherwise randomly generated observations that differed across rooms. Using only the common "landmark" locations as anchors, it was still possible to use relative representations to recover an agent's location from messages in a parallel trajectory in the other room with some success.

Summary. The results reported in Table 1 were obtained under conditions of significant uncertainty, in which messages were conditioned only on observations, without knowledge of the action that produced those observations. In this challenging setting, relative representations still enabled recovery (well above chance in all experimental conditions, and in some cases quite accurate) of one agent's maximum *a posteriori* belief about its location from those of the other agent, averaged across messages in a test sequence.[8]

In all settings, it was possible to obtain highly accurate mappings (>99% correct in most cases) by conditioning messages on actions as well as observations. This yields belief vectors sharply peaked at the hidden state corresponding to an agent's location on the map. In this regime, the reconstruction procedure acts essentially as a lookup table, as a given message $\mathbf{m}$ resembles a one-hot vector and this sparsity structure is reflected in the relative representation (which is ~ 0 everywhere except for dimensions corresponding to anchor points nearly identical to $\mathbf{m}$). The softmax weighting then simply "selects" the corresponding anchor in model B's anchor set.[9] Conditioning messages on probabilistic knowledge of actions (perhaps the most realistic scenario) can be expected to greatly improve accuracy relative to the observation-only condition, and is an interesting subject for a follow-up study.

[8] It is worth noting that this is essentially a one-of-N classification task, with effective values of N around 48 in most cases. This is because (following [16]) most experiments were performed on 6×8 rooms, and there is one "active" clone corresponding to each location in a converged CSCG.

[9] There is a variation on this in which multiple matches exist in the anchor set, but the result is the same as we then combine n identical anchor points.

6 Discussion

The "messages" used to define relative representations in the present work can be interpreted as probability distributions, but they can also be interpreted more agnostically as, simply, neuronal activity vectors. Recent work in systems neuroscience [2] has shown that it is possible to recover common abstract latent spaces from real neuronal activity profiles. As noted above, relative representations were anticipated in neuroscience by RSA, which in effect treats the neuronal responses, or computational model states, associated with certain fixed stimuli as anchor points. This technique complements others such as the analysis of attractor dynamics [26] as a tool to investigate properties of latent spaces in brains, and has been shown to be capable of revealing common latent representational structure across not only individuals, but linguistic communities [28] and even species [7,11]. Consistent with the aims of [13] and [10], this paradigm might ultimately provide fascinating future directions for brain imaging studies of navigational systems in the hippocampal-entorhinal system and elsewhere.

Relative representations generalize this paradigm to "parallel anchors", and also demonstrate the utility of high-dimensional representational similarity vectors as latent representations in their own right, which can, as demonstrated above, be used to establish zero-shot communication between distinct models.

While the conditions we constructed in our toy experiments are artificial, they have analogues in more realistic scenarios. It is plausible that animals navigating structurally homeomorphic but superficially distinct environments, for example, should learn similar cognitive maps at some level of abstraction. Something analogous to the "expansion" setting may occur across two organisms that explore the same space but (for example due to different sizes or speeds of traversal, and thus sample rates) coarse-grain it differently. The idea of landmark-based navigation is central to the SLAM paradigm generally, and the stability of landmarks across otherwise different spaces may provide a model for the ability to navigate despite changes to the same environment over time. Finally, while experiments on partially overlapping rooms seem somewhat contrived if applied naively to spatial navigation scenarios, they may be quite relevant to models of SLAM in abstract spaces [18], such as during language acquisition, where different speakers of the same language may be exposed to partially disjoint sets of stimuli, corresponding to different dialects (or in the limit, idiolects).

Crucially, the common reference frame provided by these techniques might allow for the analysis of *shared* representations, which (when derived from well-functioning systems) should embody an ideal structure that individual cognitive systems in some sense aim to approximate, allowing for comparison of individual brain-bound models against a shared, abstract ground truth. Such an abstracted "ideal" latent space could be used to measure error or misrepresentation [9], or to assess progress in developmental contexts.

7 Conclusion

In this work we have considered a toy example of the application of relative representations to graph-structured cognitive maps. The results reported here are intended mainly to illustrate concrete directions for the exploration of the latent structure of cognitive maps using relative representations, and as a proof-of-principle that the technique can be applied to the case of inferred posterior distributions over discrete latent spaces. We have also introduced a technique for reconstructing "absolute" representations from their relative counterparts without learning.

In addition to further investigating hyperparameter settings (such as choice of similarity function) to optimize performance in practical applications, future work might explore the application of relative representations to more complex models with discrete latent states, such as the discrete "world models" used in cutting-edge model-based reinforcement learning [6], or to enable belief sharing and cooperation in multi-agent active inference scenarios. Given the connection to neural self-attention described above, which has also been noted in the context of the Tolman-Eichenbaum Machine [24], it would also be intriguing to explore models in which such a translation process occurs within agents themselves, as a means of transferring knowledge across local cognitive structures.

Acknowledgements. Alex Kiefer is supported by VERSES Research. CLB is supported by BBRSC grant number BB/P022197/1 and by Joint Research with the National Institutes of Natural Sciences (NINS), Japan, program No. 0111200.

Code Availability. The CSCG implementation is based almost entirely on the codebase provided in [16]. Code for reproducing our experiments and analysis can be found at: https://github.com/exilefaker/cscg-rr.

Appendix A: Effect of Anchor Set Size on Reconstruction

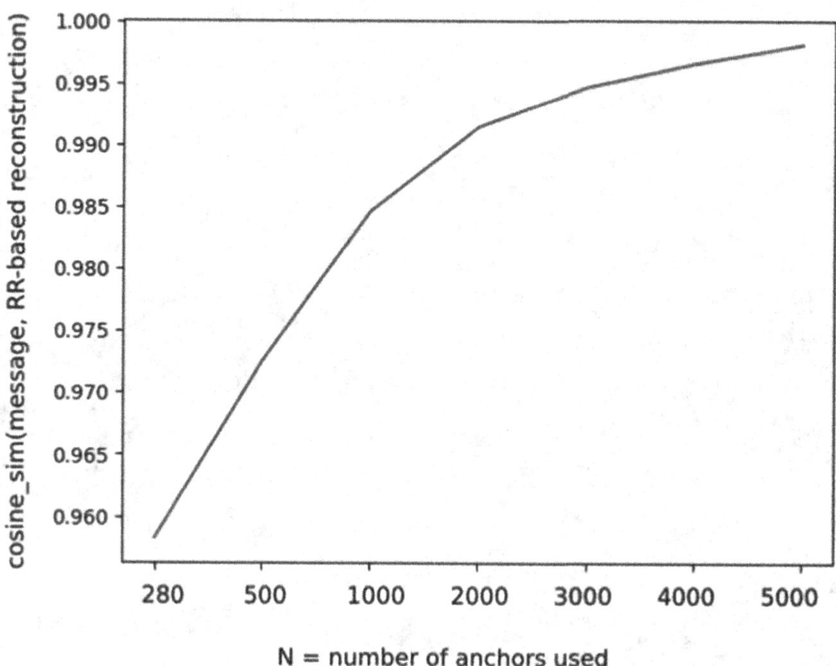

Fig. 4. Average cosine similarity ($\frac{u \cdot v}{\|u\|\|v\|}$) between ground-truth CSCG beliefs (messages) and their reconstructions from those of a distinct CSCG model trained on the same room and receiving the same sequence of observations, using the method in Eq. 2, plotted against number N of anchors used to define the relative representations. We begin by setting N to the dimensionality of the model's hidden state. The average is across all 5000 messages in a test sequence.

Appendix B: Visualizing the Correspondence of Relative Representations Across Models

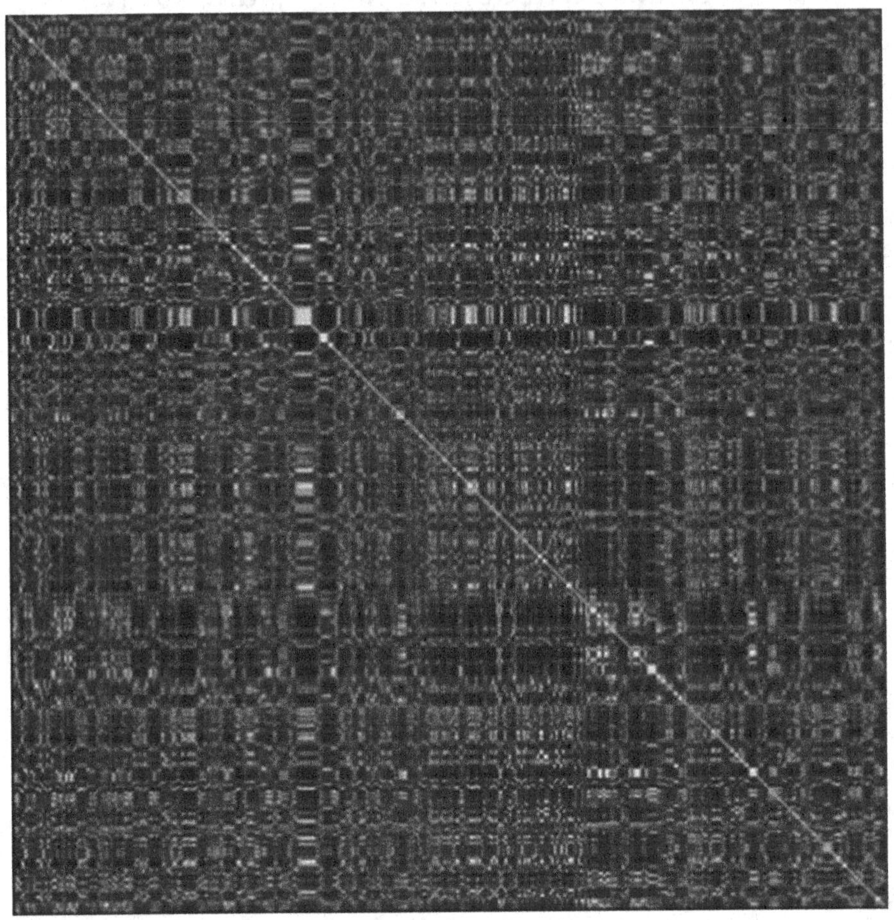

Fig. 5. Example representational similarity matrix comparing relative representations of analogous message sequences (i.e. inferred from the same observation sequence) from two distinct models trained on the same environment. This differs from the (dis)similarity matrices typically used in RSA [10], as rows and columns in this case represent distinct sets of first-order representations, i.e. cell (i, j) represents the cosine similarity between $\mathbf{r}_i^A$ and $\mathbf{r}_j^B$. Thus the diagonal symmetry illustrates the empirical equivalence of these two sets of relative representations.

Appendix C: Comparison to LLC

Locally Linear Coordination (LLC) [22] is a method for aligning the embeddings of multiple dimensionality-reducing models so that they project to the same global coordinate system. While its aims differ somewhat from the procedure outlined in the present study, LLC is also an approach to translating multiple source embeddings to a common representational format. As noted above, there is an interesting formal resemblance between the two approaches, which we explore in this Appendix.

The LLC Representation

LLC presupposes a mixture model of experts trained on N D-dimensional input datapoints $\mathcal{X} = [x_1, x_2, ..., x_N]$, in which each expert m_k is a dimensionality reducer that produces a local embedding $z_{n_k} \in \mathbb{R}^{d_k}$ of datapoint x_n. The mixture weights or "responsibilities" for the model can be derived, for example, as posteriors over each expert's having generated the data, in a probabilistic setting.

Given the local embeddings and responsibilities, LLC proposes an algorithm for discovering linear mappings $L_k \in \mathbb{R}^{d \times d_k}$ from each expert's embedding to a common (lower-dimensional) output representation $\mathcal{Y} \in \mathbb{R}^{N \times d}$, which can then be expressed as a responsibility-weighted mixture of these projections. That is to say, leaving out bias terms for simplicity: each output image y_n of datapoint x_n is computed as

$$y_n = \sum_k r_{n_k} (L_k z_{n_k}) \tag{4}$$

Crucially for what follows, with the help of a flattened (1D) index that spans the "batch" dimension N as well as the experts k, we can express this in simpler terms as $\mathcal{Y} = UL$. We define matrices $U \in \mathbb{R}^{N \times \sum_k d_k}$ and $L \in \mathbb{R}^{\sum_k d_k \times d}$ in terms of, respectively: (a) vectors u_n, where $u_{n_j} = r_{n_k} z^i_{n_k}$ (i.e. the jth element of u_n is the ith element of k's embedding of x_n scaled its responsibility term) — and (b) re-indexed, transposed columns $l_j = l^i_k$ of the L_k matrices. Intuitively, each row u_n of U concatenates the experts' responsibility-weighted embeddings $r_{n_k} z_{n_k}$ of datapoint x_n, while each of L's d columns is a concatenation of the corresponding row of the projection matrices L_k, so that the matrix product UL returns a responsibility-weighted prediction for y_n in each row (see Fig. 6).

Relationship to Our Proposal

Ignoring the motivation of dimensionality reduction which is irrelevant for present purposes, there is a precise conceptual and formal equivalence between this model and the procedure for reconstructing model B's embeddings given those of model A described above in Sect. 5.2.

Specifically, we can regard each of model A's anchor embeddings $\mathbf{e}^A_{x_k}$ as an "expert" in a fictitious mixture model, with an associated responsibility term

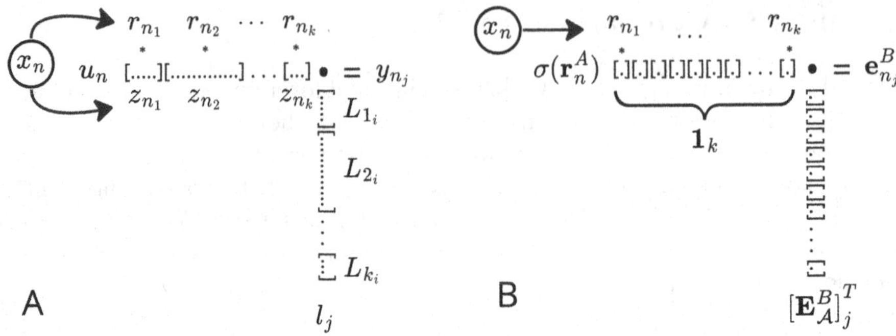

Fig. 6. Visual schematic of the computation of a single entry of the output of (A) the projection of input x_n to output y_n as in the Locally Linear Coordinates (LLC) mapping procedure; (B) the reconstruction of a latent embedding $\mathbf{e}_n^B$ in model B's embedding space given input x_n to model A. The groupings in brackets in (A) illustrate the concatenations of vector embeddings (scaled by responsibility terms r_{n_k}) in u_n, and of projection columns in l_j. $\mathbf{1}_k$ in (B) denotes a row of k 1s (where k in this case denotes the number of anchors, i.e. is set to $|\mathcal{A}|$). Each entry in the column vector $\left[\mathbf{E}_{\mathcal{A}}^B\right]_j^T$ is the jth dimension of one of model B's anchor embeddings.

measuring its fidelity to the input x_i, which in this case is given by the cosine similarity between the anchor embedding and the input embedding. Then like the rows of U, each row of $\sigma\left[\mathbf{R}_{\mathcal{X}}^A\right]$, which is a relative representation $\mathbf{r}_i^A = \mathbf{E}_{\mathcal{A}}^A \mathbf{e}_i^A$ of input i after application of the softmax, acts as a responsibility-weighted mixture of multiple "views" of the input. Similarly, since the rows of $\mathbf{E}_{\mathcal{A}}^B$ are anchor embeddings in the output space, its columns j act precisely as do the columns of L, i.e. as columns in a projection matrix, so that $\sigma[\mathbf{r}_i^A] \cdot \mathbf{E}_{\mathcal{A}j}^B$ outputs dimension j of the reconstructed target embedding $\mathbf{e}_i^B$.

There is at least one important difference between LLC and our procedure: in LLC each expert uses an internal transform to generate an input-dependent embedding, which is then scaled by its responsibility term, which also depends on the input. Reconstruction via relative representations instead employs fixed stored embeddings, so that each "expert" contributes a scalar value rather than an embedding vector to the final output. However, the expression of LLC in terms of a linear index demonstrates that this makes no essential difference mathematically (conceptually, these scalar "votes" are 1D vectors; cf. Figure 6).

The point is not that these two algorithms are doing precisely the same thing (they are not, as LLC aims to align multiple embedding spaces by deriving a mapping to a distinct common space, while our approach aims to recover the contents of one embedding space from another). The use of LLC to reconstruct input data $\mathcal{X}$ from its "global" embedding $\mathcal{Y}$ as in [22] *is* quite closely related to our procedure, however, and at this level of abstraction the approaches may be regarded as the same, with a difference in the nature of the "experts" used in the mixture model and the attendant multiple "views" of the data. The relative representation reconstruction procedure, while presumably not as expressive, may compensate to some extent for the use of scalar "embeddings" by using a large number of "experts", and has the virtue of eschewing the need for a mixture model to assign responsibilities, or indeed for multiple intermediate embedding models, to perform such a mapping.

References

1. Da Costa, L., et al.: Active inference on discrete state-spaces: a synthesis. J. Math. Psychol. **99**, 102447 (2020). ISSN: 0022-2496. https://doi.org/10.1016/j.jmp.2020.102447, https://www.sciencedirect.com/science/article/pii/S0022249620300857
2. Dabagia, M., Kording, K.P., Dyer, E.L.: Aligning latent representations of neural activity. Nat. Biomed. Eng. **7**, 337–343 (2023). https://doi.org/10.1038/s41551-022-00962-7
3. Dedieu, A., et al.: Learning higher-order sequential structure with cloned HMMs (2019). arXiv:1905.00507 [stat.ML]
4. Dimsdale-Zucker, H.R., Ranganath, C.: Chapter 27 - Representational similarity analyses: aăPractical guide for functional MRI applications. In: Manahan-Vaughan, D. (ed.) Handbook of in Vivo Neural Plasticity Techniques, vol. 28. Handbook of Behavioral Neuroscience, pp. 509–525. Elsevier (2018). https://doi.org/10.1016/B978-0-12-812028-6.00027-6, https://www.sciencedirect.com/science/article/pii/B9780128120286000276
5. George, D., et al.: Clone-structured graph representations enable flexible learning and vicarious evaluation of cognitive maps. Nat. Commun. **12**(1), 2392 (2021)
6. Hafner, D., et al.: Mastering Atari with Discrete World Models. CoRR abs/2010.02193 (2020). arXiv: 2010.02193. https://arxiv.org/abs/2010.02193
7. Haxby, J.V., Connolly, A.C., Guntupalli, J.S.: Decoding neural representational spaces using multivariate pattern analysis. Annu. Rev. Neurosci. **37**, 435–56 (2014). https://api.semanticscholar.org/CorpusID:6794418
8. Heins, C., et al.: Pymdp: a python library for active inference in discrete state spaces. CoRR abs/2201.03904 (2022). arXiv: 2201.03904, https://arxiv.org/abs/2201.03904
9. Kiefer, A., Hohwy, J.: Representation in the prediction error minimization framework. In: Robins, S.K., Symons, J., Calvo, P. (ed.), The Routledge Companion to Philosophy of Psychology, 2nd ed., pp. 384–409 (2019)
10. Kriegeskorte, N., Mur, M., Bandettini, P.A.: Representational similarity analysis connecting the branches of systems neuroscience. Front. Syst. Neurosci. **2**, 4 (2008). https://doi.org/10.3389/neuro.06.004.2008
11. Kriegeskorte, N., et al.: Matching categorical object representations in inferior temporal cortex of man and monkey. Neuron **60**, 1126–1141 (2008). https://api.semanticscholar.org/CorpusID:313180
12. Millidge, B., et al.: Universal hopfield networks: a general framework for single-shot associative memory models. In: Proceedings of the 39th International Conference on Machine Learning, vol. 162. Baltimore, Maryland, USA, pp. 15561–15583, July 2022
13. Moschella, L., et al.: Relative representations enable zero-shot latent space communication (2023). arXiv: 2209.15430 [cs.LG]
14. Pearl, J.: Reverend Bayes on inference engines: a distributed hierarchical approach. In: Proceedings of the Second AAAI Conference on Artificial Intelligence. AAAI'82. Pittsburgh, Pennsylvania: AAAI Press, pp. 133–136 (1982)

15. Ramsauer, H., et al.: Hopfield Networks is All You Need (2021). arXiv: 2008.02217 [cs.NE]
16. Rikhye, R.V., et al.: Learning cognitive maps as structured graphs for vicarious evaluation. In: bioRxiv (2020). https://doi.org/10.1101/864421. eprint: https://www.biorxiv.org/content/early/2020/06/24/864421.full.pdf. https://www.biorxiv.org/content/early/2020/06/24/864421
17. Rikhye, R.V., et al.: Memorize-generalize: an online algorithm for learning higher-order sequential structure with cloned hidden Markov Models. In: bioRxiv (2019). https://doi.org/10.1101/764456. eprint: https://www.biorxiv.org/content/early/2019/09/10/764456.full.pdf. https://www.biorxiv.org/content/early/2019/09/10/764456
18. Safron, A., Catal, O., Verbelen, T.: Generalized simultaneous localization and mapping (G-SLAM) as unification framework for natural and artificial intelligences: towards reverse engineering the hippocampal/entorhinal system and principles of high-level cognition, October 2021. https://doi.org/10.31234/osf.io/tdw82. psyarxiv.com/tdw82
19. Smith, R., Friston, K.J., Whyte, C.J.: A step-by-step tutorial on active inference and its application to empirical data. J. Math. Psychol. **107**, 102632 (2022). ISSN: 0022-2496. https://doi.org/10.1016/j.jmp.2021.102632. https://www.sciencedirect.com/science/article/pii/S0022249621000973
20. Stachenfeld, K., Botvinick, M., Gershman, S.: The hippocampus as a predictive map, July 2017. https://doi.org/10.1101/097170
21. Swaminathan, S., et al.: Schema-learning and rebinding as mechanisms of in-context learning and emergence (2023). arXiv: 2307.01201 [cs.CL]
22. Teh, Y., Roweis, S.: Automatic alignment of local representations. In: Becker, S., Thrun, S., Obermayer, K. (ed.) Advances in Neural Information Processing Systems, vol. 15. MIT Press (2002). https://proceedings.neurips.cc/paper_files/paper/2002/file/3a1dd98341fafc1dfe9bcf36360e6b84-Paper.pdf
23. Whittington, J., et al.: How to build a cognitive map. Nat. Neurosci. **25**, 1–16 (2022). https://doi.org/10.1038/s41593-022-01153-y
24. Whittington, J.C.R., Warren, J., Timothy, E.J.B.: Relating transformers to models and neural representations of the hippocampal formation. CoRR abs/2112.04035 (2021). arXiv: 2112.04035, https://arxiv.org/abs/2112.04035
25. Whittington, J.C.R., et al.: The tolman-eichenbaum machine: unifying space and relational memory through generalization in the hippocampal formation. Cell **183**(5), 1249–1263.e23 (2020). ISSN: 0092-8674. https://doi.org/10.1016/j.cell.2020.10.024, https://www.sciencedirect.com/science/article/pii/S009286742031388X
26. Wills, T.J., et al.: Attractor dynamics in the hippocampal representation of the local environment. Science **308**(5723), 873–876 (2005). https://doi.org/10.1126/science.1108905. eprint: https://www.science.org/doi/pdf/10.1126/science.1108905, https://www.science.org/doi/abs/10.1126/science.1108905
27. Winn, J., Bishop, C.M.: Variational message passing. J. Mach. Learn. Res. **6**, 661–694 (2005). ISSN: 1532–4435
28. Zinszer, B.D., et al.: Semantic structural alignment of neural representational spaces enables translation between English and Chinese words. J. Cogn. Neurosci. **28**, 1749–1759 (2016). https://api.semanticscholar.org/CorpusID:577366

Theory of Learning and Inference

Active Inference in Hebbian Learning Networks

Ali Safa[1,2,3(✉)], Tim Verbelen[4], Lars Keuninckx[1], Ilja Ocket[1], André Bourdoux[1], Francky Catthoor[1,2], Georges Gielen[1,2], and Gert Cauwenberghs[3]

[1] Imec, Leuven, Belgium
ali.safa@imec.be
[2] ESAT, KU Leuven, Leuven, Belgium
[3] University of California at San Diego, San Diego, La Jolla, USA
[4] VERSES Research Lab, Los Angeles, CA, USA

Abstract. This work studies how brain-inspired neural ensembles equipped with local Hebbian plasticity can perform active inference (AIF) in order to control dynamical agents. A generative model capturing the environment dynamics is learned by a network composed of two distinct Hebbian ensembles: a *posterior* network, which infers latent states given the observations, and a *state transition* network, which predicts the next expected latent state given current state-action pairs. Experimental studies are conducted using the Mountain Car environment from the OpenAI gym suite, to study the effect of the various Hebbian network parameters on the task performance. It is shown that the proposed Hebbian AIF approach outperforms the use of Q-learning, while *not requiring* any replay buffer, as in typical reinforcement learning systems. These results motivate further investigations of Hebbian learning for the design of AIF networks that can learn environment dynamics without the need for revisiting past buffered experiences.

Keywords: Active Inference · Hebbian Learning · Sparse Coding

1 Introduction

The study of Sparse Coding [1–4] and Predictive Coding [5–7] networks has gained much attention for understanding the mechanisms underlying learning and inference in the brain [8]. In particular, it has been shown that the learning of the *weight dictionary* used to project the input signals into sparse codes can be conducted via the biologically-plausible Hebbian learning mechanism [9], with experimental evidence behind this mechanism observed in the brain [10,11]. Hebbian learning differs from the widely-used back-propagation of error (backprop) technique due to its *local* nature [7,12,13], where the weight w_j of neuron i is modified via a combination f of the weight's input x_j and the neuron's output y_i (with η_d the learning rate parameter):

$$w_j \leftarrow w_j + \eta_d f(y_i, x_j) \tag{1}$$

C. L. Buckley et al. (Eds.): IWAI 2023, CCIS 1915, pp. 239–253, 2024.
https://doi.org/10.1007/978-3-031-47958-8_15

When applied to layers that evince some form of competition between their neurons, the Hebbian mechanism in (1) leads to the *unsupervised* learning of complementary features from the input signals [14].

At the same time, Active Inference (AIF) has gained huge interest as a *first-principle* theory, explaining how biological agents evolve and perform actions in their environment [15,16]. In recent years, the use of deep neural networks (DNNs) for parameterizing generative models has gained much attention in AIF research [17–19]. Deep AIF systems are typically composed of a *posterior* network $q_{\Phi_P}(s_l|o_{l-1}, a_{l-1})$, inferring the latent state s_l given an incoming observation-action pair $\{o_{l-1}, a_{l-1}\}$, and a *state-transition* network $p_{\Phi_S}(s_l|s_{l-1}, a_{l-1})$, predicting the next latent state s_l given the current state-action pair $\{s_{l-1}, a_{l-1}\}$ [17]. The state-transition network is used to generate the agent's roll-outs for different policies in order to compute the Expected Free Energy associated to each policy [17]. Finally, a *likelihood* network $p_{\Phi_L}(o_l|s_l)$ reconstructing the input observation o_l from the latent state s_l can also be optionally implemented [20] (not utilized in this work). Each network parameterizes its respective density function through weight tensors Φ_P, Φ_S and Φ_L.

In this work, we aim to study how AIF can be performed in Hebbian learning networks *without resorting to backprop* (as typically used in deep AIF systems). Experiments conducted in the OpenAI Mountain Car environment [21] show that the proposed Hebbian AIF approach outperforms the use of Q-learning and compares favorably to the backprop-trained Deep AIF system of [17], while *not requiring* any replay buffer, as in typical reinforcement learning systems [22]. Our derivations and experiments add to a growing number of work addressing the study of Hebbian Active Inference [23,24].

This paper is organized as follows. Background theory about Hebbian learning networks is provided in Sect. 2. Our Hebbian AIF methods are covered in Sect. 3. Experimental results are shown in Sect. 4. Conclusions are provided in Sect. 5.

2 Background Theory on Hebbian Learning Networks

Inspired by previous works that model the neural activity of biological agents through Sparse Coding [5,9] (such as in the mushroom body of an insect's brain [25]), we model each individual Hebbian Ensemble layer of our networks as an identically-distributed Gaussian likelihood model with a Laplacian prior on the neural activity c:

$$p(c|o, \Phi) \sim \exp\left(-||\Phi c - o||_2^2\right) \exp\left(-\lambda ||c||_1\right) \tag{2}$$

where o is the input of dimension N, c is the output of dimension M, Φ is the $N \times M$ weight matrix of the layer (also called *dictionary*), and λ is a hyper-parameter setting the scale of the Laplacian prior. Choosing a Laplacian prior is motivated by the fact that it promotes sparsity in the output neural code in a way similar to how sparsity is induced in networks of Spiking Leaky Integrate-and-Fire neurons, modelling cortical neural activity [9].

Under Sparse Coding (2), inference of c and learning of Φ is carried via [9]:

$$C, \Phi = \arg\min_{C,\Phi} \sum_l ||\Phi c_l - o_l||_2^2 + \lambda ||c_l||_1 \text{ with } C = \{c_l, \forall l\} \tag{3}$$

which can be solved via Proximal Stochastic Gradient Descent [26], by *alternating* between: *a)* the inference of c_l, given the current input o_l and the weight Φ and *b)* the learning of Φ, given the current c_l and o_l.

Hence, we instantiate Hebbian layers as the dynamical system given in (4), where T denotes the transpose, η_c is the coding rate, η_d is the learning rate and $\mathbf{Prox}_{\lambda||.||_1}$ is the proximal operator to the l_1 norm (non-linearity) [27]. For each input o_i, the neural and weight dynamics of the Hebbian network follows the update rules in (4) for an arbitrary number of iterations (set to 100 in this work as a good balance between speed of convergence and convergence quality), in order to infer the corresponding c_l and learn Φ [9].

$$\begin{cases} c_l \leftarrow \mathbf{Prox}_{\lambda||.||_1}\{c_l - \eta_c \Phi^T(\Phi c_l - o_l)\} \\ \Phi \leftarrow \Phi - \eta_d(\Phi c_l - o_l)c_l^T \end{cases} \tag{4}$$

with $\mathbf{Prox}_{\lambda||.||_1}$ acting as the neural non-linearity:

$$\mathbf{Prox}_{\lambda||.||_1}(c_i) = \text{sign}(c_i)\max(0, |c_i| - \eta_c\lambda), \forall i \tag{5}$$

From a neural point of view, the dynamical system of (4) can be implemented as the network architecture in Fig. 1, where all weight updates follow the standard Hebbian rule (1) [9].

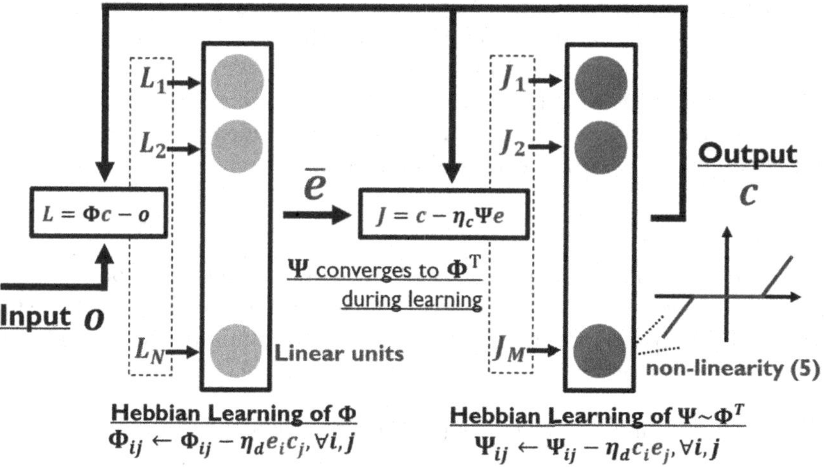

Fig. 1. ***Baseline Hebbian network architecture used in this work.*** The dynamics of the network follow (4) and minimize (3), given subsequent input vectors o. Each layer possesses its own weight matrix Φ, Ψ which evolve through Hebbian plasticity ($\Psi \sim \Phi^T$ in (4), as an independent, local set of weights).

3 Active Inference in Hebbian Learning Networks

In this Section, we show how the Hebbian network described above in Sect. 2 is utilized in order to build an AIF system. First, we describe how Variational Free Energy minimization can be performed by a cascade of two Hebbian networks: a *state-transition* network predicting the next latent states given the previous ones, and a *posterior* network providing latent states given input observations. Crucially, it is shown that Free Energy minimization necessitates top-down Hebbian learning connections from the *state-transition* network towards the *posterior* network, steering the posterior output activity towards the state-transition output during learning. Then, we show how the Expected Free Energy is computed by generating state transition roll-outs. In summary, the ensuing scheme showcased below can be regarded as learning to plan; in which the requisite inferences are amortised by Hebbian learning. Note that this learning is effectively a local scheme that eschews need for back propagation.

3.1 Minimizing the Variational Free Energy

The Variational Free Energy can be decomposed as [17,28] (where $\mathbb{E}$ denotes the expected value):

$$\mathcal{F} = \underbrace{D_{KL}[q_{\Phi_P}(s_l|o_{l-1}, a_{l-1})||p_{\Phi_S}(s_l|s_{l-1}, a_{l-1})]}_{\text{expected complexity (ambiguity)}} - \underbrace{\mathbb{E}_q[\log(p_{\Phi_L}(o_l|s_l))]}_{\text{expected accuracy (risk)}} \tag{6}$$

with the parametrized densities $q_{\Phi_P}, p_{\Phi_S}, p_{\Phi_L}$ described in Sect. 1. Since the Hebbian network architecture used in this work intrinsically provides a means to reconstruct its input x_l in (3) (i.e., *likelihood* modelling) from its produced latent code c_l in (3) (i.e., *posterior* modelling), using the *same dictionary* parameter matrix Φ in (3) that was used to generate c_l via (4), we have $\Phi_L = \Phi_P$ in (6) and we will solely use Φ_P below to denote the *posterior* weight matrix.

Under the assumption of Gaussian likelihood with identity covariance in (2), the KL divergence D_{KL} in $\mathcal{F}$ can be simplified to [29]:

$$\mathcal{F} \sim ||\Phi_S c_{S,l} - \{s_l(\Phi_P), a_l\}||_2^2 + ||\Phi_P s_l - \{o_l, a_l\}||_2^2 \tag{7}$$

where $s_l(\Phi_P)$ explicits the dependency of s_l on Φ_P (s_l being the posterior network output activity) and c_S denotes the output activity of the state-transition network given s_l. The Free Energy in (7) must be minimized with regard to the state transition weights Φ_S and the posterior weights Φ_P during learning:

$$\Phi_S, \Phi_P = \arg\min_{\Phi_S, \Phi_P} ||\Phi_S c_{S,l} - \{s_l(\Phi_P), a_l\}||_2^2 + ||\Phi_P s_l - \{o_l, a_l\}||_2^2, \forall l \tag{8}$$

This indicates that it is not only the state transition model that must be steered towards the posterior model, but also, the posterior model must be steered towards the output of the state-transition network. This effect can be achieved by re-formulating the optimization in (8) as:

$$\begin{cases} \Phi_S = \arg\min_{\Phi_S} ||\Phi_S c_{S,l} - \{s_l, a_l\}||_2^2, \forall l & \text{(a)} \\ \Phi_P = \arg\min_{\Phi_P} ||\Phi_P s_l - \{o_l, a_l\}||_2^2 + ||\Phi_P(\Phi_S c_{S,l}) - \{o_l, a_l\}||_2^2 & \text{(b)} \end{cases} \tag{9}$$

Intuitively, the right-hand term in (9 b) steers the *posterior* model towards the *state-transition* model by first re-projecting the output activity of the state-transition network c_S into the latent space as $\Phi_S c_S$ (considering Φ_S fixed). Then, minimizing $||\Phi_P(\Phi_S c_{S,l}) - \{o_l, a_l\}||_2^2$ modifies Φ_P in order to steer its posterior output s_l towards the re-projected state-transition activity $\Phi_S c_S$ (considering $\{o_l, a_l\}$ fixed).

State-Transition Model. Inspired by prior work on dictionary-based sequence modeling [30], we implement the transition model $p_{\Phi_S}(s_l|\tilde{s}_{l-1}, \tilde{a}_{l-1})$ as an *auto-regressive* Hebbian network (see Fig. 2 a), taking as input a sequence of state-and-action *history* $\tilde{s}_{l-1} = [s_{l-1}, \dots, s_{l-L_{buf}}]$, $\tilde{a}_{l-1} = [a_{l-1}, \dots, a_{l-L_{buf}}]$ and inferring the next state $\tilde{s}_l = [s_l, \dots, s_{l-L_{buf}}]$ as the re-projection of its internal sparse code c_S in the input space through the network weights Φ_S:

$$\tilde{s}_l = \Phi_S c_{S,l} \tag{10}$$

where $\Phi_{S,j}$ and $c_{S,j}$ respectively denote the weight vector and the sparse code of each layer j in the state transition network. Therefore, the state-transition network effectively projects the L_{buf} previous states (noted $\tilde{s}_{l-1}$) into a common internal sparse code c_S and reconstructs the next states $\tilde{s}_l$ by re-projection of c_S into the input space.

The state-transition network learns its weights Φ_S following (9 a) and infers its output activity c_S via sparse coding (see Sect. 2):

$$\Phi_S, c_{S,l} = \arg\min_{\Phi_S, c_{S,l}} ||\Phi_S c_{S,l} - \tilde{s}_l||_2^2 + \lambda_P ||c_{S,l}||_1, \forall l \tag{11}$$

where λ_P is a parameter that sets the strength of the *sparsity* of the *state-transition* output activity. (11) can therefore be implemented via the Sparse Coding-based Hebbian learning ensemble described in Sect. 2. This auto-regressive strategy enables the network to learn state predictions using Hebbian learning, *without* the need for non-bio-plausible back-propagation through time (BPTT) [30,31].

In order to prevent the vanishing or exploding of the state transition model when producing roll-outs further in time, we regularize the norm of the reconstructed states s_l to an arbitrary magnitude α using (12). We keep $\alpha = 5$ in our experiments in Sect. 4, giving a good balance for the dynamic range of the network output activity (adjusted empirically).

$$s_l = \alpha \frac{s_l}{||s_l||_2} \tag{12}$$

Posterior Model. Similar to the state-transition model, we use a Hebbian ensemble as posterior model, where the internal sparse code c_P (see Fig. 2 b) is identified as the hidden state $s_l \equiv c_{P,l}$ inferred by the posterior network $q_\nu(s_l|o_{l-1}, a_{l-1})$, given the observation and action pair $\{o_{l-1}, a_{l-1}\}$ in (2).

Therefore, the posterior network learns its weights Φ_P following (9 b) and infers its output activity $s_l = c_{P,l}$ via sparse coding (see Sect. 2):

$$\Phi_P, c_{P,l} = \arg\min_{\Phi_P, c_{P,l}} \underbrace{||\Phi_P c_{P,l} - \{o_l, a_l\}||_2^2 + \lambda_Q ||c_{P,l}||_1}_{\text{Standard Sparse Coding}} + \overbrace{||\Phi_P(\Phi_S c_{S,l}) - \{o_l, a_l\}||_2^2}^{\text{Top-down Connection}} \quad (13)$$

for all l, where λ_Q sets the strength of the *sparsity* of the *posterior* output activity. The left-hand *standard sparse coding* term in (13) can be implemented via the Hebbian learning ensemble described in Sect. 2, while the right-hand term in (13) can be implemented using *top-down* connections from the state-transition output activity c_S towards the posterior network, via the state-transition weight matrix Φ_S (see Fig. 2 b).

Finally, here again, we apply the regularization rule (12) to the inferred posterior state, effectively constraining s_l to lie on the α-sphere manifold.

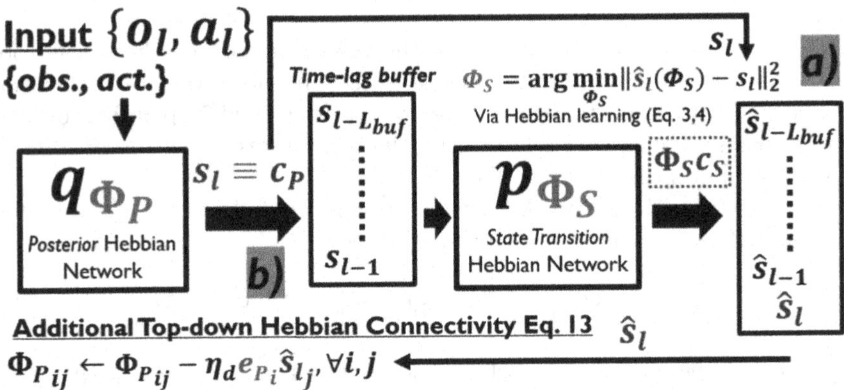

Fig. 2. ***Hebbian Active Inference Architecture.*** a) The state-transition network takes as input the L_{buf} previous latent states produced by the posterior network, and projects them onto its internal representation c_S via the learned Φ_S. When producing roll-outs, the state-transition network estimates the next state $\hat{s}_l$ by re-projecting the output activity c_S due to $[s_{l-L_{buf}}, \ldots, s_{l-1}]$ back onto the input space via (10). b) The posterior network takes observation-action pairs as input and produces latent states s (corresponding to the output in Fig. 1). In addition to the Hebbian mechanisms depicted in Fig. 1, the weights Φ_P of the posterior network are also subject to a top-down Hebbian learning mechanism for minimizing the second term in (9 b).

3.2 Minimizing the Expected Free Energy

Given a policy π, the Expected Free Energy $G(\pi)$ (EFE) can be written as [32]:

$$G(\pi) = \sum_l \mathbb{E}_{q(o_l,s_l|\pi)}[\log q(s_l|\pi) - \log p(s_l, o_l|\pi)]$$
$$= \sum_l -H\{q(s_l|\pi)\} - \mathbb{E}_{q(o_l,s_l|\pi)}[\log p(s_l, o_l|\pi)] \quad (14)$$

where H denotes the Shannon entropy. It can be seen in (14) that selecting a policy that minimizes the EFE entails the maximization of the posterior entropy (promoting exploration) and the joint posterior over the states and observations (reaching the desired goal) [32].

To reach the desired goal, we produce roll-outs of states s_l given a certain policy and approximate the risk term $-\mathbb{E}_q[\log p(s_l, o_l|\pi)]$ to be minimized as:

$$-\mathbb{E}_{q(o_l,s_l|\pi)}[\log p(s_l, o_l|\pi)] \sim ||s_l - s^*||_2^2 \quad (15)$$

where s^* is the desired state that the agent must reach, corresponding to a desired observation (e.g., the agent's position). Since the observation o_l can encompass more than just the goal to be reached (i.e., the observation o_l could be both the position and the velocity of an agent, even though the desired goal is to reach a specific position regardless of the velocity), we compute the *goal* state s^* as:

$$s^* = \arg\max_s \int_{\omega \in D_\omega} q(s|\Omega^*, \omega) d\omega \quad (16)$$

where Ω^* contains all observations that must be reached in order to attain the desired goal and ω designates all observation modalities that are *not* taking part in defining the goal that must be reached, with D_ω their domain of definition. In practice, (16) is estimated by averaging the output of the posterior network, while sweeping ω for a grid of possible values and keeping Ω^* fixed.

Regarding the exploration term in (14), our Hebbian network does not directly allow the estimation of the entropy $H\{q(o_l, s_l|\pi)\}$, since the network does not infer standard deviations as in a variational auto-encoder (VAE) [17]. We propose to replace the maximization of the entropy $H\{q(o_l, s_l|\pi)\}$ with a surrogate term, crafted to promote exploration as well. As a surrogate for $H\{q(o_l, s_l|\pi)\}$, we choose to maximize the variance (noted Var) of the state trajectory $s_l, \forall l = 1, ..., L$ along time during the roll-outs. Intuitively, a state trajectory that presents lots of variation in time will promote the exploration of new states, providing a similar qualitative effect as maximizing $H\{q(o_l, s_l|\pi)\}$. Therefore, we select the policy π such that the distance to the desired state is minimized, while achieving a state trajectory variance larger than a certain threshold t_v.

$$\pi^* = \arg\min_\pi \mathcal{G}(\pi) = \sum_{l=1}^{L} ||s_l - s^*||_2^2 \quad \text{s.t.} \quad \text{Var}(||s_l - s^*||_2^2, l = 1, ..., L) \geq t_v \quad (17)$$

Given a set of N_p policies to try, t_v can be determined in an adaptive way as follows, such that the divergence from the desired state is minimized, while ensuring the variance of counterfactual state trajectories exceeds a certain threshold:

$$t_v = \beta \times \frac{1}{2}[\max_{\pi}(\text{Var}(||s_l(\pi) - s^*||_2^2, \forall l)) + \min_{\pi}(\text{Var}(||s_l(\pi) - s^*||_2^2, \forall l))] \quad (18)$$

where β is the strength hyper-parameter (empirically set to 0.5 in our experiments reported below). β acts as the precision or inverse temperature parameter associated with prior preferences (i.e., the precision of the prediction error between predicted and desired states). It must be noted that this approach might have some limitations since it could promote a presence of noisy perturbation during state transition (as the variance in (17) represents the entropy of the *environment* vs. the entropy of the *agent's brain* in (14))

4 Experimental Results

The aim of our experimental studies is to determine *i)* how the main network hyper-parameters (number of neurons, sparsity in output activity,...) impact the success rate of the proposed Hebbian AIF system; *ii)* to what extent Hebbian AIF is robust when learning without using a replay buffer and *iii)* how Hebbian AIF compares to Q-learning (which uses *dense rewards* versus *unsupervised learning* in Hebbian AIF).

4.1 Mountain Car Environment

We perform experiments in the *Mountain Car* environment from the OpenAI gym suite [21]. In this task, a car starts at a *random* position at the bottom of a hill and is expected to reach the top of a mountain within 200 time steps. The agent is subject to gravity and cannot reach the goal trivially, just by accelerating towards it. Rather, the agent must learn to gain momentum before accelerating towards the goal.

In this environment, the x-axis position x and the velocity v_x of the car constitute the input observations to the Hebbian AIF network. Before feeding the observation tuple (x, v_x) to our Hebbian network, we normalize (x, v_x) using (19) in order to equalize the dynamic range of the position and velocity signals:

$$\begin{cases} x \leftarrow \frac{x-\mu_x}{\sigma_x} \\ v_x \leftarrow \frac{v_x-\mu_{v_x}}{\sigma_{v_x}} \end{cases} \quad (19)$$

where (μ_x, σ_x) and $(\mu_{v_x}, \sigma_{v_x})$ denote the mean and standard deviation of the position and velocity signals respectively (estimated during random environment runs).

We use an action space constituted by two discrete actions: *accelerate to the left* and *accelerate to the right*. In addition, each action is repeated for 10 consecutive time steps once selected during the Expected Free Energy minimization in (17).

In order to compute the Expected Free Energy, we generate roll-outs of $L = 200$ time step predictions for 100 different random policies $\pi^j, j = 1, ...100$ with equal probability of selecting the *accelerate to the left* or the *accelerate to the right* actions.

As learning rate for the Hebbian learning mechanism (4), we use $\eta_d = 10^{-4}$ with a *decay rate* of 0.8 applied at the end of each *successful* episode, i.e. if the episode terminates successfully, $\eta_d \leftarrow \eta_d \times 0.8$ (else no decay is applied on η_d). All weights are initialized randomly from a normal distribution with standard deviation 0.01.

In the remainder of this Sections, we perform all our experiments using a 10-fold validation approach, by reporting the success rate curves as averages over 10 different runs (with 35 episodes per runs), with different random network initializations. For each run, we compute the success rate curve using a moving average window of size 5, and report the mean success rate curve by averaging over the 10 runs, alongside with its standard deviation (see e.g. Fig. 3). We will now study the impact of the various network hyper-parameters on the achieved success rates so as to give a *complete account* of their effect during model tuning.

4.2 Impact of the Number of Neurons in the Posterior and State Transition Networks

Figure 3 and 4 show the effect of sweeping the number of coding neurons M_Q and M_P in both the *posterior* and *state-transition* networks. Figure 3 shows that for $M_Q < 8$, the success rate is sub-optimal, but reaches a steady plateau around $M_Q = 8$ (orange curve in Fig. 3). Then, as M_Q is increased for $M_Q > 8$, the success rate becomes sub-optimal again, with dips in the performance along the episodes (e.g., red curve in Fig. 3). This phenomenon can be explained as follows: for $M_Q < 8$, the posterior network does not have enough parameters to capture the input dynamics into its latent space and *under-fits*, while for $M_Q > 8$, the posterior network starts over-fitting, reducing the success rate again[1]. Regarding the *state-transition* network, Fig. 4 shows that the higher the number of neurons M_P, the flatter the success rate curves become, leading to higher performance. The state transition network does not seem to over-fit as M_P is increased (for $\lambda_P = 10^{-4}$ kept fixed). Rather, Fig. 4 indicates that a higher state-transition

[1] Note that we are using our Hebbian scheme to amortize variational inference. This means we are optimizing amortization (encoding) parameters, not the (decoding) parameters of the generative model. An alternative approach would be to treat the model parameters as random variables and derive the update rules for minimizing variational free energy, in the form of a Hebbian update. In this instance, over-fitting would be precluded because of the complexity term in (6). However, this would not be amortization; this would be an implementation of AIF as described in [34].

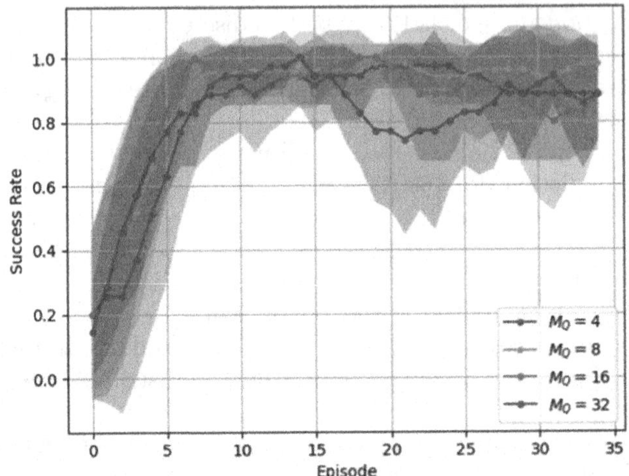

Fig. 3. Impact on the success rate when changing the *number of neurons* M_Q in the *posterior network*.

network capacity is beneficial for capturing important dynamics in the latent space, at the output of the posterior network.

4.3 Impact of the Sparsity of the Output Activity in the Posterior and State Transition Networks

Figure 5 and 6 show the effect of sweeping the *sparsity-defining* hyper-parameters λ_Q and λ_P in both the *posterior* and *state-transition* networks. For the posterior network, Fig. 5 shows that the success rate performance initially grows as λ_Q is increased from $\lambda_Q = 10^{-6}$ to $\lambda_Q = 10^{-5}$. Doing so, the non-linearity of the posterior network is increased, better capturing observation features into its latent space. Then, as λ_Q grows past $\lambda_Q = 10^{-4}$, the success rate degrades again, indicating a too strong posterior network non-linearity.

Regarding the *state-transition* network, Fig. 6 a) shows that the lower λ_P, the higher the success rate becomes. This suggests that making the state-transition network *more linear* (i.e., lower λ_P) better captures the dynamics of the latent space produced by the posterior network (other parameters kept fixed).

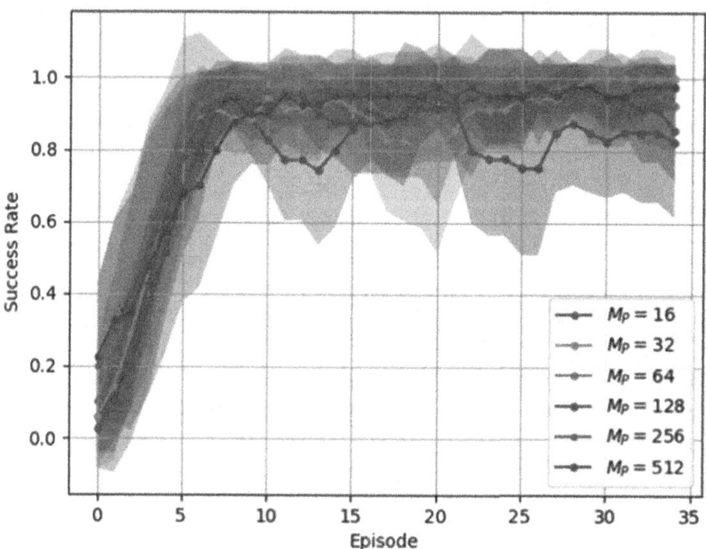

Fig. 4. Impact on the success rate when changing the *number of neurons* M_P in the *state transition network*.

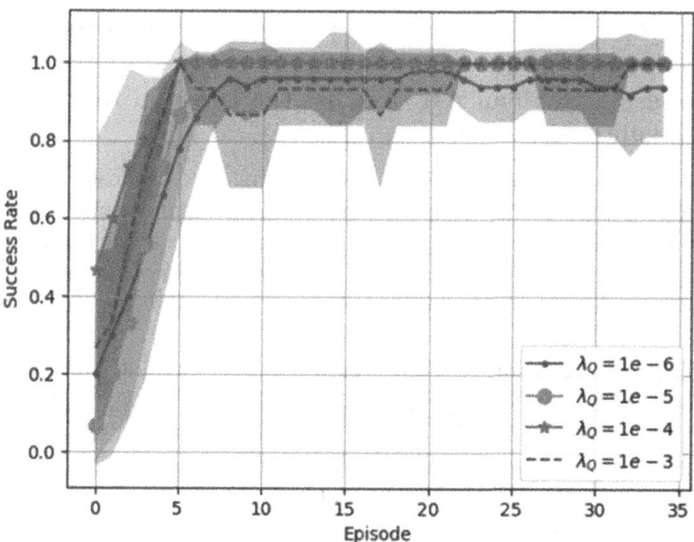

Fig. 5. Impact on the success rate when changing the *sparsity hyper-parameter* λ_Q in the *posterior network*.

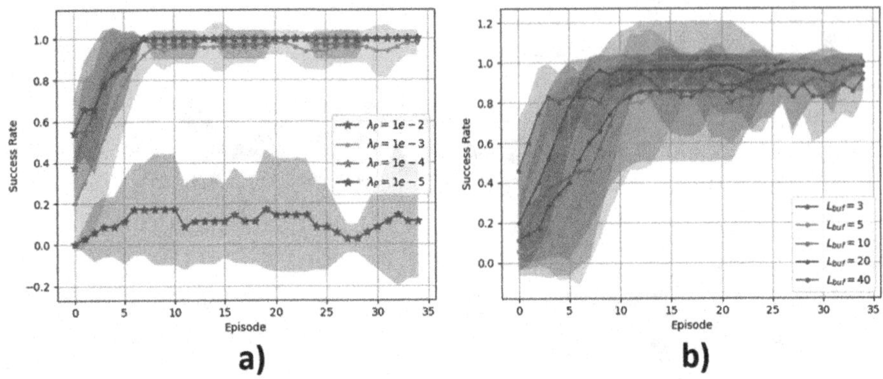

Fig. 6. a) Impact on the success rate when changing the *sparsity* λ_P in the *state transition network*. b) Success rate when changing the *time-lag buffer length* L_{buf}.

4.4 Impact of the Time-Lag Buffer Length on Task Performance

Figure 6 b) shows how the length L_{buf} of the time-lag buffer impacts the achieved success rate. Initially, as L_{buf} increases, the success rate increases as well, due to an increased availability of past latent states used by the state-transition network to estimate the next expected state. Then, as L_{buf} is further increased for $L_{buf} > 20$, the success rate drops again due to the addition of latent states *from deep in the past* that are less useful for estimating the present dynamics.

4.5 Comparing Hebbian AIF Against the Use of a Replay Buffer and Against Q-Learning

Figure 7 compares the success rate obtained using our proposed Hebbian AIF system against *a)* the use of a replay buffer during learning and *b)* the use of a Q-learning agent. Experience replay is done by saving the history of observation-action pairs in a buffer after each episode. After the end of the episode, a past experience is randomly selected and used to train the Hebbian AIF system for one episode.

Regarding the Q-learning setup, we use a standard Q-table learning approach [22], with the python implementation proposed in [33].

Figure 7 shows that our Hebbian AIF system converges much faster than the Q-learning system and behaves in a comparable manner to the Hebbian AIF setup with a *replay buffer*. Indeed, the Q-learning agent needs two orders of magnitude more training episodes in order to converge, despite the fact that it utilizes the *dense rewards* provided by the Mountain Car environment [21] (vs. *unsupervised* learning for Hebbian AIF)[2]. This confirms prior observations

[2] Although we have referred to the AIF scheme as unsupervised, there is an implicit constraint on behavior that is implemented, in this instance, by the goal states

about the efficient convergence of AIF systems, due to their ability to learn a generative model of environment dynamics used to select actions during learning (vs. supervised learning of a Q-table) [17].

Finally, it is interesting to note that, compared to the Deep AIF results *reported in* [17] (using a fully-connected 2-hidden-layer network trained through backprop), the Hebbian AIF system proposed in this work eventually reaches ~100% success rate (see red curve in Fig. 6 a) while the system in [17] reaches ~95%, motivating further investigations of Hebbian learning for AIF systems.

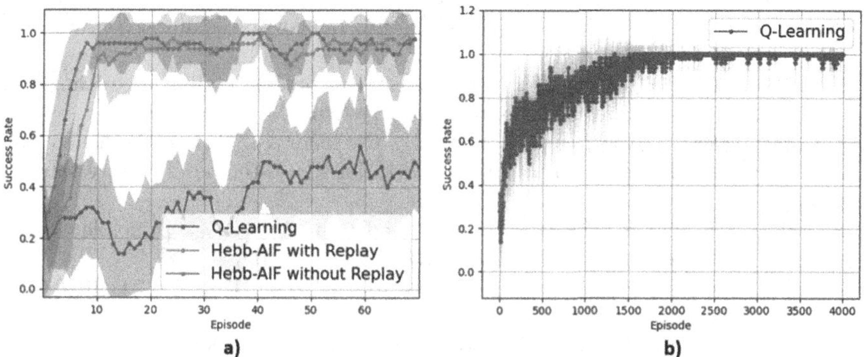

Fig. 7. Hebbian AIF versus the use of a *replay buffer* and *Q-learning* (a). Q-learning needs two orders of magnitude more episodes in order to converge (b).

5 Conclusion

This paper has investigated how neural ensembles equipped with local Hebbian plasticity can perform active inference for the control of dynamical agents. First, a Hebbian network architecture performing joint dictionary learning and sparse coding has been introduced for implementing both the posterior and the state-transition models forming our generative Active Inference system. Then, it has been shown how Free Energy minimization can be performed by the proposed Hebbian AIF system. Finally, extensive experiments for parameter exploration and benchmarking have been performed to study the impact of the network parameters on the task performance. Experimental results on the Mountain Car environment show that the proposed system outperforms the use of Q-learning, while not requiring the use of a replay buffer during learning, motivating future investigations of using Hebbian learning for designing active inference systems.

in (16). In AIF, goal-directed behavior emerges from inferring the right courses of action that lead to preferred outcomes. In amortized AIF, this planning as inference is learned; as we have demonstrated. In contrast, reinforcement learning ignores inference and simply learns rewarded behaviors, which can take a very long time - because there is no learning of a generative model, or the constraints that it affords.

Acknowledgement. This research was partially funded by a Long Stay Abroad grant from the Flemish Fund of Research - Fonds Wetenschappelijk Onderzoek (FWO) - grant V413023N. This research received funding from the Flemish Government under the "Onderzoeksprogramma Artificiële Intelligentie (AI) Vlaanderen" programme.

References

1. Olshausen, B.A., Field, D.J.: Sparse coding with an overcomplete basis set: a strategy employed by V1? Vis. Res. **37**(23), 3311–3325 (1997)
2. Fang, M.S., Mudigonda, M., Zarcone, R., Khosrowshahi, A., Olshausen, B.: Learning and inference in sparse coding models with Langevin dynamics. Neural Comput. **34**(8), 1676–1700 (2022)
3. Lee, H., Battle, A., Raina, R., Ng, A.: Efficient sparse coding algorithms. In: Advances in Neural Information Processing Systems. MIT Press (2006)
4. Safa, A., Ocket, I., Bourdoux, A., Sahli, H., Catthoor, F., Gielen, G.: A new look at spike-timing-dependent plasticity networks for spatio-temporal feature learning (2022)
5. Friston, K., Kiebel, S.: Predictive coding under the free-energy principle. Philos. Trans. R. Soc. Lond. Ser. B Biol. Sci. **364**, 1211–21 (2009)
6. Friston, K.: Does predictive coding have a future? Nat. Neurosci. **21**, 1019–1021 (2018). https://doi.org/10.1038/s41593-018-0200-7
7. Zahid, U., Guo, Q., Fountas, Z.: Predictive coding as a neuromorphic alternative to backpropagation: a critical evaluation (2023)
8. Olshausen, B., Field, D.: Emergence of simple-cell receptive field properties by learning a sparse code for natural images. Nature **381**, 607–609 (1996)
9. Safa, A., Ocket, I., Bourdoux, A., Sahli, H., Catthoor, F., Gielen, G.: Event camera data classification using spiking networks with spike-timing-dependent plasticity. In: 2022 International Joint Conference on Neural Networks (IJCNN), pp. 1–8 (2022)
10. Bi, G., Poo, M.: Synaptic modifications in cultured hippocampal neurons: dependence on spike timing, synaptic strength, and postsynaptic cell type. J. Neurosci. **18**(24), 10464–10472 (1998)
11. Rao, R., Ballard, D.: Predictive coding in the visual cortex: a functional interpretation of some extra-classical receptive-field effects. Nat. Neurosci. **2**(1), 79–87 (1999)
12. Safa, A., Ocket, I., Bourdoux, A., Sahli, H., Catthoor, F., Gielen, G.G.E.: STDP-driven development of attention-based people detection in spiking neural networks. IEEE Transactions on Cognitive and Developmental Systems (2022). https://doi.org/10.1109/TCDS.2022.3210278
13. Neftci, E., Das, S., Pedroni, B., Kreutz-Delgado, K., Cauwenberghs, G.: Event-driven contrastive divergence for spiking neuromorphic systems. Front. Neurosci. **7** (2014)
14. Krotov, D., Hopfield, J.J.: Unsupervised learning by competing hidden units. Proc. Natl. Acad. Sci. **116**(16), 7723–7731 (2019)
15. Parr, T., Pezzulo, G., Friston, K.: Active Inference: The Free Energy Principle in Mind, Brain, and Behavior. The MIT Press, Cambridge (2022)
16. Isomura, T., Shimazaki, H., Friston, K.J.: Canonical neural networks perform active inference. Commun. Biol. **5**, 55 (2022). https://doi.org/10.1038/s42003-021-02994-2

17. Çatal, O., Wauthier, S., De Boom, C., Verbelen, T., Dhoedt, B.: Learning generative state space models for active inference. Front. Comput. Neurosci. **14** (2020)
18. Ueltzhöffer, K.: Deep active inference. Biol. Cybern. **112**, 547–573 (2018). https://doi.org/10.1007/s00422-018-0785-7
19. Fountas, Z., Sajid, N., Mediano, P., Friston, K.: Deep active inference agents using Monte-Carlo methods. In: Advances in Neural Information Processing Systems, pp. 11662–11675. Curran Associates Inc. (2020)
20. Van de Maele, T., Verbelen, T., Çatal, O., De Boom, C., Dhoedt, B.: Active vision for robot manipulators using the free energy principle. Front. Neurorobotics **15** (2021)
21. Brockman, G., et al.: Openai gym (2016). arXiv preprint arXiv:1606.01540
22. Sutton, R., Barto, A.: Reinforcement Learning: An Introduction. The MIT Press, Cambridge (2018)
23. Ororbia, A.G., Mali, A.: Backprop-free reinforcement learning with active neural generative coding. In: Proceedings of the AAAI Conference on Artificial Intelligence, vol. 36, no. 1, pp. 29–37 (2022)
24. Ororbia, A., Mali, A.: Active predicting coding: brain-inspired reinforcement learning for sparse reward robotic control problems. In: IEEE International Conference on Robotics and Automation (ICRA) (2023)
25. Liang, Y., et al.: Can a fruit fly learn word embeddings? In: International Conference on Learning Representations (2021)
26. Ablin, P., Moreau, T., Massias, M., Gramfort, A.: Learning step sizes for unfolded sparse coding. In: Advances in Neural Information Processing Systems. Curran Associates Inc. (2019)
27. Lin, T.H., Tang, P.T.P.: Sparse Dictionary Learning by Dynamical Neural Networks. Presented at the (2019)
28. Friston, K.: The free-energy principle: a unified brain theory? Nat. Rev. Neurosci. **11**, 127–138 (2010). https://doi.org/10.1038/nrn2787
29. Hershey, J.R., Olsen, P.A.: Approximating the Kullback Leibler divergence between gaussian mixture models. In: 2007 IEEE International Conference on Acoustics, Speech and Signal Processing - ICASSP '07, Honolulu, HI, USA, 2007, pp. IV-317–IV-320 (2007). https://doi.org/10.1109/ICASSP.2007.366913
30. Kim, E., Lawson, E., Sullivan, K., Kenyon, G.: Spatiotemporal sequence memory for prediction using deep sparse coding. In: Proceedings of the 7th Annual Neuro-Inspired Computational Elements Workshop. Association for Computing Machinery (2019)
31. Werbos, P.J.: Backpropagation through time: what it does and how to do it. Proc. IEEE **78**(10), 1550–1560 (1990). https://doi.org/10.1109/5.58337
32. Schwartenbeck, P., FitzGerald, T., Dolan, R., Friston, K.: Exploration, novelty, surprise, and free energy minimization. Front. Psychol. **4** (2013)
33. https://gist.github.com/gkhayes/3d154e0505e31d6367be22ed3da2e955 . Accessed 1 May 2023
34. Friston, K.: Hierarchical models in the brain. PLoS Comput. Biol. **4**(11), 1–24 (2008)

Learning One Abstract Bit at a Time Through Self-invented Experiments Encoded as Neural Networks

Vincent Herrmann[1(✉)], Louis Kirsch[1], and Jürgen Schmidhuber[1,2]

[1] IDSIA/USI/SUPSI, Lugano, Switzerland
{vincent.herrmann,louis.kirsch,juergen}@idsia.ch
[2] AI Initiate, King Abdullah University of Science and Technology (KAUST), Thuwal, Saudi Arabia

Abstract. There are two important things in science: (A) Finding answers to given questions, and (B) Coming up with good questions. Our artificial scientists not only learn to answer given questions, but also continually invent new questions, by proposing hypotheses to be verified or falsified through potentially complex and time-consuming experiments, including thought experiments akin to those of mathematicians. While an artificial scientist expands its knowledge, it remains biased towards the simplest, least costly experiments that still have surprising outcomes, until they become boring. We present an empirical analysis of the automatic generation of interesting experiments. In the first setting, we investigate self-invented experiments in a reinforcement-providing environment and show that they lead to effective exploration. In the second setting, pure thought experiments are implemented as the weights of recurrent neural networks generated by a neural experiment generator. Initially interesting thought experiments may become boring over time.

Keywords: Reinforcement Learning · Exploration

1 Introduction and Previous Work

It has been pointed out that there are two important things in science: (A) Finding answers to given questions, and (B) Coming up with good questions, e.g., [2,30,31,42,60,63,65,68]. (A) is arguably just the standard problem of computer science. But how to implement the creative part (B) in artificial systems through reinforcement learning (RL), gradient-based artificial neural networks (NNs), and other machine learning methods?

For at least three decades, work on artificial scientists equipped with artificial curiosity and creativity has been published that addresses this question, e.g., [33, 38,40,42,48,53,57,60,70,72,73]. One early such work is the intrinsic motivation-based **adversarial system** from 1990 [38,42]. It is an artificial Q&A system designed to invent and answer questions. For that, it uses two artificial NNs. The first NN is called the controller C. C probabilistically generates outputs that

C. L. Buckley et al. (Eds.): IWAI 2023, CCIS 1915, pp. 254–274, 2024.
https://doi.org/10.1007/978-3-031-47958-8_16

may influence an environment. The second NN is called the world model M. It predicts the environmental reactions to C's outputs. Using gradient descent, M minimizes its error, thus becoming a better predictor. But in a zero-sum game, the reward-maximizing C tries to find sequences of output actions that maximize the error of M. M's loss is the gain of C (like in the later application of artificial curiosity called GANs [10,64], but also for the more general cases of sequential data and RL [20,74,80]).

C is asking questions through its action sequences: What happens if I do that? M is learning to answer those questions. C is motivated to come up with questions where M does not yet know the answer and loses interest in questions with known answers.

This type of Q&A system helps to understand the world, which is necessary for planning [38,39,42] and may boost external reward [2,31,40,50,52,58]. Clearly, the adversarial approach makes for a fine exploration strategy in many deterministic environments. **In stochastic environments, however, it might fail.** C might learn to focus on those parts of the environment where M can always get high prediction errors due to randomness, or due to computational limitations of M. For example, an agent controlled by C might get stuck in front of a TV screen showing highly unpredictable white noise, e.g., [2,57]. Therefore, in stochastic environments, C's reward should not be the errors of M, but (an approximation of) the *first derivative* of M's errors across subsequent training iterations, that is, M's **learning progress or improvements** [40,54]. As a consequence, despite M's high errors in front of a noisy TV screen, C won't get rewarded for getting stuck there, simply because M's errors won't improve. Both the totally predictable and the fundamentally unpredictable will get boring.

This simple insight led to lots of follow-up work [57]. For example, one particular RL approach for artificial curiosity in stochastic environments was published in 1995 [72]. A simple M learned to predict or estimate the probabilities of the environment's possible responses, given C's actions. After each interaction with the environment, C's intrinsic reward was the KL-Divergence [25] between M's estimated probability distributions before and after the resulting new experience—the **information gain** [72]. This was later also called *Bayesian Surprise* [19]. Compare earlier work on information gain [66] and its maximization *without* RL & NNs [6].

In the general RL setting where the environment is only partially observable [61, Sec. 6], C and M may greatly profit from a memory of previous events [38,39,43]. Towards this end, both C and M can be implemented as LSTMs [7,12,16,61] or Transformers [28,75].

The better the predictions of M, the fewer bits are required to encode the history H of observations because short codes can be used for observations that M considers highly probable [17,83]. That is, the learning progress of M has a lot to do with the concept of *compression progress* [53,55–57]. But it's not quite the same thing. In particular, it does not take into account the bits of information needed to specify M. A more general approach is based on algorithmic information theory, e.g., [22,26,51,69,78,79]. Here C's intrinsic reward is indeed based on **algorithmic compression progress** [53,55–57] based on some coding scheme for the weights of the model network, e.g., [8,15,23,24,46,47,71], and also a coding scheme for the history of all observations so far, given the model [15,17,34,53,78,83]. Note that the history of science is a history of compression progress through incremental discovery of simple laws that govern seemingly complex observation sequences [53,55–57].

In early systems, the questions asked by C were restricted in the sense that they always referred to all the details of future inputs, e.g., pixels [38,42]. That's why in 1997, a more general adversarial RL machine was built that could ignore many or all of these details and ask **arbitrary abstract questions** with computable answers [48–50]. Example question: if we run this policy (or program) for a while until it executes a special interrupt action, will the internal storage cell number 15 contain the value 5, or not? Again there are two learning, reward-maximising adversaries playing a zero-sum game, occasionally betting on different yes/no outcomes of such computational experiments. The winner of such a bet gets a reward of 1, the loser −1. So each adversary is motivated to come up with questions whose answers surprise the other. And both are motivated to avoid seemingly trivial questions where both already agree on the outcome, or seemingly hard questions that none of them can reliably answer for now. This is the approach closest to what we will present in the following sections.

All the systems above (now often called CM systems [62]) actually maximize the sum of the standard external rewards (for achieving user-given goals) and the intrinsic rewards. **Does this distort the basic RL problem?**

It turns out not so much. Unlike the external reward for eating three times a day, the curiosity reward in the systems above is ephemeral, because once something is known, there is no additional intrinsic reward for discovering it again. That is, the external reward tends to dominate the total reward. In totally learnable environments, in the long run, the intrinsic reward even *vanishes* next to the external reward. Which is nice, because in most RL applications we care only for the external reward.

RL Q&A systems of the 1990s did not **explicitly, formally enumerate their questions.** But the more recent PowerPlay framework (2011) [60,70] does. Let us step back for a moment. What is the set of all formalisable questions? How to decide whether a given question has been answered by a learning machine? To define a question, we need a computational procedure that takes a solution candidate (possibly proposed by a policy) and decides whether it is an answer to the question or not. PowerPlay essentially enumerates the set of all such procedures (or some user-defined subset thereof), thus enumerating all possible questions or problems. **It searches for the simplest question that the current policy cannot yet answer but can quickly *learn* to answer *without* forgetting the answers to previously answered questions.** What is the simplest such Q&A to be added to the repertoire? It is the cheapest one—the one that is found first. Then the next trial starts, where new Q&As may build on previous Q&As.

In our empirical investigation of Sect. 3, we will revisit the above-mentioned concepts of complex computational experiments with yes/no outcomes, focusing on two settings: (1) The generation of experiments driven by model prediction error in a deterministic reinforcement-providing environment, and (2) An approach where C (driven by information gain) generates pure thought experiments in form of weight matrices of RNNs.

2 Self-invented Experiments Encoded as Neural Networks

We present a CM system where C can design essentially arbitrary computational experiments (including thought experiments) with binary yes/no outcomes. Experiments may run for several time steps. However, C will prefer simple experiments whose outcomes still surprise M, until they become boring.

In general, both the controller C and the model M can be implemented as (potentially multi-dimensional) LSTMs [11]. At each time step $t = 1, 2, \ldots,$ C's input includes the current sensory input vector $in(t)$, the external reward vector $R_e(t)$, and the intrinsic curiosity reward $R_i(t)$. C may or may not interact directly with the environment through action outputs. How does C ask questions and propose experiments? C has an output unit called the START unit. Once it becomes active (>0.5), C uses a set of extra output units for producing the *weight matrix or program* θ of a separate RNN or LSTM called E (for Experiment), in fast weight programmer style [4,9,18,21,36,37,41,44,45].

E takes sensory inputs from the environment and produces actions as outputs. It also has two additional output units, the HALT unit [59] and the RESULT unit. Once the weights θ are generated at time step t', E is tested in a trial, interacting with some environment. Once E's HALT unit exceeds 0.5 in a later time step t'', the current experiment ends. That is, the experiment computes its own runtime [59]. The experimental outcome $r(t'')$ is 1 if the activation $result(t'')$ of E's RESULT unit exceeds 0.5, and 0 otherwise. At time t', so before the experiment is being executed, M has to compute its output $pr(t') \in [0, 1]$ from θ (and the history of C's inputs and actions up to t', which includes all previous experiments their outcomes). Here, $pr(t')$ models M's (un)certainty that the final binary outcome of the experiment will be 1 (YES) or 0 (NO). Then the experiment is run.

In short, C is proposing an experimental question in form of θ that will yield a binary answer (unless some time limit is reached). M is trying to predict this answer before the experiment is executed. Since E is an RNN and thus a general computer whose weight matrix can implement any program executable on a traditional computer [67], any computable experiment with a binary outcome can be implemented in its weight matrix (ignoring storage limitations of finite RNNs or other computers). That is, by generating an appropriate weight matrix θ, C can ask any scientific question with a computable solution. In other words, C can propose any scientific hypothesis that is experimentally verifiable or falsifiable.

At t'', M's previous prediction $pr(t')$ is compared to the later observed outcome $r(t'')$ of C's experiment (which spans $t'' - t'$ time steps), and C's intrinsic curiosity reward $R_i(t'')$ is proportional to M's surprise. To calculate it, we interpret $pr(t')$ as M's estimated probability of $r(t'')$, given the history of observations so far. Then we train M by gradient descent (with regularization to avoid overfitting) for a fixed amount of time to improve all of its previous predictions including the most recent one. This yields an updated version of M called M^*.

In general, M^* will compute a different prediction $PR(t')$ of $r(t'')$, given the history up to $t' - 1$. At time t'', the contribution $R_{IG}(t'')$ to C's curiosity reward is proportional to the apparent resulting information gain, the KL-divergence

$$R_{IG}(t'') \sim D_{KL}\big(PR(t')||pr(t')\big).$$

If M had a confident belief in a particular experimental outcome, but this belief gets shattered in the wake of C's experiment, there will be a major surprise and a big insight for M, as well as lots of intrinsic curiosity reward for C. On the other hand, if M was quite unsure about the experimental outcome, and remains quite unsure afterwards, then C's experiment can hardly surprise M and C will fail to profit much. C is motivated to propose *interesting* hypotheses or experiments that violate M's current deep beliefs and expand its horizon. An alternative intrinsic curiosity reward would be based on compression progress [53,55–57].

Note that the entire experimental protocol is the responsibility of θ. Through θ, E must initialize the experiment (e.g., by resetting the environment or moving the agent to some start position if that is important to obtain reliable results), then run the experiment by executing a sequence of computational steps or actions, and translate the incoming data sequence into some final abstract binary outcome YES or NO.

C is motivated to design experimental protocols θ that surprise M. C will get bored by experiments whose outcomes are predicted by M with little confidence (recall the noisy TV), as well as by experiments whose outcomes are correctly predicted by M with high confidence. *C will get rewarded for surprising experiments whose outcomes are incorrectly predicted by M with high confidence.*

A negative reward per time step encourages C to be efficient and lazy and come up with simple and fast still surprising experiments. If physical actions in the environment cost much more energy (resulting in immediate negative reward) than E's internal computations per time step, C is motivated to propose a θ defining a "thought experiment" requiring only internal computations, without executing physical actions in the (typically non-differentiable) environment. In fact, due to C's bias towards the computationally cheapest and least costly experiments that are still surprising to M, most of C's initial experiments may be thought experiments. Hence, since C, E and M are differentiable, not only M but also C may be often trainable by backpropagation [4] rather than the generally slower policy gradient methods [1,29,77,81]. Of course, this is only true if the reward function is also differentiable with respect to C's parameters.

3 Experimental Evaluation

Here we present initial studies of the automatic generation of interesting experiments encoded as NNs. We evaluate these systems empirically and discuss the associated challenges. This includes two setups: (1) Adversarial intrinsic reward encourages experiments executed in a differentiable environment through sequences of continuous control actions. We demonstrate that these experiments aid the discovery of goal states in a sparse reward setting. (2) Pure thought experiments encoded as RNNs (without any environmental interactions) are guided by an information gain reward.

Together, these two setups cover the important aspects discussed in Sect. 2: the use of abstract experiments with binary outcomes as a method for curious exploration, and the creation of interesting pure thought experiments encoded as RNNs. We leave the integration of both setups into a single system (as described in Sect. 2) for future work.

3.1 Generating Experiments in a Differentiable Environment

Reinforcement learning (RL) usually involves exploration in an environment with non-differentiable dynamics. This requires RL methods such as policy gradients [82]. To simplify our investigation and focus solely on the generation of self-invented experiments, we introduce a fully differentiable environment that allows for computing analytical policy gradients via backpropagation. This does not limit the generality of our approach, as standard RL methods can be used instead.

Our continuous force field environment is depicted in Fig. 1. The agent has to navigate through a 2D environment with a fixed external force field. This force field can have different levels of complexity. The states in this environment are the position and velocity of the agent. The agent's actions are real-valued force vectors applied to itself. To encourage laziness and a bias towards simple experiments, each time step is associated with a small negative reward (-0.1). A sparse large reward (100) is given whenever the agent gets very close to the goal state. We operate in the single life setting without episodic resets. Additional information about the force field environment can be found in Appendix A. Since the environment is deterministic, it is sufficient for C to generate experiments whose results the current M cannot predict.

Method. Algorithm 1 and Fig. 2 summarize the process for generating a sequence of interesting abstract experiments with binary outcomes. The goal is to test the following three hypotheses:

- Generated experiments implement exploratory behavior, facilitating the reaching of goal states.
- If there are negative rewards in proportion to the runtime of experiments, then the average runtime will increase over time, as the controller will find it harder and harder to come up with new short experiments whose outcomes the model cannot yet predict.

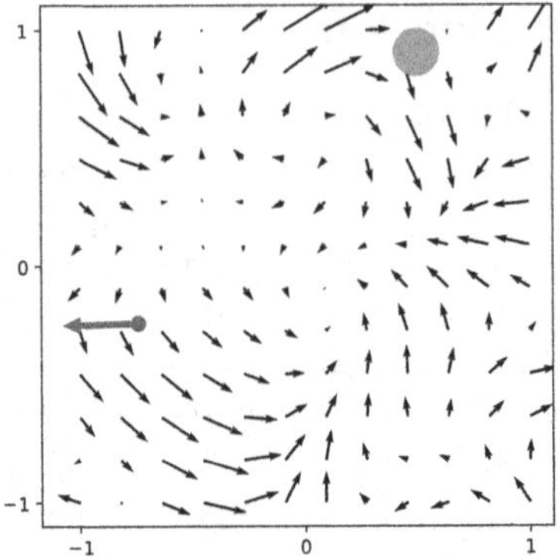

Fig. 1. A differentiable force field environment. The agent (red) has to navigate to the goal state (yellow) while the external force field exerts forces on the agent. (Color figure online)

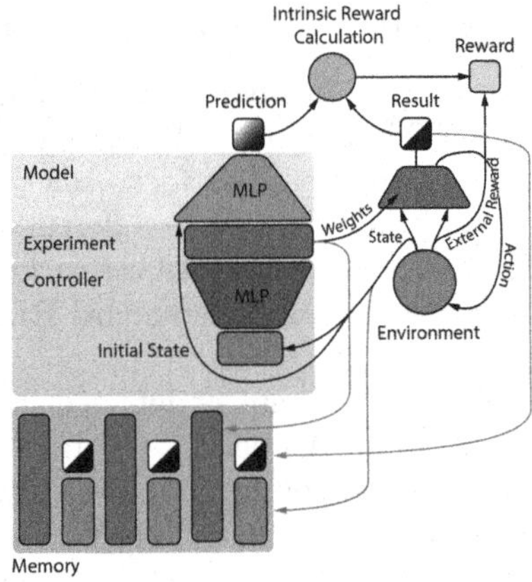

Fig. 2. Generating self-invented experiments in a differentiable environment. A controller C_ϕ is motivated to generate experiments E_θ that still surprise the model $M_\mathbf{w}$. After execution in the environment, the experiments and their binary results are stored in memory. The model is trained on the history of previous experiments.

- As the model learns to predict the yes/no results of more and more experiments, it becomes harder for the controller to create experiments whose outcomes surprise the model.

The generated experiments have the form $E_\psi(s) = (a, \hat{r})$, where E_ψ is a linear feedforward network with parameters ψ, s is the environment state, a are the actions and $\hat{r} \in [0, 1]$ is the experimental result. Both s and a are real-valued vectors.

Instead of a HALT unit, a single scalar $\tau \in \mathbb{R}^+$ determines the number of steps for which an experiment will run. To further simplify the setup, the experiment network is a feedforward NN without recurrence. To make the experimental result differentiable with respect to the runtime parameter, τ predicts the mean of a Gaussian distribution with fixed variance over the number of steps. The actual result $\tilde{r}$ is the expectation of the result unit $\hat{r}$ over the distribution defined by τ (more details on this can be found in Appendix A.1). The binarized result r has the value 1 if $\tilde{r} > 0.5$, and 0 otherwise. The parameters θ of the experiment are the network parameters ψ together with the runtime parameter τ, i.e. $\theta := (\psi, \tau)$.

For a given starting state s, the controller C_ϕ generates experiments: $C_\phi(s) = \theta$. C_ϕ is a multi-layer perceptron (MLP) with parameters ϕ, and θ denotes the parameters of the generated experiment. The model $M_\mathbf{w}$ is an MLP with parameters $\mathbf{w}$. It makes a prediction $M_\mathbf{w}(s, \theta) = \hat{o}$, with $\hat{o} \in [0, 1]$, for an experiment defined by the starting state s and the parameters θ.

During each iteration of the algorithm, C_ϕ generates an experiment based on the current state s of the environment. This experiment is executed until the cumulative halting probability defined by the generated τ exceeds a certain threshold (e.g., 99%). The starting state s, experiment parameters θ and binary result r are saved in a memory buffer $\mathcal{D}$ of experiments. Every state encountered during the experiment is saved to the state memory buffer $\mathcal{B}$.

After the experiment execution, the model $M_\mathbf{w}$ is trained for a fixed number of steps of stochastic gradient descent (SGD) to minimize the loss

$$\mathcal{L}_M = \mathbb{E}_{(s,\theta,r)\sim\mathcal{D}}[\text{bce}(M_\mathbf{w}(s, \theta), r)], \tag{1}$$

where bce is the binary cross-entropy loss function.

The third and last part of each iteration is the training of the controller C_ϕ. The loss that is being minimized via SGD is

$$\mathcal{L}_C = \mathbb{E}_{s\sim\mathcal{B}}[-\text{bce}\big(M_\mathbf{w}(s, C_\phi(s)), \tilde{r}(C_\phi(s), s)\big) - R_e(C_\phi(s), s)]. \tag{2}$$

The function $\tilde{r}$ maps the experiment parameters and starting state to the continuous result of the experiment. The function R_e maps the experiment parameters and starting state to the external reward. Note that gradient information will flow back from $\tilde{r}$ and R to ϕ through the execution of the experiment in the differentiable environment. The first term corresponds to the intrinsic reward for the controller, which encourages it to generate experiments whose outcomes $M_\mathbf{w}$ cannot predict. The second term is the external reward from the environment, which punishes long experiments. Since the reward for reaching the goal is sparse and not differentiable with respect to the experiment's actions, no information about the goal state reaches C_ϕ through the gradient.

Results and Discussion. To investigate our first hypothesis, Fig. 3a shows the cumulative number of times a goal state was reached during an experiment, adjusted by the number of environment interactions of each experiment. Specifically, it shows $h(j) = \sum_{k=1}^{j} \frac{g_k}{n_k}$, where $j = 1, 2, \ldots$ is the index of the generated experiment, g_k is 1 if the goal state was reached during the kth experiment and 0 otherwise, and n_k is the runtime of the kth experiment. Our method,

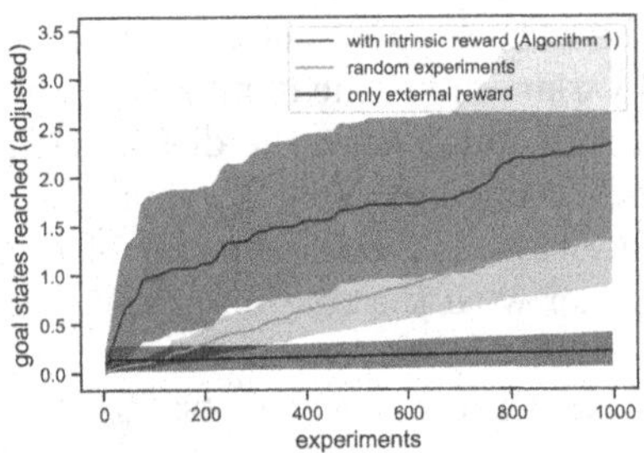

(a) Number of times the goal state was reached, adjusted by the number of environment interactions. Experiments generated with adversarial intrinsic reward benefit exploration more than random experiments. Without intrinsic motivation, the agent usually fails to reach any goal states in the sparse reward setting. Mean with bootstrapped 95% confidence intervals across 30 seeds.

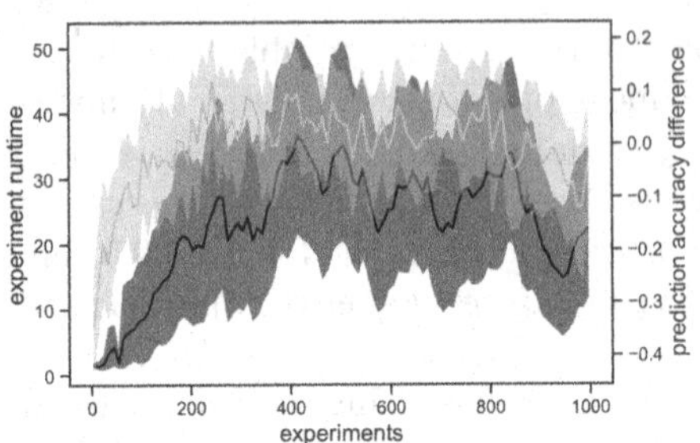

(b) Blue: the average runtime of experiments generated by C_ϕ. Purple: the difference between result prediction accuracy of the current $M_\mathbf{w}$ for the generated experiment and the average prediction accuracy of the current $M_\mathbf{w}$ for random experiments. As it gets harder for C_ϕ to generate experiments θ that are surprising to $M_\mathbf{w}$ (hard to predict), the runtime increases and the experiments tend to be harder to predict than the average randomly drawn experiment. Mean with bootstrapped 95% confidence intervals across 30 seeds.

Fig. 3. Experiments in the differentiable force field environment

as described above and in Algorithm 1, reaches the most goal states per environment interaction. Purely random experiments also discover goal states, but less frequently. Note that such random exploration in parameter space has been shown to be a powerful exploration strategy [32,35,76]. The average runtime of the random experiments is 50 steps, compared to 22.9 for the experiments generated by C_ϕ. To rule out a potential unfair bias due to different runtimes, Fig. 6 in the Appendix shows an additional baseline of random experiments with an average runtime of 20 steps, leading to results very similar to those of longer running random experiments. If we remove the intrinsic adversarial reward, the controller is left only with the external reward. This means that there is no bce term in Eq. 2. It is not surprising that in this setting, C_ϕ fails to generate experiments that discover goal states, since the gradient of $\mathcal{L}_C$ contains no information about the sparse goal reward.

Figure 3b addresses our second and third hypotheses. C_ϕ indeed tends to prolong experiments as $M_\mathbf{w}$ has been trained on more experiments, even though experiments with long runtimes are discouraged through the punitive external reward. Our explanation for this is that it becomes harder with time for C_ϕ to come up with short experiments for which $M_\mathbf{w}$ cannot yet accurately predict the correct results. This is supported by the fact that the prediction accuracy of $M_\mathbf{w}$ for newly generated experiments goes up. Specifically, Fig. 3b shows the difference between prediction accuracy of the current $M_\mathbf{w}$ for the newly generated experiment and the expected prediction accuracy of the current $M_\mathbf{w}$ for experiments randomly sampled from a simple prior. This accounts for the general gain of $M_\mathbf{w}$'s prediction accuracy over the course of training. It can be seen that in the beginning, C_ϕ is successful at creating adversarial experiments that surprise

$M_{\mathbf{w}}$. With time, however, it fails to continue doing so and is forced to create longer experiments to challenge $M_{\mathbf{w}}$.

3.2 Pure RNN Thought Experiments

The previous experimental setup uses feedforward NNs as experiments and an intrinsic reward function that is differentiable with respect to the controller's weights. This section investigates a complementary setup: interesting pure thought experiments (with no environment interactions) are generated in the form of RNNs without any inputs, driven by an intrinsic curiosity reward based on information gain which we treat as non-differentiable.

Algorithm 1. Adversarial yes/no experiments in a differentiable environment

Input: Randomly initialized differentiable Controller $C_\phi : \mathrm{S} \rightarrow \Theta$, randomly initialized differentiable Model $M_{\mathbf{w}} : \mathrm{S} \times \Theta \rightarrow \mathbb{R}$, empty experiment memory $\mathcal{D}$, empty state memory $\mathcal{B}$, set of random initial experiments $\mathcal{E}_{\text{init}}$, Differentiable environment
Output: An experiment memory populated with (formerly) interesting experiments

for $\theta \in \mathcal{E}_{\text{init}}$ **do**
  $s \leftarrow$ current environment state
  Execute the experiment parametrized by θ in the environment, obtain binary result r
  Save the tuple (s, θ, r) to $\mathcal{D}$
  Save all encountered states during the experiment to $\mathcal{B}$
end for
repeat
  $s \leftarrow$ current environment state
  $\theta \leftarrow C_\phi(s)$
  Execute the experiment parametrized by θ in the environment, obtain binary result r
  Save tuple (s, θ, r) to $\mathcal{D}$
  $\hat{s} \leftarrow$ current environment state
  for some steps **do**
    Sample tuple (s, θ, r) from $\mathcal{D}$
    Update the model using SGD: $\nabla_{\mathbf{w}} \mathrm{bce}(M_{\mathbf{w}}(s, \theta), r)$
  end for
  for some steps **do**
    Sample starting state s from $\mathcal{B}$
    Set environment to state s
    Execute the experiment parametrized by $C_\phi(s)$ in the environment, obtain continuous result $\tilde{r}$ and external reward R_e
    Update the controller using SGD: $\nabla_\phi\big(- \mathrm{bce}(M_{\mathbf{w}}(s, C_\phi(s)), \tilde{r}) - R_e\big)$
  end for
  Set environment to state $\hat{s}$
until no more interesting experiments are found

Method. In many ways, this new setup (depicted in Fig. 4 and described in Algorithm 2 in the Appendix) is similar to the one presented in Sect. 3.1. In what follows, we highlight the important differences.

An experiment E_θ is an RNN of the form $(h_{t+1}, r_{t+1}, \gamma_{t+1}) = E_\theta(h_t)$, where h_t is the hidden state vector, $r_t \in \{0, 1\}$ is the binary result at experiment time step t, and $\gamma_t \in [0, 1]$ is the HALT unit. The result r of E_θ is the r_t for the experiment step t where γ_t first is larger than 0.5. Since there is no external environment and the experiments are independent of each other, the model $M_\mathbf{w}$ is again a simple MLP with parameters $\mathbf{w}$. It takes only the experiment parameters θ as input and makes a result prediction $\hat{o} = M_\mathbf{w}(\theta), \hat{o} \in [0, 1]$.

As mentioned above, here we treat the intrinsic reward signal as non-differentiable. This means that—in contrast to the method presented in Sect. 3.1—the controller cannot receive information about $M_\mathbf{w}$ from gradients that are backpropagated through the model. Instead, it has to infer the learning behavior of $M_\mathbf{w}$ from the history ω of previous experiments and intrinsic rewards to come up with new surprising experiments. The controller C_ϕ is now an LSTM that is trained by DDPG [27] and generates new experiments solely based on the history of past experiments: $C_\phi(\omega) = \theta$. The history ω is a sequence of tuples (θ_i, r_i, R_i), where $i = 1, 2, \ldots$ is the index of the experiment. It contains experiments up to the last one that has been executed. More details on the training of $M_\mathbf{w}$ and the algorithm can be found in Appendix B.

For these pure thought experiments, we use a reward based on information gain. Let $\mathbf{w}$ be M's weights at certain point in time. Then a new experiment with parameters θ is generated, executed and saved to the buffer. On this buffer $\mathcal{D}$, which now includes θ, M is trained for a fixed number of SGD steps to obtain new weights $\mathbf{w}^*$. Then, the information gain reward associated with experiment θ is

$$R_{IG}(\theta, \mathbf{w}, \mathbf{w}^*) = \frac{1}{|\mathcal{D}|} \sum_{\tilde{\theta} \in \mathcal{D}} D_{KL}(M_{\mathbf{w}^*}(\tilde{\theta}) || M_\mathbf{w}(\tilde{\theta})), \tag{3}$$

where we interpret the output of the model as a Bernoulli distribution.

Results and Discussion. Figure 5 shows the information gain reward associated with each new experiment that C_ϕ generates. We observe that, after a short initial phase, the intrinsic information gain reward steadily declines. This is similar to what we observe for the prediction accuracy in Sect. 3.1: it becomes harder for the controller to generate experiments that surprise the model. It should be mentioned that this is a natural effect, since—as the model is trained on more and more experiments—every new additional experiment contributes on average less to the model's change during training, and thus is associated with less information gain reward. An interesting, albeit minor, effect shown in Fig. 5 is that also in this setup, the average runtime of the generated experiments increases slightly over time, even though there is no negative reward for longer thought experiments. For shorter experiments, however, it is apparently easier for the model to learn to predict the results. Hence, at least in the beginning,

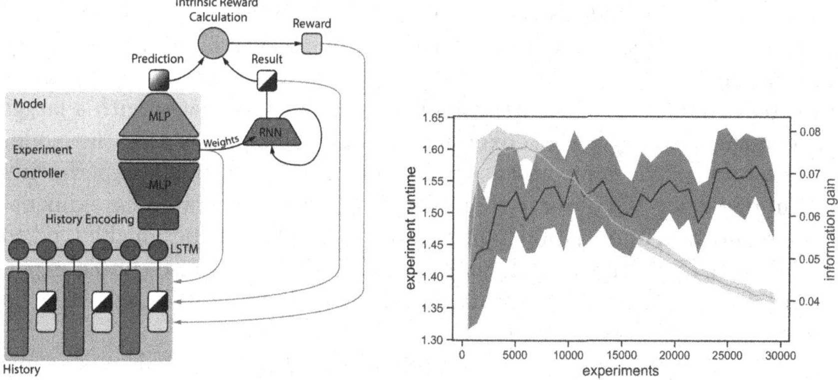

Fig. 4. Generating abstract thought experiments encoded as RNNs. The model is trained to predict the results of previous experiments. The controller generates new interesting thought experiments (without environment interactions) based on the history of previous experiments, their results and rewards.

Fig. 5. Empirical results for pure thought experiments encoded as RNNs. Blue: the average runtime of each experiment generated by C_ϕ. Purple: information gain reward (Eq. 3) for C_ϕ associated with each experiment. Mean with bootstrapped 95% confidence intervals across 20 seeds.

they yield more learning progress and more information gain. Later, however, longer experiments become more interesting.

In comparison to the experiments generated in Sect. 3.1, the present ones have a much shorter runtime. This is a side-effect of the experiments being RNNs with a HALT unit; for randomly initialized experiments, the average runtime is approximately 1.6 steps.

4 Conclusion and Future Work

We extended the neural Controller-Model (CM) framework through the notion of arbitrary self-invented computational experiments with binary outcomes: experimental protocols are essentially programs interacting with the environment, encoded as the weight matrices of RNNs generated by the controller. The model has to predict the outcome of an experiment based solely on the experiment's parameters. By creating experiments whose outcomes surprise the model, the controller curiously explores its environment and what can be done in it. Such a system is analogous to a scientist who designs experiments to gain insights about the physical world. However, an experiment does not necessarily involve actions taken in the environment: it may be a pure thought experiment akin to those of mathematicians.

We provide an empirical evaluation of two simple instances of such systems, focusing on different and complementary aspects of the idea. In the first setup, we show that self-invented abstract experiments encoded as feedforward networks interacting with a continuous control environment facilitate the discovery of rewarding goal states. Furthermore, we see that over time the controller is forced to create longer experiments (even though this is associated with a larger negative external reward) as short experiments start failing to surprise the model. In the second setup, the controller generates pure abstract thought experiments in the form of RNNs. We observe that over time, newly generated experiments result in less intrinsic information gain reward. Again, later experiments tend to have slightly longer runtime. We hypothesize that this is because simple experiments initially lead to a lot of information gain per time interval, but later do not provide much insight anymore.

These two empirical setups should be seen as initial steps towards more capable systems such as the one proposed in Sect. 2. Scaling these methods to more complex environments and the generation of more sophisticated experiments, however, is not without challenges. Direct generation and interpretation of NN weights may not be very effective for large and deep networks. Previous work [3] already combined hypernetworks [13] and policy fingerprinting [5,14] to generate and evaluate policies. Similar innovations will facilitate the generation of abstract self-invented experiments beyond the small scale setups presented in this paper.

A Experiments in the Force Field Environment

The force field of the environment is based on a 2D grid of randomly sampled force vectors. To get a continuous force field, bicubic interpolation between the vectors of the grid is used. Hence, the resolution of the grid influences the complexity of the force field (higher resolution $\rightarrow$ more intricate force field). In all experiments, the grid resolution is sampled uniformly from $\{(3,3),(5,5),(7,7)\}$. The random seed of each run affects both the force field and the position of the goal state. This means that every run has its own unique environment.

A.1 Experiment Execution

Let $\hat{r}_t \in [0,1]$ be the value of the result node at step t of the experiment whose runtime is determined by the parameter $\tau \in [0,100]$. The maximum runtime is fixed to 100 steps. A distribution over experiment steps t is defined by τ as follows: $p_\tau(t) = \frac{\exp(-0.5(t-\tau)^2)}{\sum_{u=1}^{100}\exp(-0.5(u-\tau)^2)}$.

The continuous result of the experiment is the expectation of the result unit over this distribution: $\tilde{r} = \mathbb{E}_{t\sim p_\tau}\hat{r}_t$. The binary result of the experiment r is the boolean value $\tilde{r} > 0.5$.

A.2 Hyperparameters for the Force Field Experiments

Table 1 shows the hyperparameters for Algorithm 1. The output nodes of C_ϕ that generate the parameters ψ of the experiment network have a tanh output nonlinearity and are then scaled to the predefined range. The output node that generates τ is clipped to the range $[0, 100]$.

The experiment parameters for random baselines are generated as $\psi = 2\tanh(v)$, where $v \sim \mathcal{N}(0, 4I)$. The runtime parameter τ is sampled uniformly from the allowed range. The hyperparameters for the model are the same as in Table 1. The baseline with only external reward also uses the hyperparameters of Table 1. The difference is that in this setting, the loss of the C_ϕ is simply $\mathcal{L}_C = \mathbb{E}_{s \sim \mathcal{B}}[-R(C_\phi(s), s)]$ instead of the one defined in Eq. 2.

Table 1. Hyperparameters for Algorithm 1

Hyperparameter	Value
hidden layers $M_{\mathbf{w}}$	[128, 128, 128, 128]
hidden layers C_ϕ	[128, 128, 128, 128]
training steps per iteration $M_{\mathbf{w}}$	100
training steps per iteration C_ϕ	100
learning rate $M_{\mathbf{w}}$	0.0003
learning rate C_ϕ	0.0003
weight decay $M_{\mathbf{w}}$	0.01
weight decay C_ϕ	0.01
experiment parameter range	[−2, 2]
noise input nodes C_ϕ	8
environment grid resolutions	[(3, 3), (5, 5), (7, 7)]
number of iterations	1000
number of initial experiments in $\mathcal{E}_{\text{init}}$	100

A.3 Additional Results

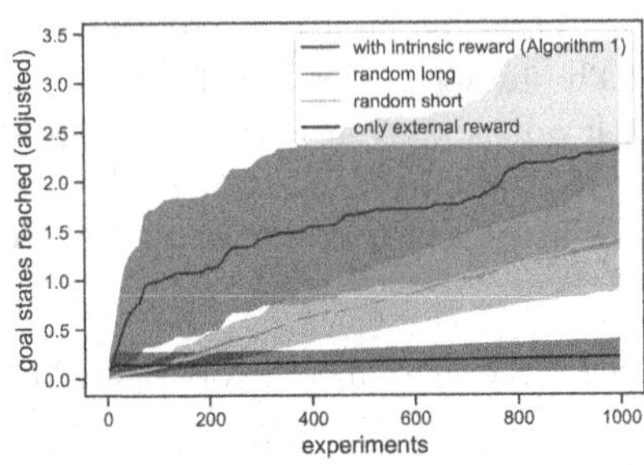

Fig. 6. Similar to Fig. 3a, but with an additional baseline of short random experiments with an average runtime of 20 steps.

To account for a potential bias due to experimental runtime, Fig. 6 shows the adjusted number of goal states for a baseline of shorter random experiments.

B Pure Thought Experiments

Algorithm 2 summarizes the method described in Sect. 3.2. In this setup, the model $M_{\mathbf{w}}$ is trained to minimize the following loss:

$$\mathcal{L}_M = \mathbb{E}_{(\theta,r)\sim\mathcal{D}}[\text{bce}(M_{\mathbf{w}}(\theta), r)]. \tag{4}$$

Efficient approximation of the policy gradients for the controller is achieved through an actor-critic method, specifically DDPG [27]. The controller C_ϕ has an additional LSTM encoder that generates a vector-sized representation of the history ω of previous experiments, their results and the reward associated with them. The actor is an MLP that receives as input the history representation created by the LSTM and generates the weights of an experiment RNN, whereas the critic receives both a history representation and experiment weights as input, and outputs a scalar reward estimation. Actor and critic share the same LSTM history encoder and take alternating gradient descent steps during training. The input to the LSTM history encoder is the sequence ω of the last 1000 that have been executed.

The experiment RNNs E_θ used in this empirical evaluation have 3 hidden units and no inputs. The initial hidden state h_0 is treated as part of the parameters θ and is thus also generated by C_ϕ. Random experiments are sampled the same way as described in Sect. A.2. All other hyperparameters are listed in Table 2.

Table 2. Hyperparameters for Algorithm 2

Hyperparameter	Value
hidden layers $M_{\mathbf{w}}$	[128, 128, 128, 128]
hidden layers C_ϕ LSTM	[64]
hidden layers C_ϕ MLP	[128, 128, 128, 128]
training steps per iteration $M_{\mathbf{w}}$	50
training steps per iteration C_ϕ	10
learning rate $M_{\mathbf{w}}$	0.0001
learning rate C_ϕ	0.0001
weight decay $M_{\mathbf{w}}$	0.01
weight decay C_ϕ	0.01
experiment parameter range	[–3, 3]
number of iterations	30000
number of initial experiments in $\mathcal{E}_{\text{init}}$	100

Algorithm 2. Pure thought experiments encoded by RNNs

Input: Randomly initialized differentiable Controller $C_\phi : \Omega \rightarrow \Theta$, where Ω is the set of sequences of the form $(\theta_i, r_i, R_i, \theta_{i+1}, r_{i+1}, R_{i+1}, \ldots)$, randomly initialized differentiable Model $M_{\mathbf{w}} : \Theta \rightarrow \mathbb{R}$, empty sequential experiment memory $\mathcal{D}$, set of random initial experiments $\mathcal{E}_{\text{init}}$

Output: An experiment memory populated with (formerly) interesting pure thought experiments

for $\theta \in \mathcal{E}_{\text{init}}$ **do**
  Execute the RNN thought experiment parametrized by θ, obtain binary result r
  Save the tuple (θ, r) to $\mathcal{D}$
  Train $M_{\mathbf{w}}$ on data from $\mathcal{D}$ for a fixed number of steps minimizing Equation 4 to obtain updated weights $\mathbf{w}^*$
  Calculate the intrinsic reward $R_i = R_{IG}(\theta, \mathbf{w}, \mathbf{w}^*)$ (Equation 3)
  $\mathbf{w} \leftarrow \mathbf{w}^*$
  Save R_i to $\mathcal{D}$
end for
repeat
  $\omega \leftarrow$ sequence of the last experiments from $\mathcal{D}$
  $\theta \leftarrow C_\phi(\omega)$
  Execute the RNN thought experiment parametrized by θ, obtain binary result r
  Train $M_{\mathbf{w}}$ on data from $\mathcal{D}$ for a fixed number of steps to obtain updated weights $\mathbf{w}^*$
  Calculate the intrinsic reward $R_i = R_{IG}(\theta, \mathbf{w}, \mathbf{w}^*)$
  $\mathbf{w} \leftarrow \mathbf{w}^*$
  Save R_i to $\mathcal{D}$
  Train C_ϕ for a fixed number of steps with DDPG to maximize the expected intrinsic reward
until no more interesting experiments are found

References

1. Berner, C., et al.: Dota 2 with large scale deep reinforcement learning. CoRR abs/1912.06680 (2019). http://arxiv.org/abs/1912.06680
2. Burda, Y., Edwards, H., Pathak, D., Storkey, A., Darrell, T., Efros, A.A.: Large-scale study of curiosity-driven learning. Preprint arXiv:1808.04355 (2018)
3. Faccio, F., Herrmann, V., Ramesh, A., Kirsch, L., Schmidhuber, J.: Goal-conditioned generators of deep policies. arXiv preprint arXiv:2207.01570 (2022)
4. Faccio, F., Kirsch, L., Schmidhuber, J.: Parameter-based value functions. Preprint arXiv:2006.09226 (2020)
5. Faccio, F., Ramesh, A., Herrmann, V., Harb, J., Schmidhuber, J.: General policy evaluation and improvement by learning to identify few but crucial states. arXiv preprint arXiv:2207.01566 (2022)
6. Fedorov, V.V.: Theory of Optimal Experiments. Academic Press, Cambridge (1972)
7. Gers, F.A., Schmidhuber, J., Cummins, F.: Learning to forget: continual prediction with LSTM. Neural Comput. **12**(10), 2451–2471 (2000)
8. Gomez, F., Koutník, J., Schmidhuber, J.: Compressed network complexity search. In: Coello, C.A.C., Cutello, V., Deb, K., Forrest, S., Nicosia, G., Pavone, M. (eds.) PPSN 2012. LNCS, vol. 7491, pp. 316–326. Springer, Heidelberg (2012). https://doi.org/10.1007/978-3-642-32937-1_32
9. Gomez, F., Schmidhuber, J.: Evolving modular fast-weight networks for control. In: Duch, W., Kacprzyk, J., Oja, E., Zadrożny, S. (eds.) ICANN 2005. LNCS, vol. 3697, pp. 383–389. Springer, Heidelberg (2005). https://doi.org/10.1007/11550907_61
10. Goodfellow, I., et al.: Generative adversarial nets. In: Advances in Neural Information Processing Systems (NIPS), pp. 2672–2680, December 2014
11. Graves, A., Fernández, S., Schmidhuber, J.: Multi-dimensional recurrent neural networks. In: Proceedings of the 17th International Conference on Artificial Neural Networks, September 2007
12. Graves, A., Liwicki, M., Fernandez, S., Bertolami, R., Bunke, H., Schmidhuber, J.: A novel connectionist system for improved unconstrained handwriting recognition. IEEE Trans. Pattern Anal. Mach. Intell. **31**(5) (2009)
13. Ha, D., Dai, A., Le, Q.V.: Hypernetworks. arXiv preprint arXiv:1609.09106 (2016)
14. Harb, J., Schaul, T., Precup, D., Bacon, P.L.: Policy evaluation networks. arXiv preprint arXiv:2002.11833 (2020)
15. Hochreiter, S., Schmidhuber, J.: Flat minima. Neural Comput. **9**(1), 1–42 (1997)
16. Hochreiter, S., Schmidhuber, J.: Long short-term memory. Neural Comput. **9**(8), 1735–1780 (1997). based on TR FKI-207-95, TUM (1995)
17. Huffman, D.A.: A method for construction of minimum-redundancy codes. Proc. IRE **40**, 1098–1101 (1952)
18. Irie, K., Schlag, I., Csordás, R., Schmidhuber, J.: Going beyond linear transformers with recurrent fast weight programmers. Adv. Neural Inf. Process. Syst. **34**, 7703–7717 (2021)
19. Itti, L., Baldi, P.F.: Bayesian surprise attracts human attention. In: Advances in Neural Information Processing Systems (NIPS), vol. 19, pp. 547–554. MIT Press, Cambridge, MA (2005)
20. Kaelbling, L.P., Littman, M.L., Moore, A.W.: Reinforcement learning: a survey. J. AI Res. **4**, 237–285 (1996)
21. Kirsch, L., Schmidhuber, J.: Meta learning backpropagation and improving it. Adv. Neural Inf. Process. Syst. **34**, 14122–14134 (2021)

22. Kolmogorov, A.N.: Three approaches to the quantitative definition of information. Probl. Inf. Transm. **1**, 1–11 (1965)
23. Koutník, J., Gomez, F., Schmidhuber, J.: Evolving neural networks in compressed weight space. In: Proceedings of the 12th Annual Conference on Genetic and Evolutionary Computation, pp. 619–626 (2010)
24. Koutník, J., Cuccu, G., Schmidhuber, J., Gomez, F.: Evolving large-scale neural networks for vision-based reinforcement learning. In: Proceedings of the Genetic and Evolutionary Computation Conference (GECCO), pp. 1061–1068. ACM, Amsterdam, July 2013
25. Kullback, S., Leibler, R.A.: On information and sufficiency. Ann. Math. Stat. 79–86 (1951)
26. Li, M., Vitányi, P.M.B.: An Introduction to Kolmogorov Complexity and its Applications (2nd edition). Springer, New York (1997). https://doi.org/10.1007/978-1-4757-2606-0
27. Lillicrap, T.P., et al.: Continuous control with deep reinforcement learning. arXiv preprint arXiv:1509.02971 (2015)
28. Micheli, V., Alonso, E., Fleuret, F.: Transformers are sample efficient world models. arXiv preprint arXiv:2209.00588 (2022)
29. Andrychowicz, O.M., et al.: Learning dexterous in-hand manipulation. Int. J. Robot. Res. **39**(1), 3–20 (2020)
30. Oudeyer, P.-Y., Baranes, A., Kaplan, F.: Intrinsically motivated learning of real-world sensorimotor skills with developmental constraints. In: Baldassarre, G., Mirolli, M. (eds.) Intrinsically Motivated Learning in Natural and Artificial Systems, pp. 303–365. Springer, Heidelberg (2013). https://doi.org/10.1007/978-3-642-32375-1_13
31. Pathak, D., Agrawal, P., Efros, A.A., Darrell, T.: Curiosity-driven exploration by self-supervised prediction. In: Proceedings of the IEEE Conference on Computer Vision and Pattern Recognition Workshops, pp. 16–17 (2017)
32. Plappert, M., et al.: Parameter space noise for exploration. arXiv preprint arXiv:1706.01905 (2017)
33. Ramesh, A., Kirsch, L., van Steenkiste, S., Schmidhuber, J.: Exploring through random curiosity with general value functions. In: Oh, A.H., Agarwal, A., Belgrave, D., Cho, K. (eds.) Advances in Neural Information Processing Systems (2022)
34. Rissanen, J.: Modeling by shortest data description. Automatica **14**, 465–471 (1978)
35. Rückstieß, T., Felder, M., Schmidhuber, J.: State-dependent exploration for policy gradient methods. In: Daelemans, W., Goethals, B., Morik, K. (eds.) ECML PKDD 2008. LNCS (LNAI), vol. 5212, pp. 234–249. Springer, Heidelberg (2008). https://doi.org/10.1007/978-3-540-87481-2_16
36. Schlag, I., Irie, K., Schmidhuber, J.: Linear transformers are secretly fast weight programmers. In: International Conference on Machine Learning, pp. 9355–9366. PMLR (2021)
37. Schlag, I., Schmidhuber, J.: Learning to reason with third order tensor products. In: Advances in Neural Information Processing Systems (NIPS), pp. 9981–9993 (2018)
38. Schmidhuber, J.: Making the world differentiable: On using fully recurrent self-supervised neural networks for dynamic reinforcement learning and planning in non-stationary environments. Technical report FKI-126-90 (1990). http://people.idsia.ch/~juergen/FKI-126-90_(revised)bw_ocr.pdf, Tech. Univ. Munich

39. Schmidhuber, J.: An on-line algorithm for dynamic reinforcement learning and planning in reactive environments. In: Proceedings of the IEEE/INNS International Joint Conference on Neural Networks, San Diego, vol. 2, pp. 253–258 (1990)
40. Schmidhuber, J.: Curious model-building control systems. In: Proceedings of the International Joint Conference on Neural Networks, Singapore, vol. 2, pp. 1458–1463. IEEE press (1991)
41. Schmidhuber, J.: Learning temporary variable binding with dynamic links. In: Proceedings of the International Joint Conference on Neural Networks, Singapore, vol. 3, pp. 2075–2079. IEEE (1991)
42. Schmidhuber, J.: A possibility for implementing curiosity and boredom in model-building neural controllers. In: Meyer, J.A., Wilson, S.W. (eds.) Proc. of the International Conference on Simulation of Adaptive Behavior: From Animals to Animats, pp. 222–227. MIT Press/Bradford Books (1991)
43. Schmidhuber, J.: Reinforcement learning in Markovian and non-Markovian environments. In: Lippman, D.S., Moody, J.E., Touretzky, D.S. (eds.) Advances in Neural Information Processing Systems, vol. 3 (NIPS 3), pp. 500–506. Morgan Kaufmann (1991)
44. Schmidhuber, J.: Learning to control fast-weight memories: an alternative to recurrent nets. Neural Comput. **4**(1), 131–139 (1992)
45. Schmidhuber, J.: On decreasing the ratio between learning complexity and number of time-varying variables in fully recurrent nets. In: Proceedings of the International Conference on Artificial Neural Networks, Amsterdam, pp. 460–463. Springer (1993)
46. Schmidhuber, J.: Discovering solutions with low Kolmogorov complexity and high generalization capability. In: Prieditis, A., Russell, S. (eds.) Machine Learning: Proceedings of the Twelfth International Conference, pp. 488–496. Morgan Kaufmann Publishers, San Francisco, CA (1995)
47. Schmidhuber, J.: Discovering neural nets with low Kolmogorov complexity and high generalization capability. Neural Netw. **10**(5), 857–873 (1997)
48. Schmidhuber, J.: What's interesting? Technical report IDSIA-35-97, IDSIA (1997). ftp://ftp.idsia.ch/pub/juergen/interest.ps.gz; extended abstract in Proc. Snowbird'98, Utah, 1998; see also [50]
49. Schmidhuber, J.: Artificial curiosity based on discovering novel algorithmic predictability through coevolution. In: Angeline, P., Michalewicz, Z., Schoenauer, M., Yao, X., Zalzala, Z. (eds.) Congress on Evolutionary Computation, pp. 1612–1618. IEEE Press (1999)
50. Schmidhuber, J.: Exploring the predictable. In: Ghosh, A., Tsuitsui, S. (eds.) Advances in Evolutionary Computing, pp. 579–612. Springer, Berlin, Heidelberg (2003). https://doi.org/10.1007/978-3-642-18965-4_23
51. Schmidhuber, J.: Hierarchies of generalized Kolmogorov complexities and nonenumerable universal measures computable in the limit. Int. J. Found. Comput. Sci. **13**(4), 587–612 (2002)
52. Schmidhuber, J.: Overview of artificial curiosity and active exploration, with links to publications since 1990 (2004). http://www.idsia.ch/~juergen/interest.html
53. Schmidhuber, J.: Developmental robotics, optimal artificial curiosity, creativity, music, and the fine arts. Connect. Sci. **18**(2), 173–187 (2006)
54. Schmidhuber, J.: Simple algorithmic principles of discovery, subjective beauty, selective attention, curiosity & creativity. In: Corruble, V., Takeda, M., Suzuki, E. (eds.) DS 2007. LNCS (LNAI), vol. 4755, pp. 26–38. Springer, Heidelberg (2007). https://doi.org/10.1007/978-3-540-75488-6_3

55. Schmidhuber, J.: Compression progress: the algorithmic principle behind curiosity and creativity (with applications of the theory of humor) (2009). 40 min video of invited talk at Singularity Summit 2009, New York City: http://www.vimeo.com/7441291. 10 min excerpts: http://www.youtube.com/watch?v=Ipomu0MLFaI
56. Schmidhuber, J.: Driven by compression progress: a simple principle explains essential aspects of subjective beauty, novelty, surprise, interestingness, attention, curiosity, creativity, art, science, music, jokes. In: Pezzulo, G., Butz, M.V., Sigaud, O., Baldassarre, G. (eds.) ABiALS 2008. LNCS (LNAI), vol. 5499, pp. 48–76. Springer, Heidelberg (2009). https://doi.org/10.1007/978-3-642-02565-5_4
57. Schmidhuber, J.: Formal theory of creativity, fun, and intrinsic motivation (1990–2010). IEEE Trans. Auton. Ment. Dev. **2**(3), 230–247 (2010). https://doi.org/10.1109/TAMD.2010.2056368
58. Schmidhuber, J.: Overviews of artificial curiosity/creativity and active exploration (with links to publications since 1990) (2012). http://www.idsia.ch/~juergen/interest.html, http://www.idsia.ch/~juergen/creativity.html
59. Schmidhuber, J.: Self-delimiting neural networks. Technical report. IDSIA-08-12, arXiv:1210.0118v1 [cs.NE], The Swiss AI Lab IDSIA (2012)
60. Schmidhuber, J.: PowerPlay: Training an increasingly general problem solver by continually searching for the simplest still unsolvable problem. Front. Psychol. (2013). https://doi.org/10.3389/fpsyg.2013.00313, (Based on arXiv:1112.5309v1 [cs.AI], 2011)
61. Schmidhuber, J.: Deep learning in neural networks: an overview. Neural Netw. **61**, 85–117 (2015). https://doi.org/10.1016/j.neunet.2014.09.003, published online 2014; 888 references; based on TR arXiv:1404.7828 [cs.NE]
62. Schmidhuber, J.: On learning to think: algorithmic information theory for novel combinations of reinforcement learning controllers and recurrent neural world models. Preprint arXiv:1511.09249 (2015)
63. Schmidhuber, J.: Artificial Curiosity & Creativity Since 1990–91. https://people.idsia.ch/~juergen/artificial-curiosity-since-1990.html (AI Blog, 2021), https://people.idsia.ch/~juergen/artificial-curiosity-since-1990.html
64. Schmidhuber, J.: Generative adversarial networks are special cases of artificial curiosity (1990) and also closely related to predictability minimization (1991). Neural Networks (2020)
65. Schmidhuber, J.: Learning one abstract bit at a time through self-invented experiments. Unpublished Tech Report, IDSIA & NNAISENSE (2020)
66. Shannon, C.E.: A mathematical theory of communication (parts I and II). Bell Syst. Tech. J. **XXVII**, 379–423 (1948)
67. Siegelmann, H.T., Sontag, E.D.: Turing computability with neural nets. Appl. Math. Lett. **4**(6), 77–80 (1991)
68. Singh, S., Barto, A.G., Chentanez, N.: Intrinsically motivated reinforcement learning. In: Advances in Neural Information Processing Systems (NIPS), vol. 17. MIT Press, Cambridge, MA (2005)
69. Solomonoff, R.J.: A formal theory of inductive inference. Part I. Inf. Control **7**, 1–22 (1964)
70. Srivastava, R.K., Steunebrink, B.R., Schmidhuber, J.: First experiments with PowerPlay. Neural Netw. **41**, 130–136 (2013). https://doi.org/10.1016/j.neunet.2013.01.022, http://www.sciencedirect.com/science/article/pii/S0893608013000373, special Issue on Autonomous Learning

71. van Steenkiste, S., Koutník, J., Driessens, K., Schmidhuber, J.: A wavelet-based encoding for neuroevolution. In: Proceedings of the Genetic and Evolutionary Computation Conference 2016, pp. 517–524. GECCO '16, ACM, New York, NY, USA (2016)
72. Storck, J., Hochreiter, S., Schmidhuber, J.: Reinforcement driven information acquisition in non-deterministic environments. In: Proceedings of the International Conference on Artificial Neural Networks, Paris, vol. 2, pp. 159–164. EC2 & Cie (1995)
73. Sun, Y., Gomez, F., Schmidhuber, J.: Planning to be surprised: optimal Bayesian exploration in dynamic environments. In: Proceedings of the Fourth Conference on Artificial General Intelligence (AGI), Google, Mountain View, CA (2011)
74. Sutton, R., Barto, A.: Reinforcement Learning: An Introduction. MIT Press, Cambridge, MA (1998)
75. Vaswani, A., et al.: Attention is all you need. Adv. Neural Inf. Process. Syst. 5998–6008 (2017)
76. Vemula, A., Sun, W., Bagnell, J.: Contrasting exploration in parameter and action space: a zeroth-order optimization perspective. In: The 22nd International Conference on Artificial Intelligence and Statistics, pp. 2926–2935. PMLR (2019)
77. Vinyals, O., et al.: Grandmaster level in StarCraft II using multi-agent reinforcement learning. Nature **575**(7782), 350–354 (2019)
78. Wallace, C.S., Boulton, D.M.: An information theoretic measure for classification. Comput. J. **11**(2), 185–194 (1968)
79. Wallace, C.S., Freeman, P.R.: Estimation and inference by compact coding. J. R. Stat. Soc. Ser. B **49**(3), 240–265 (1987)
80. Wiering, M., van Otterlo, M.: Reinforcement Learning. Springer, Berlin, Heidelberg (2012). https://doi.org/10.1007/978-3-642-27645-3
81. Wierstra, D., Foerster, A., Peters, J., Schmidhuber, J.: Recurrent policy gradients. Log. J. IGPL **18**(2), 620–634 (2010)
82. Williams, R.J.: Simple statistical gradient-following algorithms for connectionist reinforcement learning. Mach. Learn. **8**, 229–256 (1992)
83. Witten, I.H., Neal, R.M., Cleary, J.G.: Arithmetic coding for data compression. Commun. ACM **30**(6), 520–540 (1987)

Probabilistic Majorization of Partially Observable Markov Decision Processes

Tom Lefebvre(✉)

Faculty of Engineering, Ghent University, Ghent, Belgium
tom.lefebvre@ugent.be
https://dynamics.ugent.be

Abstract. Markov Decision Processes (MDPs) are wielded by the Reinforcement Learning and control community as a framework to bestow artificial agents with the ability to make autonomous decisions. Control as Inference (CaI) is a tangent research direction that aims to recast optimal decision making as an instance of probabilistic inference, with the dual hope to incite exploration and simplify calculations. Active Inference (AIF) is a sibling theory conforming to similar directives. Notably, AIF also entertains a procedure for per- and proprio-ception, which is currently lacking from the CaI theory. Recent work has established an explicit connection between CaI and Markov Decision Processes (MDPs). In particular, it was shown that the CaI policy can be iterated recursively, ultimately retrieving the associated MDP policy. In this work, such results are generalized to Partially Observable Markov Decision Processes, that – apart from a procedure to make optimal decisions – now also entertains a procedure for model based per- and proprio-ception. By extending the theory of CaI to the context of optimal decision making under partial observability, we mean to further our understanding of and illuminate the relationship between these different frameworks.

1 Introduction

The Reinforcement Learning and control community at large is concerned with automated decision making or control system synthesis. To that end, the community often relies on the framework of Markov Decision Processes (MDPs) or Stochastic Optimal Control (SOC). These synonymous frameworks synthesize policy makers or controllers by minimising the expected cost over a(n) (in)finite decision or control horizon [19,23]. In a model-based setting, probabilistic models are utilised to assess the uncertain (future) behaviour of the system. The solution is provided by a deterministic function known as the optimal policy or control. Though theoretically appealing, often these solutions can only be attained at the result of complex calculations directly pursuing the deterministic result.

An intriguing question that has been pursued by several authors is whether the complexity of these calculations can be alleviated by drawing on the probabilistic setting that is already utilised to model the system [2,7–10,21,22]. It has

C. L. Buckley et al. (Eds.): IWAI 2023, CCIS 1915, pp. 275–288, 2024.
https://doi.org/10.1007/978-3-031-47958-8_17

been argued that it is sometimes easier to approach a problem probabilistically, even if the problem and the eventual result are deterministic [6,17].

In control, these endeavours let to a paradigm that is referred to as Control as Inference (CaI) [1,13]. Here one attempts to recast optimal control as an inference process. To that end, the framework entertains an extension of the standard Markov Chain generative model that is used otherwise in optimal decision making, i.e. MDPs. To attain an MDP, a Markov Chain is equipped with a utility function[1] – in other words a cost function – associating value to particular decisions. In CaI, rather than through a utility function, value is encoded through an auxiliary set of exogenous observation variables whose (future) values are assumed fixed and indicate that an optimal decision has been made. Thence, the control system is inferred by calculating the probability of making a decision at present time assuming (future) optimally has been achieved[2].

By a specific choice of the auxiliary emission model, the framework resumes close analogies with the theory of optimal control. Recent work has established an explicit connection between CaI and MDPs [12]. Particularly, it has been shown that the two main governing problems in CaI majorize conventional optimal control problems. This observation and the particular structure of the associated solutions, then invites to establish a fixed point iteration, maintaining probabilistic controllers, but whose stationary point eventually coincides with the deterministic control. This result characterizes CaI by its computational implications rather than by it efficacy to incite explorative behavioural tendencies.

Active Inference (AIF) is another framework that casts planning as an inference problem and which leverages approximate inference tools to solve this problem. An interesting comparison between AIF and CaI was made by Milledge et al. [14], the key difference being identified by the way in which value is encoded in the generative model. Whereas CaI extends a veridical generative model with exogenous optimisation variables, AIF encodes value into the generative model itself directly. Exploration in the context of CaI manifests as entropy-maximization, whereas exploration in the context of AIF is said to be goal-directed through maximization of an expected information gain [14].

Distinctively, compared to the thus far existing literature on CaI, AIF also entertains a procedure for per- and proprio-ception. To make way for a more nuanced comparison between CaI and AIF, we establish an explicit connection between CaI and Partially Observable Markov Decision Processes (POMDPs). We will show that the probabilistic fixed point iterations applying to MDPs extend to POMDPs and shall make a first attempt at interpreting these results.

1.1 Notation

With notation, $\underline{x}_t = \{x_0, \ldots, x_t\}$, and, $\overline{x}_t = \{x_t, \ldots, x_T\}$, we refer to the leading or trailing part of a time series or sequence. The index, t, refers to the final or

[1] Applying to the whole history of the system.

[2] Ergo by conditioning the present action on future auxiliary observation variables. The exact technical details somewhat deviate from this verbal exposition, however it succeeds elegantly at capturing the gist of the idea.

initial time instance of the corresponding subsequence. We silently assume that a complete sequence starts at time $t = 0$ and ends at time $t = T$.

2 Background

The present paper and the results it communicates are rather technical in nature. It is, therefore, essential to clearly sketch the mathematical stage upon which our results are founded. No attempt is made at achieving vigorous mathematical rigor.

We take interest in solving the following stochastic optimal control problem

$$\min_{\underline{u}_{T-1}} \int \underline{R}_T(\underline{x}_T, \underline{u}_{T-1}) p(\underline{x}_T | \underline{u}_{T-1}) \mathrm{d}\underline{x}_T \tag{1}$$

The integrand is defined as an accumulated cost (i.e. the utility function)

$$\underline{R}_T(\underline{x}_T, \underline{u}_{T-1}) = r_T(x_T) + \sum_{t=0}^{T-1} r_t(x_t, u_t) \tag{2}$$

For the generative model, $p(\underline{x}_T | \underline{u}_{T-1})$, we adopt a (Hidden) Markov Chain configuration depicted in Fig. 1. Conventionally, here $\underline{x}_T$ denotes a sequence of (hidden) state variables, $\underline{y}_T$ denotes a sequence of measurement or observation variables, and, $\underline{u}_{T-1}$ denotes a sequence of arbitrary control inputs.

So, next to the state variable sequence, $\underline{x}_T$, that is already represented in the control problem, there is also a measurement sequence, $\underline{y}_T$. This implies that the generative model is truly characterised by the following joint density.

$$p(\underline{x}_T, \underline{y}_T | \underline{u}_{T-1}) \tag{3}$$

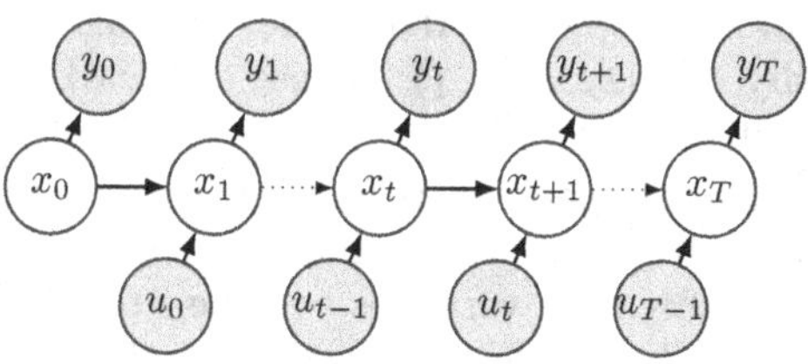

Fig. 1. Probabilistic graph model of a Hidden Markov Chain.

Without loss of generality we can make the presence of the measurement sequence explicit in the optimal control objective

$$\min_{\underline{u}_{T-1}} \int \underline{R}_T(\underline{x}_T, \underline{u}_{T-1}) p(\underline{x}_T, \underline{y}_T | \underline{u}_{T-1}) \mathrm{d}\underline{x}_T \mathrm{d}\underline{y}_T \tag{4}$$

The goal is now to find the optimal control, $\underline{u}_T^*$, as the argument that minimizes (1). Here our notation falls somewhat short because the optimal control or *agent* is in fact a function that may depend on several variables of the generative model. The variables that the agent uses, depends on the information that we grant it access to. This information can be reflected explicitly in the way the generative model is decomposed.

There are two main strategies to decompose the generative model.

1. The *causal* decomposition

$$p(\underline{x}_T, \underline{y}_T | \underline{u}_{T-1}) = \prod_{t=0}^{T} p(x_t|x_{t-1}, u_{t-1}) p(y_t|x_t) \tag{5}$$

which depends on the state transition density model, $p(x_t|x_{t-1}, u_{t-1})$, and, the emission density model, $p(y_t|x_t)$. In optimal decision making, it is typically assumed that the agent has access to these models as part of the generative model it entertains.

2. The *evidential* decomposition

$$p(\underline{x}_T, \underline{y}_T | \underline{u}_{T-1}) = \prod_{t=0}^{T} p(y_t|\underline{y}_{t-1}, \underline{u}_{t-1}) p(x_t|\underline{y}_t, \underline{u}_{t-1}) \tag{6}$$

which depends on the output transition density model, $p(y_t|\underline{y}_{t-1}, \underline{u}_{t-1})$, and, the Bayesian belief density, $p(x_t|\underline{y}_t, \underline{u}_{t-1})$. The latter can be calculated using the Bayesian filtering equations [18]. The former is governed by

$$p(y_{t+1}|\underline{y}_t, \underline{u}_t) = \int p(y_{t+1}|x_{t+1}) p(x_{t+1}|x_t, u_t) p(x_t|\underline{y}_t, \underline{u}_{t-1}) \mathrm{d}x_t \mathrm{d}x_{t+1} \tag{7}$$

As we mentioned, the decomposition strategy shall be determinative with regard to what information the controller is granted access to. By construction, the information will coincide with that required by the transition density, governing the dynamics. In case of MDPs, the dynamics are governed by the state transition density, $p(x_t|x_{t-1}, u_{t-1})$. In case of POMDPs, the dynamics are governed by the output transition density, $p(y_t|\underline{y}_{t-1}, \underline{u}_{t-1})$.

As such, the causal decomposition is useful when we grant the control system access to the state variable x_t to compute u_t. This setting is characteristic of MDPs. Then the measurement variables become irrelevant and can be disregarded altogether. The evidential decomposition is useful in the setting where we deny the control system access to the state and only present it with a real-time measurement, y_t, and, a memory, storing the variables $\underline{y}_t$ and $\underline{u}_{t-1}$. This setting is characteristic of POMDPs and will be the setting that enjoys our interest in the remainder of this paper.

For notational convenience, the historical variables $\underline{y}_t$ and $\underline{u}_{t-1}$ that are available at time t can be concatenated in a variable, w_t. Note that the variable, w_t, contains all the information required to calculate the Bayesian state belief function, $p(x_t|w_t)$, so that often no distinction is made between the two. Substituting the new variable, w_t, into our earlier definitions, the output transition density simplifies to $p(y_{t+1}|w_t, u_t)$. This substitution is particularly relevant because it neatly distinguishes between the decisions that have been taken, $\underline{u}_{t-1}$, and the decision, u_t that needs to be taken at the present time t.

Adopting notation w_t and substitution of the evidential decomposition in (4) then yields a problem formulation tailored to the POMDP setting

$$\min_{\underline{u}_{T-1}} \int \sum_{t=0}^{T} \int r(x_t, u_t) p(x_t|w_t) \mathrm{d}x_t \prod_{t=0}^{T} p(y_t|w_{t-1}, u_{t-1}) \mathrm{d}\underline{y}_T \tag{8}$$

It is well known that optimal control problems exhibit a so called optimal substructure and can be treated by means of dynamic programming by consequence. To expose the substructure in the present setting, it is crucial that we recognize that a decision at time t can rely on the information in w_t. This variable however contains the earlier decision variables $\underline{u}_{t-1}$. As such it appears the older decision variables affect the present decision variable, apparently destroying the optimal substructure. Fortunately, once the decision has been made, u_t, becomes a *regular* variable that we can no longer optimize. Therefore, when treating the problem, the earlier decision variables contained in w_t should not be treated in the same manner as the optimization variables u_t.

Once convinced by this last observation, it is easily verified that the optimal control is governed by the following backward recursion.

$$\begin{aligned} V_t(w_t) &= \min_{\overline{u}_t} \int \overline{R}_t(\overline{x}_t, \overline{u}_t) p(\overline{x}_t, \overline{y}_{t+1}|w_t, \overline{u}_t) \mathrm{d}\overline{x}_t \mathrm{d}\overline{y}_{t+1} \\ &= \min_{u_t} \int r_t(x_t, u_t) p(x_t|w_t) \mathrm{d}x_t + \int V_{t+1}(w_{t+1}) p(y_{t+1}|w_t, u_t) \mathrm{d}y_{t+1} \end{aligned} \tag{9}$$

This defines the standard Bellman equation for POMDPs [20]. Retrospectively, Eq. (9) also illustrates why we may disregard the older decision variables, $\underline{u}_{t-1}$, when taking the present decision, u_t. This is because the present decision is only affected by the present belief, $p(x_t|w_t)$, – directly or through (7) – not by the particular values that determine w_t itself[3].

3 Control as Inference

Control as Inference (CaI) is a paradigm within optimal control theory which attempts to cast optimal decision making as an inference problem. The premise is that, if successful, this would allow to bring to bear a wide range of inference techniques to alleviate treatment of difficult optimal control problems.

There exist several angles to arrive at the framework [9,12,13]. Though, before we engage in further discussion, it is necessary to expand upon the problem formulation that has been established thus far. To that end we generalize

[3] In fact, the set populated by w_t is surjective to the set populated with belief functions, $p(x_t|w_t)$, defined on the state-space.

the generative model explicitly annotating that the model is *parametrized* by a policy sequence, $\underline{\pi}_{T-1}$[4]. We further assume that the policy sequence is populated by policy densities, $\pi_t(u_t|w_t)$, conditioning the probability of taking some decision, u_t, onto the information that is contained in w_t.

$$p(\underline{x}_T, \underline{u}_{T-1}, \underline{y}_T; \underline{\pi}_{T-1}) \tag{10}$$

In the extended formulation, the evidential decomposition is given by

$$p(\underline{x}_T, \underline{u}_{T-1}, \underline{y}_T; \underline{\pi}_{T-1}) = \prod_{t=0}^{T} p(x_t|w_t)\pi_t(u_t|w_t)p(y_t|w_{t-1}, u_{t-1}) \tag{11}$$

This first intervention places the control variables on an equal footing with the other variables.

3.1 Encoding Value

Second, we need a mechanism that introduces the notion of value in the generative model [14]. To that end we introduce an auxiliary set of binary measurements variables, $\underline{z}_T$. It is presumed that all of these auxiliary variables have assumed the value 1 with probability proportional to the negative exponential transform of the cost rate in (2)[5]. We further write z_t when we mean $z_t = 1$.

$$p(z_t|x_t, u_t) \propto e^{-r_t(x_t, u_t)} \tag{12}$$

Introduction of these variables into the generative model results into the graphical depiction in Fig. 2. The associated joint density follows

$$p(\underline{x}_T, \underline{u}_{T-1}, \underline{y}_T, \underline{z}_T; \underline{\pi}_{T-1}) \propto p(\underline{x}_T, \underline{u}_{T-1}, \underline{y}_T; \underline{\pi}_{T-1})e^{-\underline{R}_T(\underline{x}_T, \underline{u}_{T-1})} \tag{13}$$

Because the variables $\underline{z}_T$ have a fixed value, often the density in the right-hand side of Eq. (13) is referred to as the *desired* joint density,

$$p^*(\underline{x}_T, \underline{u}_{T-1}, \underline{y}_T; \underline{\pi}_{T-1}) \tag{14}$$

dropping the ubiquitous dependency on $\underline{z}_T$.

[4] Note that we could have introduced this formulation at the very beginning and optimized for π_t rather than u_t. Formally this is equivalent since the set of all densities also contains the set of all deterministic functions. Moreover, this would have saved us from the trouble explaining why the decision variables contained in w_t are treated differently then the decision variable u_t. Now it is clear this is because we do not optimize the decision variable, u_t, itself but rather the policy, π_t.

[5] It is rather difficult to give a convincing justification for this model. Rather it should be understood as a technical trick.

3.2 Inferring Policies

At this stage, alternative strategies can be traced to extract a policy. The strategies are equivalent in the sense that, eventually, they arrive at the same principle problems and inference mechanism [12]. In this work we follow the approach used in [12], referred to as probabilistic (optimal) control [9,10].

Probabilistic control interprets CaI as a density matching problem, defining the optimal control as that sequence that makes the generative model closest to the desired generative model, p^*. Put differently, the goal of an optimal policy sequence, $\underline{\pi}_{T-1}$, is to induce a density that exhibits the same statistics as the desired density, p^*. The only remaining question is how we quantify the proximity between two densities. Therefore we rely on the information-theoretic projection strategies known as the *information* (I) and *moment* (M) projection [3,16][6].

1. *information* projected probabilistic optimal control problem

$$\underline{\pi}^{\bullet}_{T-1} = \min_{\underline{\pi}_{T-1}} \mathbb{D}\left[p(\underline{x}_T, \underline{u}_{T-1}, \underline{y}_T; \underline{\pi}_{T-1}) \middle\| p^*(\underline{x}_T, \underline{u}_{T-1}, \underline{y}_T; \underline{\rho}_{T-1})\right] \tag{15}$$

2. *moment* projected probabilistic optimal control problem

$$\underline{\pi}^{\star}_{T-1} = \min_{\underline{\pi}_{T-1}} \mathbb{D}\left[p^*(\underline{x}_T, \underline{u}_{T-1}, \underline{y}_T; \underline{\rho}_{T-1}) \middle\| p(\underline{x}_T, \underline{u}_{T-1}, \underline{y}_T; \underline{\pi}_{T-1})\right] \tag{16}$$

Note that in either case, the desired generative model depends on some arbitrary policy sequence, $\underline{\rho}_{T-1}$. These policy sequences can be interpreted as our agent's prior belief about the policy before encoding value into the optimal policy belief sequences, $\underline{\pi}^{\bullet}_{T-1}$, or $\underline{\pi}^{\star}_{T-1}$, respectively.

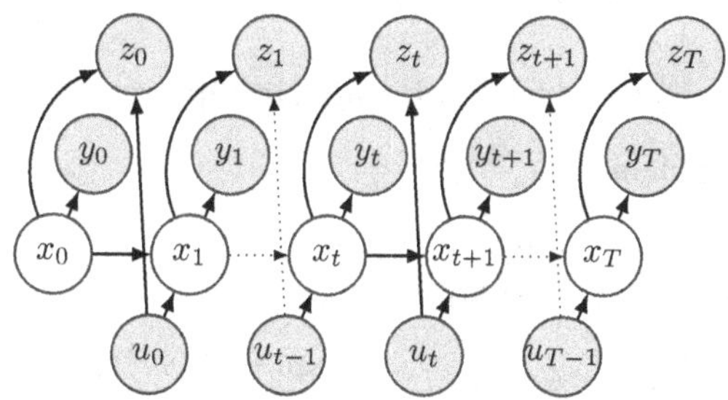

Fig. 2. Probabilistic graph model of the extended Hidden Markov Chain.

[6] Both projection strategies rely on the relative entropy or Kullback-Leibler divergence, $\mathbb{D}[\pi||\rho]$. The relative entropy is a divergence and not a distance and thus asymmetric in its arguments. Therefore the I-projection and the M-projection do not yield the same projection [3,15]. They are either *mode seeking* or *covering* for π. As a result the I-projection will underestimate the support of ρ and vice versa.

4 Probabilistic Majorization of Optimal Decision Making

To establish our main results it is required that we give a brief introduction of the Majorizing-Minimizing (MM) principle which shall proof essential to interpret and wield the solutions of problems (15) and (16).

4.1 The Majorizing-Minimizing Principle

The MM principle aims to to convert hard optimization problems into sequences of simple ones [11]. When the goal is to minimize the objective, say $\min_\theta f(\theta)$, the MM principle requires to majorize the objective function, $f(\theta)$, with a surrogate, $g(\theta, \theta')$, anchored at the current iterate, θ'. Majorization of an objective imposes two requirements on the surrogate: (1) the tangency condition, and, (2) the domination condition, with $a > 0$ and b independent of θ

$$\begin{aligned} f(\theta') &= a \cdot g(\theta', \theta') + b \\ f(\theta) &\leq a \cdot g(\theta, \theta') + b \end{aligned} \tag{17}$$

The surrogate can then be used as a proxy for the true objective to obtain a new iterate through the following fixed point iteration

$$\theta^* \leftarrow \arg\min_\theta g(\theta, \theta^*) \tag{18}$$

By definition, the iteration drives the objective function downhill. Strictly speaking, the descent property depends only on decreasing $g(\theta, \theta')$, not on strictly minimizing it. Under appropriate regularity conditions, an MM approach is guaranteed to converge to a stationary point of the objective function.

$$f(\theta'') \leq a \cdot g(\theta'', \theta') + b \leq a \cdot g(\theta', \theta') + b = f(\theta') \tag{19}$$

4.2 Information Projected Probabilistic Optimal Control

First let us treat problem (15).

Lemma 1. *Consider the I-projection in (15) and let p^* be defined as in (13). Then the probabilistic optimal control is given by*

$$\pi_t^\bullet(u_t|w_t) = \rho_t(u_t|w_t) \frac{\exp(-Q_t^\bullet(w_t, u_t))}{\exp(-V_t^\bullet(w_t)}$$

The functions $V_t^\bullet$ and $Q_t^\bullet$ are generated recursively in a backward manner

$$V_t^\bullet(w_t) = -\log \int \exp(-Q_t^\bullet(w_t, u_t))\rho_t(u_t|w_t)du_t$$

and

$$Q_t^\bullet(w_t, u_t) = \int r_t(x_t, u_t)p(x_t|w_t)dx_t + \int V_{t+1}^\bullet(w_{t+1})p(y_{t+1}|w_t, u_t)dy_{t+1}$$

The proof is analogous to the proof of Lemma 1 in [12].

Now, so far not much attention was given to the choice of the prior policy, $\underline{\rho}_{T-1}$, rather than that it is arbitrary in some sense. Further, given the structure of the probabilistic control policies, $\pi_t^\bullet$, an evident question is to ask what happens if we were to iterate the solutions? It turns out that the answer contains the key to understanding how the CaI paradigm relates to conventional optimal control theory.

The relation is established by the following proposition

Proposition 1. *Objective (15) majorizes objective (1).*

The proof is analogous to the proof of Proposition 3 in [12].

Then, by merit of the MM principle, the following fixed point iteration converges to the optimal control as defined by the argument of problem (1).

$$\underline{\pi}_{T-1}^\bullet \leftarrow \arg\min_{\underline{\pi}_{T-1}} \mathbb{D}\left[p(\underline{x}_T, \underline{u}_{T-1}, \underline{y}_T; \underline{\pi}_{T-1}) \middle\| p^*(\underline{x}_T, \underline{u}_{T-1}, \underline{y}_T; \underline{\pi}_{T-1}^\bullet)\right] \tag{20}$$

4.3 Moment Projected Probabilistic Optimal Control

Second we shift attention to problem (16).

Lemma 2. *Consider the M-projection in (16) and let p^* be defined as in (13). Then the probabilistic optimal control is given by*

$$\pi_t^\star(u_t|w_t) = \rho_t(u_t|w_t)\frac{\exp(-Q_t^\star(w_t, u_t))}{\exp(-V_t^\star(w_t))}$$

The functions $V_t^\star$ and $Q_t^\star$ are generated recursively in a backward manner

$$V_t^\star(w_t) = -\log\int \exp(-Q_t^\star(w_t, u_t))\rho_t(u_t|w_t)\,du_t$$

and

$$\begin{aligned} Q_t^\star(w_t, u_t) = &-\log\int \exp(-r_t(x_t, u_t))p(x_t|w_t)\,dx_t \\ &-\log\int \exp(-V_{t+1}^\star(w_{t+1}))p(y_{t+1}|w_t, u_t)\,dy_{t+1} \end{aligned}$$

The proof is analogous to the proof of Proposition 1 in [12].

First note that its solution is governed by a similar, though not equivalent, backward recursion. Especially, the definition of the corresponding Q-function is distinct. This has already one direct and significant implication. One easily verifies that the value function is governed by a path integral

$$V_t^\star(w_t) = -\log\int e^{-\overline{R}_t(\overline{x}_t, \overline{u}_t)}p(\overline{x}_t, \overline{u}_t, \overline{y}_{t+1}|w_t; \overline{\rho}_t)\,d\overline{x}_t d\overline{u}_t d\overline{y}_{t+1} \tag{21}$$

Clearly, we may now also set out and attempt to establish a similar result as in Proposition 1. Though it cannot be that (16) majorizes the same objective as (15). This simple observation warrants further exploration. To that end it is required that we engage in a different line of inquiry.

4.4 Risk Sensitive Optimal Control and Estimation

Let us reconsider the desired joint density and marginalize out all variables but the auxiliary measurement sequence, $\underline{z}_T$. This function then reads as the likelihood of the measurement sequence $\underline{z}_T$. Now also recall that the generative model has been parametrised by the policy sequence, $\underline{\pi}_T$. Consequently, we can establish a Maximum Likelihood Estimation (MLE) problem

$$\max_{\underline{\pi}_{T-1}} \log p(\underline{z}_T; \underline{\pi}_{T-1}) \tag{22}$$

where

$$p(\underline{z}_T; \underline{\pi}_{T-1}) \propto \int e^{-\underline{R}_T(\underline{x}_T, \underline{u}_{T-1})} p(\underline{x}_T, \underline{u}_{T-1}, \underline{y}_T; \underline{\pi}_T) \mathrm{d}\underline{x}_T \mathrm{d}\underline{u}_{T-1} \mathrm{d}\underline{y}_T \tag{23}$$

It is interesting to note that problem (23) corresponds exactly with the definition of a Risk Sensitive Optimal Control (RSOC) problem. RSOC is an extension of the optimal control framework using an exponential utility function rather than a linear one. Such an exponential utility function puts less (or more) emphasis on the successful histories. We refer to the body of work in [23].

This brief investigation has to two interesting implications. First, we can establish a similar relation between the RSOC problem in (23) and M-projected optimal control problem in (16).

Proposition 2. *Objective (16) majorizes objective (23).*

The proof is analogous to the proof of Proposition 4 in [12].

Again, by merit of the MM principle, the following fixed point iteration converges to the optimal control as defined by the argument of problem (22).

$$\underline{\pi}^\star_{T-1} \leftarrow \arg\min_{\underline{\pi}_{T-1}} \mathbb{D}\left[p^*(\underline{x}_T, \underline{u}_{T-1}, \underline{y}_T; \underline{\pi}^\star_{T-1}) \middle\| p(\underline{x}_T, \underline{u}_{T-1}, \underline{y}_T; \underline{\pi}_{T-1})\right] \tag{24}$$

Second, it turns out that the RSOC problem can be viewed as a MLE problem. Treatment of MLE problems associated to generative models with a similar complexity as is presently the case, are usually treated by means of the Expectation-Maximization (EM) algorithm, which is a specialization of the MM principle to probabilistic graph models. Treatment of the MLE in (22) with the EM principle results into the following fixed point iteration.

$$\underline{\pi}^\star_{T-1} \leftarrow \arg\min_{\underline{\pi}_{T-1} \in \mathcal{P}} \mathbb{D}\left[p(\underline{x}_T, \underline{u}_{T-1}, \underline{y}_T | \underline{z}_T; \underline{\pi}^\star_{T-1}) \middle\| p(\underline{x}_T, \underline{u}_{T-1}, \underline{y}_T; \underline{\pi}_{T-1})\right] \tag{25}$$

One easily verifies that the fixed point iteration in (25) is equivalent to that in (24). Therefore, the solution given in Lemma 2 extends to problem (25). Furthermore, one verifies that the solution of (25), after substituting $\underline{\rho}_{T-1}$ for $\underline{\pi}^\star_{T-1}$, is given alternatively by the marginalised Bayesian smoother [18], making this the sole expression that can be evaluated by means of conventional inference.

$$\pi^\star_t(u_t|w_t) = \frac{p(w_t, u_t | \underline{z}_T; \underline{\rho}_{T-1})}{p(w_t | \underline{z}_T; \underline{\rho}_{T-1})} \tag{26}$$

5 Discussion

We give first a resume of what has been presented.

The goal of this paper was to give a technical exposé of the theory of CaI under the restriction of partial observability. Therewith it serves the dual purpose of, (1) extending the present literature on CaI to POMDPs – which remained limited to MDPs – and, (2) providing ground to further illuminate the close analogies that exist with the theory of AIF. To that end we have taken the route of probabilistic control, that formulates CaI as a distribution matching problem. The optimal probabilistic control policy is defined as that which makes the generative model as close as possible to some desired generative model. Here, one interpretation is that the notion of value in the desired generative model is encoded by means of an auxiliary sequence of exogenous observation variables. Depending on the information-theoretic projection method pursued to quantify proximity in density spaces, the resulting problems are then shown to either majorize the SOC or RSOC problems. Both results imply at a fixed point iteration that maintains probabilistic controllers that eventually collapse on the associated deterministic optimal control.

Next we briefly discuss these results in light of (1) calculation, (2) exploration and finally (3) Active Inference.

(1) We argue that one of the main advantages of the CaI framework is computational. Remark that the present work associates CaI irrevocably with classical optimal control theory. Rather than viewing (15) and (16) as stand-alone problems, we belief they should be viewed within context of the fixed point iterations – put differently, we belief the intermediate solutions have limited value on their own. The benefit of the present over the classical problems is that they can be solved explicitly yielding backward induction rules for the policy. As opposed to the backward induction rules of classical optimal control theory, they have been stripped from any, difficult to evaluate, optimisation operators (i.e. arg min). Instead, any quantity of interest, i.e. the V- and Q-functions, may be calculated by evaluating an expectation operator with respect to the prior model, possibly by approximation – though it is recognized that the resulting procedures will remain challenging to practice in general. Commenting further on the fixed point iterations, it is possible to interpret the policy sequences from the frequentist point of view instead of the Bayesian. Technically such an interpretation is irrelevant. Though here we argue that one may interpret each intermediate iterate policy sequence as a set of belief functions that express our uncertainty about the underlying deterministic solution. Put differently, the sequences give expression to our epistemic uncertainty about the deterministic solution. Finally, we argue that problem (16) claims a special place due to the technical equivalence between the MLE and RSOC problem. As a direct result, it follows that the probabilistic control itself (26) can be evaluated by means of the Bayesian smoother, which itself is a well-established problem with many known numerical treatments.

(2) In principal, exploration is not required in the context of MDPs because the framework presupposes exact knowledge of the generative model. Hence our statement, that CaI is characterised by its computational advantages rather than its capacity to incite purposeful exploration. As noted by Millidge [14] and others, explorative behavioural tendencies obtained through CaI on MDPs boils down to (naive) entropy maximization of the policy. In the context of POMDPs however goal-directed exploration is imminent. Any POMDP agent will determine, based on the generative model it started with, whether it is useful to explore for the sake of exploration, i.e. to reduce its uncertainty about its own state and that of the world, or, whether it is more beneficial to pursue value by minimizing the objective – even though the agent may not be too certain about its state and that of the world at that given time. These two behavioural tendencies are balanced out automatically. Explorative behavioural tendencies are often associated to random tendencies but we do not think this is necessarily so. The decision maker can be certain about its decision. The fact that it may incite behaviour that appears 'random' is a result of the observation it makes next, which itself is indeed subject to uncertainty. This mechanism makes it effectively appear so that the decision maker entertains some randomness in its decisions. That being said, 'exploration' in terms of entropy maximization on POMDP may very well exhibit all the attributes we would like it to exhibit.

(3) These final comments lead us to a comparison between CaI and AIF. A first comparison between CaI and AIF was made by Millidge et al. [14]. Their conclusion was that CaI retains a degree of freedom over AIF that entertains a non-veridical generative model that is biased towards the agent's preferences. With AIF, the same model that is used to 'truthfully' infer the present state, is also used to express behavioural preferences, inevitable leading to a conflict of interest. Recent AIF extensions are embracing strategies to encode value without impeding perception [4,5]. These strategies are in close agreement with the information projected optimal control strategy in (15). Consider the following control objective that is standard in AIF. Here, q_π denotes the variational posterior, and, $\tilde{p}$, usually corresponds with the 'biased' model prior, $p(\underline{x}_T, \underline{u}_{T-1}, \underline{y}_T)$ [4,14] – more generally it can be interpreted as the desired model [5].

$$\mathbb{E}_{q_\pi(\underline{x}_T, \underline{u}_{T-1}, \underline{y}_T)} \left[\log q_\pi(\underline{x}_T, \underline{u}_{T-1}) - \log \tilde{p}\right] \tag{27}$$

Then a first distinction between AIF and the present CaI theory, is that here we rely on exact Bayesian inference (filtering) to obtain the state belief, $p(x_t|w_t)$. If we were to adopt the same strategy in AIF this would imply that

$$q_\pi \leftarrow p(\underline{x}_T, \underline{u}_{T-1}, \underline{y}_T; \underline{\pi}_T) \tag{28}$$

Further, assuming the generative model is unbiased, we have to come up with a different desired model, $\tilde{p}$. To that end, let us substitute one of the following desired models (adopting notation from Sect. 3)

$$\begin{aligned}\tilde{p} &\leftarrow p^*(\underline{x}_T, \underline{u}_{T-1}; \underline{\pi}_T) \\ \tilde{p} &\leftarrow p^*(\underline{x}_T, \underline{u}_{T-1}, \underline{y}_T; \underline{\pi}_T)\end{aligned} \tag{29}$$

Substituting the first model, one easily verifies that problem (27) and (15) are equivalent. If instead we substitute the second model, one verifies problem (27) reduces to (15) subtracting an additional term, referred to as the 'ambiguity' [5].

$$(15) - \mathbb{E}_{p(\underline{x}_T, \underline{u}_{T-1}, \underline{y}_T; \underline{\pi}_T)} \left[\log p(\underline{y}_T | \underline{x}_T)\right] \tag{30}$$

To interpret this term, we remark that the same effect can be obtained within the context of CaI by using an alternative cost function definition, in particular

$$r_t(x_t, u_t, y_t) \leftarrow r_t(x_t, u_t) - \log p(y_t | x_t) \tag{31}$$

Quantitatively, this is equivalent to seeking out 'likely' observations.

All of the presented results support the observations that CaI on POMDPs and AIF are very similar frameworks. This of course raises the question which framework is to be preferred. This and other related questions are topics for future research.

References

1. Abdolmaleki, A., Springenberg, J., Tassa, Y., Munos, R., Heess, N., Riedmiller, M.: Maximum a posteriori policy optimisation. In: International Conference on Learning Representations (2018). https://openreview.net/forum?id=S1ANxQW0b
2. Attias, H.: Planning by probabilistic inference. In: International Workshop on Artificial Intelligence and Statistics, pp. 9–16. PMLR (2003)
3. Bishop, C.M., Nasrabadi, N.M.: Pattern Recognition and Machine Learning, vol. 4, no. 4, p. 738. Springer, New York (2006)
4. Da Costa, L., Parr, T., Sajid, N., Veselic, S., Neacsu, V., Friston, K.: Active inference on discrete state-spaces: a synthesis. J. Math. Psychol. **99**, 102447 (2020)
5. Da Costa, L., Sajid, N., Parr, T., Friston, K., Smith, R.: Reward maximization through discrete active inference. Neural Comput. **35**(5), 807–852 (2023). https://doi.org/10.1162/neco_a_01574
6. Hennig, P., Osborne, M., Girolami, M.: Probabilistic numerics and uncertainty in computations. Proc. R. Soc. A Math. Phys. Eng. Sci. **471**(2179), 20150142 (2015)
7. Hoffmann, C., Rostalski, P.: Linear optimal control on factor graphs-a message passing perspective—. IFAC-PapersOnLine **50**(1), 6314–6319 (2017)
8. Kappen, H.J., Gómez, V., Opper, M.: Optimal control as a graphical model inference problem. Mach. Learn. **87**(2), 159–182 (2012). https://doi.org/10.1007/s10994-012-5278-7
9. Kárný, M.: Towards fully probabilistic control design. Automatica **32**(12), 1719–1722 (1996)
10. Kárný, M., Guy, T.V.: Fully probabilistic control design. Syst. Control Lett. **55**(4), 259–265 (2006)
11. Lange, K.: MM optimization algorithms. SIAM (2016)
12. Lefebvre, T.: A review of probabilistic control and majorization of optimal control (2022). https://doi.org/10.48550/ARXIV.2205.03279

13. Levine, S.: Reinforcement learning and control as probabilistic inference: tutorial and review. arXiv preprint arXiv:1805.00909 (2018)
14. Millidge, B., Tschantz, A., Seth, A.K., Buckley, C.L.: On the relationship between active inference and control as inference. In: IWAI 2020. CCIS, vol. 1326, pp. 3–11. Springer, Cham (2020). https://doi.org/10.1007/978-3-030-64919-7_1
15. Murphy, K.P.: Probabilistic Machine Learning: An Introduction. MIT Press, Cambridge (2022)
16. Murphy, K.P.: Probabilistic Machine Learning: Advanced Topics. MIT Press, Cambridge (2023)
17. Oates, C.J., Sullivan, T.J.: A modern retrospective on probabilistic numerics. Stat. Comput. **29**(6), 1335–1351 (2019). https://doi.org/10.1007/s11222-019-09902-z
18. Särkkä, S.: Bayesian Filtering and Smoothing, no. 3. Cambridge University Press, Cambridge (2013)
19. Sutton, R.S., Barto, A.G.: Reinforcement Learning: An Introduction. MIT Press, Cambridge (2018)
20. Thrun, S.: Probabilistic robotics. Commun. ACM **45**(3), 52–57 (2002)
21. Toussaint, M.: Robot trajectory optimization using approximate inference. In: Proceedings of the 26th Annual International Conference on Machine Learning, pp. 1049–1056 (2009)
22. Toussaint, M., Storkey, A.: Probabilistic inference for solving discrete and continuous state Markov decision processes. In: Proceedings of the 23rd International Conference on Machine Learning, pp. 945–952 (2006)
23. Whittle, P.: Optimal Control: Basics & Beyond. Wiley, Chichester (1996)

Author Index

C. L. Buckley et al. (Eds.): IWAI 2023, CCIS 1915, pp. 289–290, 2024.
https://doi.org/10.1007/978-3-031-47958-8

Printed in the United States
by Baker & Taylor Publisher Services